Pepperdine Papers on Linear Temporal Logic

J. Stanley Warford
Computer Science Department
Pepperdine University
Malibu, CA 90263

David Vega*
The Aerospace Corporation
El Segundo, CA 90245

Scott M. Staley
Ford Motor Company Research Labs (retired)
Dearborn, MI 48124

May 24, 2021
Version: 1.0

---

*Research supported by Tooma Undergraduate Research Fellowship Program, Pepperdine University, Summer 2009 and academic year 2009-10.

Paperback copyright © 2021 by Warford, Vega and Staley.

All rights reserved. We hereby grant permission for this publication to be photo-copied for personal or classroom use. No part of this publication may be stored in a retrieval system or transmitted in any form or by any means other than paper copies without the prior written permission of the authors. Paperback available in any quantity, print-on-demand, from lulu.com. Please contact any author or Julee S. Staley (julees831@gmail.com) for details.

Production managed by Julee S. Staley.
Printed in the United States of America.
First Printing, May 2021
ISBN 978-1-6671-4094-0

Contents

Editor's Introduction   7

A Calculational Deductive System for Linear Temporal Logic (Complete)   9
J. Stanley Warford, David Vega, and Scott M. Staley

1 Introduction   10
    1.1 Axiomatic Logic Systems 10
    1.2 Propositional Logic Systems 11
    1.3 Linear Temporal Logic Systems 12

2 Background   13
    2.1 Calculational Deductive Systems 13
       2.1.1 Propositional and Temporal Operators 13
       2.1.2 Inference Rules 14
       2.1.3 Proof Technique Metatheorems 16
       2.1.4 Predicate Calculus 18
    2.2 Linear Temporal Logic 20
       2.2.1 Models and Anchored Sequences 20
       2.2.2 The Temporal Operators 21
       2.2.3 Duality 27

3 The Calculational Temporal System   27
    3.1 Next 28
    3.2 Until 30
    3.3 Eventually 37
    3.4 Always 41
    3.5 Temporal Deduction 51
    3.6 Always, Continued 53
    3.7 Proof Metatheorems 71
    3.8 Always, Continued 73
    3.9 Wait 91

4 Comparison with Previous Work   121
    4.1 Modal Logic Systems 121
    4.2 Calculational Modal Logic Systems 122
    4.3 Survey of LTL Deductive Systems 122

5 Conclusion   125

6 Acknowledgements   125

7	References	125

Theorems from CDS4LTL   **129**
J. Stanley Warford, David Vega, and Scott M. Staley

A Calculational Deductive System for
Linear Temporal Logic: Additional Theorems   **145**
Scott M. Staley

   1 Preliminaries   **145**

   2 Additional Theorems   **145**
      2.1 Temporal Modus Ponens and Modus Tollens 146
      2.2 Theorems from StackExchange and Spot 149
      2.3 Induction 153
      2.4 Absorption 159
      2.5 Frame Rules 160
      2.6 Duality 162
      2.7 Next $\bigcirc$ 165
      2.8 Until $\mathcal{U}$ 166
            2.8.1 In-state Expansion of $\mathcal{U}$ 170
            2.8.2 Nested Insertion 171
      2.9 Eventually $\Diamond$ 173
      2.10 Always $\square$ 175
      2.11 Always Eventually $\square\Diamond$ and its Dual $\Diamond\square$ 177
      2.12 Wait $\mathcal{W}$ 180
      2.13 Variations of the Dummett Formula 188
      2.14 Proof of the Dummett Formula 192

   3 Theorem Counterexamples   **193**

   4 Left-over Lemmas   **195**

   5 Summary   **198**

   6 Acknowledgements   **198**

   7 References   **198**

Theorems from CDS4LTL (Expanded)   **201**
J. Stanley Warford, David Vega, and Scott M. Staley

Using ACL2 to Confirm Theorems in CDS4LTL   **223**
Scott M. Staley

1	**Preface to this Paper (June 2020)**	**223**
2	**Preliminaries**	**224**
3	**True and False**	**224**
4	**Propositional Operators**	**226**
5	**Propositional Calculus Proofs**	**228**
6	**Linear Temporal Logic Operators**	**230**
	6.1 Next	233
	6.2 Until	234
	6.3 Eventually	235
	6.4 Always	237
	6.5 Always-Eventually	243
	6.6 Wait	246
7	**CDS4LTL Proofs**	**248**
8	**Summary**	**249**
9	**Acknowledgements**	**250**
10	**References**	**250**
	Book Acknowledgements	**251**
	CDS4LTL Research Team	**251**

Editor's Introduction

This book was motivated by the June 2020 publication of our survey paper, in ACM Computing Surveys, on proving theorems in linear temporal logic using a Calculational Deductive System (see https://doi.org/10.1145/3387109). The ACM Computing Surveys paper is freely available and can be downloaded at the previous link. That paper could also be used, on its own, as a student edition of a book supporting teaching LTL at the university level. Students could attempt proofs before an example valid proof, taken from the first paper in this collection, is made available by the instructor.

In the ACM Computing Surveys paper we claim to have proved all the theorems presented there using an extension of the calculational deductive system of Gries and Schneider. The first article in this collection is the Pepperdine technical report that preceded the submission of the ACM paper to the peer review process. This technical report, not constrained by a page limit, provides a proof of every theorem in the ACM paper. The early drafts of this report were used in teaching CoSc450, a senior-level computer science course at Pepperdine University that includes concurrent programming in Java. This article could be used as the instructor edition of a book supporting teaching LTL along with the student edition described above.

In the second article we provide, mainly for the benefit of students in Math220 (Formal Methods), Math221 (Discrete Structures), and CoSc450 (Programming Paradigms) at Pepperdine University, an equations sheet which collects the theorems of Gries and Schneider's propositional calculus along with our LTL theorems. This has proven over the years to be a valuable tool for constructing proofs in propositional calculus and LTL by hand.

The third article in the collection presents additional theorems proven during and since the publication of the ACM Computing Surveys paper. Some of the theorems are just lemmas that might be helpful in constructing more significant theorems. Others are theorems that show new results or provide formal proof, in the calculational deductive style, of well known literature theorems like the Prior formula, the Lemmon formula and the Dummett formula.

The next article provides the student of LTL with an updated equation sheet that includes the addition of all 111 of the S-labeled equations from the previous article.

The final article in this collection documents work that was done to use the ACL2 automated theorem proving software from the University of Texas-Austin to construct proofs confirming the correctness of the CDS4LTL theorems. In this work, which stopped in December 2015, over 150 CDS4LTL theorems had been proven using the ACL2 theorem prover. In addition, all the theorems of Gries and Schneider's propositional calculus on the equation sheet were also proven using ACL2.

In very recent developments, John Wiegley, Principal Engineer at DFINITY, has completed a project in which he used the Coq theorem prover to confirm every theorem and its accompanying proof presented in the ACM Computing Surveys' paper. This gives us additional confidence in the theorems and proofs of that paper, and is consistent with the results we produced earlier using the ACL2 theorem prover. We expect John will publish his results in the near future. In addition, during his work with Coq, John also found a proof of our axiom (56) $\mathcal{U}$ induction, thereby demoting it to a theorem and eliminating an axiom from the system. John also supplied theorem (238) and provided an alternate proof for (58) $\Diamond$ induction, that does not use Duality. We have included these new results in an updated CDS4LTL technical report which is the first article of this collection.

J. Stanley Warford
David Vega
Scott M. Staley

August 8, 2021

A Calculational Deductive System for Linear Temporal Logic (Complete)

J. Stanley Warford
Computer Science Department
Pepperdine University
Malibu, CA 90263

David Vega [*]
The Aerospace Corporation
El Segundo, CA 90245

Scott M. Staley
Ford Motor Company Research Labs (retired)
Dearborn, MI 48124

Abstract

This paper surveys the linear temporal logic (LTL) literature and presents all the LTL theorems from the survey, plus many new ones, in a calculational deductive system. Calculational deductive systems, developed by Dijkstra and Scholten and extended by Gries and Schneider, are based on only four inference rules – Substitution, Leibniz, Equanimity, and Transitivity. Inference rules in the older Hilbert-style systems, notably modus ponens, appear as theorems in this calculational deductive system. This paper extends the calculational deductive system of Gries and Schneider to LTL, using only the same four inference rules. Since there are no space limitations in this technical report, every theorem in this report has been proved using the calculational logic.

---

[*]Research supported by Tooma Undergraduate Research Fellowship Program, Pepperdine University, Summer 2009 and academic year 2009-10.

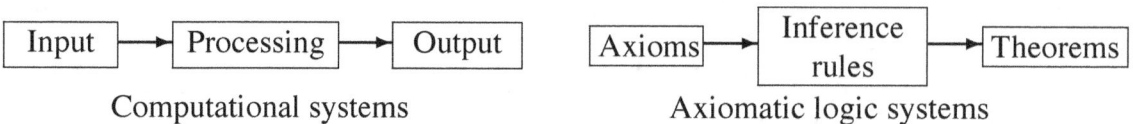

Figure 1: Computational systems and axiomatic logic systems.

1 Introduction

Linear temporal logic (LTL) has application to proof of correctness for concurrent programs. Many concurrent programs, such as operating systems and embedded systems that control physical equipment, are nonterminating by design. Consequently, proof techniques that depend on proving the correctness of postconditions on program termination do not apply. LTL, on the other hand, can be used to prove desirable program traits such as freedom from deadlock.

Most treatments of LTL consist of cursory introductions in one or two chapters of graduate-level textbooks. [2, 20, 21, 24] While many LTL theorems are common in the different treatments, each treatment has theorems that are unique to it. This survey is a comprehensive collection of all the LTL theorems that we have found in the literature, together with many new theorems, all of which are presented in an axiomatic logic system. It serves as an introduction to LTL and should be accessible with a prerequisite only of the standard propositional and predicate logic at the undergraduate level.

1.1 Axiomatic Logic Systems

All axiomatic logic systems have three components – inference rules, axioms, and theorems. Both inference rules and axioms are assumed. Theorems are proved from axioms using inference rules. From a computational systems perspective, the inference rules process axioms as input and produce theorems as output. Figure 1 shows the parallel between traditional computational systems and axiomatic logic systems. In the same way that a processor executes program statements with input to produce output, a proof uses inference rules with axioms to produce theorems.

In a conventional computational system, placement of the hardware/software boundary is a design decision. Any given computational task can be implemented either in hardware or in software. The tradeoff in such systems is usually between speed of execution and flexibility. Usually, a task implemented in hardware executes faster than if it is implemented in software. However, once implemented in hardware a task is more difficult to modify or extend than if it is implemented in software. One goal of RISC design is to simplify the hardware by moving tasks from hardware to software. For example, CISC machines provide complex addressing modes with hardware circuits to compute array cell addresses. The equivalent address computation is done in software in a RISC machine.

A similar design decision exists in axiomatic logic systems with the placement of the inference rule/axiom boundary. It is possible to have two different logic systems produce equivalent sets of theorems but with different sets of inference rules and axioms. What is an inference rule in one system might be a corresponding theorem or axiom in the other. The tradeoff is more subjective in logic systems, as there is apparently no metric of goodness that can be quantified as objectively as can the speed of execution in computational systems. It can be harder to prove that an inference rule is sound than it is to prove that an axiom is sound. Deductive systems often arrange for fewer inference rules to make the soundness proof easier.

This paper presents a logic system that places the boundary between inference rules and axioms to minimize the number of inference rules. We maintain that the primary advantage of such a system is a human one. That is, manual proofs in such systems are easier to understand and to design than in other systems.

1.2 Propositional Logic Systems

Propositional calculus is a formal system of logic based on the unary operator negation ¬, the binary operators conjunction ∧, disjunction ∨, implies ⇒ (also written →), and equivalence ≡ (also written ↔), variables (lowercase letters p, q, ...), and the constants *true* and *false*. Hilbert-style logic systems, $\mathcal{H}$, are the deductive logic systems traditionally used in mathematics to describe the propositional calculus. Typical of such descriptions with applications to computer science is the text by Ben-Ari [3]. A key feature of such systems is their multiplicity of inference rules and the importance of modus ponens as one of them.

In the late 1980's, Dijkstra and Scholten [7], and Feijen [11] developed a method of proving program correctness with a new logic based on an equational style. This equational deductive system, $\mathcal{E}$, is the basis of books by Kaldewaij [19] and Cohen [5]. In contrast to $\mathcal{H}$ systems, $\mathcal{E}$ has only four inference rules – Substitution, Leibniz, Equanimity, and Transitivity. In $\mathcal{E}$, modus ponens plays a secondary role. It is not an inference rule, nor is it assumed as an axiom, but instead is proved as a theorem from the axioms using the inference rules.

Gries and Schneider [12, 16] show that $\mathcal{E}$, also known as a *calculational* system, has several advantages over traditional logic systems. The primary advantage of $\mathcal{E}$ over $\mathcal{H}$ systems is that the calculational system has only four proof rules, with inference rule Leibniz as the primary one. Roughly speaking, Leibniz is "substituting equals for equals," hence the moniker *equational* deductive system. In contrast, $\mathcal{H}$ systems rely on a more extensive set of inference rules. We find proofs in $\mathcal{E}$ easy to understand and to teach, because the substitution of equals for equals is common in elementary algebraic manipulations.

Another major advantage of $\mathcal{E}$ over $\mathcal{H}$ systems is the sequential format of its proof syntax. Proofs in $\mathcal{H}$ systems have a bottom-up tree structure, which is sequentialized with multiple references to previously numbered lines. For example, a proof of formula f_2 might begin by establishing the validity of a formula f_1 on lines 1 through 4. Then, on lines 5 through 9, it might establish the validity of $f_1 \Rightarrow f_2$. Then, on line 10, it would refer back to lines 4 and 9 and invoke modus ponens to establish the validity of f_2.

In contrast, proofs in $\mathcal{E}$ have a top-down structure and proceed sequentially with each step self-contained. There is no need to number the lines in a proof in $\mathcal{E}$ because reference is never made to a previous intermediate step of the proof. Instead, each line depends only on the immediately preceding line by invoking a previously-proved theorem or an axiom.

There is an analogy between the proof style of $\mathcal{H}$ systems versus the proof style of $\mathcal{E}$, and the unstructured *"goto"* style of programming versus structured programming. In the same way that the goto statement can produce spaghetti code that is more difficult to understand than structured code, proofs in $\mathcal{H}$ systems are more difficult to understand than proofs in $\mathcal{E}$. It is perhaps not coincidental that Dijkstra, who ignited the goto controversy with his famous CACM letter [6], was the prime developer of $\mathcal{E}$.

Proof syntax is no guarantee of clarity. In the same way that a well-written assembly language program can be easier to understand than a poorly-written program in a structured high-order language, a well-written proof in $\mathcal{H}$ can be easier to understand than a poorly-written proof in $\mathcal{E}$.

We agree with Gries and Schneider [13] that, "We need a style of logic that can be used as a tool in every-day work. In our experience, an equational logic, which is based on equality and Leibniz's rule for substitution of equals for equals, is best suited for this purpose." These advantages of $\mathcal{E}$ over $\mathcal{H}$ systems are primarily *human* advantages, not necessarily machine advantages. That is, the motivation behind this work is based on teaching and human understanding, as opposed to machine theorem provers or proof assistants.

In 1994, Gries and Schneider published *A Logical Approach to Discrete Math* (LADM) [13], in which they first develop $\mathcal{E}$ for propositional and predicate calculus, and then extend it to a theory of sets, a theory of sequences, relations and functions, a theory of integers, recurrence relations, modern algebra, and a theory of graphs. Using calculational logic as a tool, LADM brings all the advantages of $\mathcal{E}$ to these additional knowledge domains. The treatment is in marked contrast to the traditional one exemplified by the classic undergraduate text by Rosen [23].

1.3 Linear Temporal Logic Systems

Linear temporal logic describes how the truth values of propositions change over time. It extends the propositional operators with the unary operators *next* $\bigcirc$, *eventually* $\Diamond$, and *always* $\Box$, and the binary operators *until* $\mathcal{U}$ and *wait* $\mathcal{W}$. Most treatments of linear temporal logic use $\mathcal{H}$ systems instead of $\mathcal{E}$. Typical are Ben-Ari [4], Emerson [10], Kröger and Merz [20], and Manna and Pnueli [21]. Each of these authors describes the semantics of the above temporal operators and provides an axiomatization for linear temporal logic. One characteristic of these $\mathcal{H}$ systems is the introduction of temporal inference rules along with temporal axioms from which temporal theorems are proved. Section 4 summarizes these systems and compares them with this work.

Our institution has used LADM at the introductory undergraduate level since its publication, and the calculational proof style has consequently permeated the computer science curriculum. A problem arises, however, when those who are schooled in $\mathcal{E}$ study concurrency and need the correctness proof tools of linear temporal logic. Schneider [24] appears to be the only treatment of linear temporal logic that uses $\mathcal{E}$. Although this graduate-level text presents a calculational deductive system, the only appearance of a calculational proof of a temporal logic theorem is a single example. Likewise, Baier and Katoen [2] have a single calculational proof of a linear temporal logic theorem in their chapter on linear temporal logic.

To solve the problem of teaching linear temporal logic at the undergraduate level, this paper presents a comprehensive linear temporal logic system suitable for those versed in the calculational deductive system of LADM. In the same way that LADM brings the advantages of $\mathcal{E}$ to set theory and other mathematical domains, this paper brings the advantages of $\mathcal{E}$ to linear temporal logic.

This axiomatization follows the spirit of $\mathcal{E}$ in its design to minimize the number of inference rules. A unique characteristic of this system is the absence of temporal inference rules. It extends the propositional calculus of LADM using only the same four inference rules of $\mathcal{E}$ along with additional temporal axioms. In our judgment, the absence of temporal inference rules brings the same clarity to linear temporal logic that $\mathcal{E}$ brings to the propositional calculus.

The paper also adds to the linear temporal logic literature by proving many previously unpublished theorems. It is comprehensive, as we have tried to include all known linear temporal theorems described in the literature. Since space limitations do not apply to a technical report, we give a proof of every theorem in this paper. And, we have proved every theorem with $\mathcal{E}$.

Section 2 describes the deductive axioms and the proof rules for $\mathcal{E}$. It also defines the syntax and semantics of linear temporal logic. Section 3 presents the calculational deductive system for linear temporal logic. Section 4 summarizes previous linear temporal logic axiomatization systems and compares them with the current work.

2 Background

The first section below summarizes the calculational system $\mathcal{E}$ from LADM [13]. The summary is minimal, and assumes the reader is familiar with the propositional and predicate calculus. The second section introduces temporal logic and assumes no prior familiarity with it. The paper can serve as an introduction to linear temporal logic.

2.1 Calculational Deductive Systems

2.1.1 Propositional and Temporal Operators

Expressions are the basis of propositional calculus in the calculational system. Propositional theorems are simply boolean expressions that are true in all states. The definition of an expression has four parts:

$[x := e]$ (textual substitution)	Highest precedence
$\neg \quad \circ \quad \diamond \quad \square$	
$\mathcal{U} \quad \mathcal{W}$	
$=$ (conjunctional)	
$\vee \quad \wedge$	
$\Rightarrow \quad \Leftarrow$	
$\equiv$ (associative)	Lowest precedence

Figure 2: Precedence of the propositional and temporal logic operators.

- A constant or variable is an expression.

- If E is an expression, then (E) is an expression.

- If $\triangleright$ is a unary prefix operator and E is an expression, then $\triangleright E$ is an expression with operand E.

- If $\star$ is a binary infix operator and D and E are expressions, then $D \star E$ is an expression with operands D and E.

By convention, upper-case letters (*e.g.* $X, Y, \ldots$) represent expressions and lower-case letters (*e.g.* $x, y, \ldots$) represent variables. In the propositional calculus, the constants are *true* and *false*.

Figure 2 is the table of precedences. Textual substitution has the highest precedence. All the unary operators have the next highest precedence. They are necessarily right associative. For example, $\neg \circ \neg p$ means $\neg(\circ(\neg p))$. In this system, two binary operators that have the same precedence require parentheses to disambiguate. As in LADM, conjunction $\wedge$ and disjunction $\vee$ have the same precedence so that $p \wedge q \vee r$ must be disambiguated as either $(p \wedge q) \vee r$ or $p \wedge (q \vee r)$. This contrasts with many systems in which conjunction has higher precedence than disjunction.

Also consistent with the calculational system of LADM but different from most other deductive logic systems is the difference between operators equals $=$ and equivales $\equiv$. Equals applies to any mathematical type including, *e.g.*, boolean, natural number, and set. Equivales applies only to boolean and is commonly denoted $\leftrightarrow$ in other systems. Another difference is that equals is conjunctive, while equivales is associative. For example, the expression $p = q = r$ has conjunctive meaning $(p = q) \wedge (q = r)$, while the expression $p \equiv q \equiv r$ can be taken as either $(p \equiv q) \equiv r$ or $p \equiv (q \equiv r)$. Associativity of equivales is the first axiom in the calculational deductive system of LADM.

2.1.2 Inference Rules

The inference rules for the calculational deductive system are Substitution, Leibniz, Equanimity, and Transitivity.

(I1) **Substitution:** $\dfrac{E}{E[z := F]}$

(I2) **Leibniz:** $\dfrac{X = Y}{E[z := X] = E[z := Y]}$

(I3) **Equanimity:** $\dfrac{X,\quad X = Y}{Y}$

(I4) **Transitivity:** $\dfrac{X = Y,\quad Y = Z}{X = Z}$

where the square bracket in $E[z := F]$ indicates textual substitution of expression F for variable z everywhere z occurs in expression E. In a typical proof, Substitution and Leibniz are explicit, while Equanimity and Transitivity are implicit.

Substitution allows the generalization of a single theorem to represent an infinite number of theorems. For example, because $p \Rightarrow \textit{false} \equiv \neg p$ is a theorem, then, with $p := p \wedge q$, the expression $(p \wedge q) \Rightarrow \textit{false} \equiv \neg(p \wedge q)$ is also a theorem.

Roughly speaking, Leibniz allows for the substitution of equals for equals in a proof step. The general form of a proof step is

$$\begin{aligned}&E[z := X]\\=\ &\langle X = Y\rangle\\&E[z := Y]\end{aligned}$$

where the expression enclosed in angle brackets $\langle\ \rangle$, called the "hint", is the justification for the step.

An example of a proof step from the proof of theorem (166) in Section 3.8 is

$$\begin{aligned}&\Box(\Box p \wedge \Diamond q) \Rightarrow \Box\Diamond(p \wedge q)\\=\ &\langle(99)\ \text{Distributivity of}\ \Box\ \text{over}\ \wedge\rangle\\&\Box\Box p \wedge \Box\Diamond q \Rightarrow \Box\Diamond(p \wedge q)\end{aligned}$$

This proof step uses the previously proved theorem (99) Distributivity of $\Box$ over $\wedge$, which is $\Box(p \wedge q) \equiv \Box p \wedge \Box q$. The justification in the hint $X = Y$ comes from inference rule Substitution, with the textual substitution of $\Box p$ for p and $\Diamond q$ for q in (99) as follows

$$(\Box(p \wedge q) \equiv \Box p \wedge \Box q)[p, q := \Box p, \Diamond q]:\quad \Box(\Box p \wedge \Diamond q) \equiv \Box\Box p \wedge \Box\Diamond q$$

The expressions in Leibniz for the step are

$$\begin{aligned}E:\quad &z \Rightarrow \Box\Diamond(p \wedge q)\\X:\quad &\Box(\Box p \wedge \Diamond q)\\Y:\quad &\Box\Box p \wedge \Box\Diamond q\end{aligned}$$

The textual substitutions are

$$E[z := X]: \quad \Box(\Box p \land \Diamond q) \Rightarrow \Box \Diamond (p \land q)$$
$$E[z := Y]: \quad \Box \Box p \land \Box \Diamond q \Rightarrow \Box \Diamond (p \land q)$$

The proof of a theorem consists of showing the equivalence of that theorem to a previously proved theorem through a sequence of the proof steps. For example, here is a one-step proof of (3) Linearity: $\bigcirc p \equiv \neg \bigcirc \neg p$ in Section 3.1.

Proof:

$$\begin{aligned}
& \bigcirc p \equiv \neg \bigcirc \neg p \\
= \quad & \langle (3.11)\ \neg p \equiv q \equiv p \equiv \neg q \text{ with } p, q := \bigcirc \neg p, \bigcirc p, \\
& \neg \bigcirc \neg p \equiv \bigcirc p \equiv \bigcirc \neg p \equiv \neg \bigcirc p \rangle \\
& \neg \bigcirc p \equiv \bigcirc \neg p \quad -(1)\ \text{Self-dual} \quad \blacksquare
\end{aligned}$$

In a proof hint, numeric references that contain a decimal point, such as (3.11) above, refer to a theorem in $\mathcal{E}$ from LADM. Equanimity is implicit in the proof. Because $\neg \bigcirc p \equiv \bigcirc \neg p$ (*i.e.* X) is a previous theorem, and $\neg \bigcirc p \equiv \bigcirc \neg p$ is equivalent to $\bigcirc p \equiv \neg \bigcirc \neg p$ (*i.e.* $X = Y$), by equanimity $\bigcirc p \equiv \neg \bigcirc \neg p$ (*i.e.* Y) is proved.

Transitivity of equality allows a derivation to be given as a sequence of equivalent expressions, which, at the end, proves the equivalence of the first expression in the sequence with the last expression in the sequence. For example, here is a two-step proof of (22) Idempotency of $\mathcal{U}$, $p\,\mathcal{U}\,p \equiv p$ in Section 3.2.

Proof:

$$\begin{aligned}
& p\,\mathcal{U}\,p \equiv p \\
= \quad & \langle (10)\ \text{Expansion of } \mathcal{U} \rangle \\
& p \lor (p \land \bigcirc (p\,\mathcal{U}\,p)) \equiv p \\
= \quad & \langle (3.43\text{b})\ \text{Absorption}, p \lor (p \land q) \equiv p \text{ with } q := \bigcirc (p\,\mathcal{U}\,p) \rangle \\
& p \equiv p \quad -(3.5)\ \text{Reflexivity of } \equiv \quad \blacksquare
\end{aligned}$$

Transitivity of equality is implicit in the proof. Because $p\,\mathcal{U}\,p \equiv p$ is equivalent to $p \lor (p \land \bigcirc (p\,\mathcal{U}\,p)) \equiv p$ (*i.e.* $X = Y$), and $p \lor (p \land \bigcirc (p\,\mathcal{U}\,p)) \equiv p$ is equivalent to $p \equiv p$ (*i.e.* $Y = Z$), by transitivity $p\,\mathcal{U}\,p \equiv p$ is equivalent to $p \equiv p$ (*i.e.* $X = Z$).

2.1.3 Proof Technique Metatheorems

The logic system $\mathcal{E}$ of LADM [13] has 13 axioms for the propositional calculus from which theorems are deduced with the above inference rules in the calculational style. The system also contains a number of metatheorems based on properties of equivalence and implication, which allow the proof style to be extended. Here are four of the proof technique metatheorems.

(4.4) **Deduction (assume conjuncts of antecedent):**
To prove $P_1 \wedge P_2 \Rightarrow Q$, assume P_1 and P_2, and prove Q
You cannot use textual substitution in P_1 or P_2.

(4.7) **Mutual implication:** To prove $P \equiv Q$, prove $P \Rightarrow Q$ and $Q \Rightarrow P$

(4.7.1) **Truth implication:** To prove P, prove $\mathit{true} \Rightarrow P$

(4.12) **Contrapositive:** To prove $P \Rightarrow Q$, prove $\neg Q \Rightarrow \neg P$

The validity of each metatheorem is established from the theorems of the propositional calculus and the inference rules. Deduction is established by showing that any deductive proof has an equivalent calculational proof. Mutual implication is based on (3.80), Truth implication is based on (3.73), and Contrapositive is based on (3.61).

(3.80) **Mutual implication:** $(p \Rightarrow q) \wedge (q \Rightarrow p) \equiv (p \equiv q)$

(3.73) **Left identity of $\Rightarrow$:** $\mathit{true} \Rightarrow p \equiv p$

(3.61) **Contrapositive:** $p \Rightarrow q \equiv \neg q \Rightarrow \neg p$

The proof format is extended to the case where the theorem to be proved is of the form $P \equiv Q$. For theorems of this form, the proof may begin with the left hand side and show equivalence to the right hand side through a sequence of proof steps. This proof style is established by showing that such a proof is equivalent to a calculational proof and is based on (3.5).

(3.5) **Reflexivity of $\equiv$:** $p \equiv p$

For example, the following proof is the preferred style for the previous proof of (22) Idempotency of $\mathcal{U}$.
Proof:

$\quad p \, \mathcal{U} \, p$
$= \quad \langle (10) \text{ Expansion of } \mathcal{U} \rangle$
$\quad p \vee (p \wedge \circ (p \, \mathcal{U} \, p))$
$= \quad \langle (3.43b) \text{ Absorption, } p \vee (p \wedge q) \equiv p \text{ with } q := \circ (p \, \mathcal{U} \, p) \rangle$
$\quad p \quad \blacksquare$

Gries and Schneider also extend the proof format to incorporate implication using its transitive properties with itself and with equivales. Instead of proving a theorem of the form $P \Rightarrow Q$ to be equivalent to a previously proved theorem, P can be shown to imply Q, or Q can be shown to follow from P. The following mutual transitivity theorems justify this extension.

(3.82) **Transitivity:**
(a) $(p \Rightarrow q) \wedge (q \Rightarrow r) \Rightarrow (p \Rightarrow r)$
(b) $(p \equiv q) \wedge (q \Rightarrow r) \Rightarrow (p \Rightarrow r)$
(c) $(p \Rightarrow q) \wedge (q \equiv r) \Rightarrow (p \Rightarrow r)$

An example is a proof of (46) Weakening of $\Diamond$, $p \Rightarrow \Diamond p$ in Section 3.3.

Proof:

$$\begin{array}{rl} & \Diamond p \\ = & \langle (45) \text{ Expansion of } \Diamond \rangle \\ & p \vee \circ \Diamond p \\ \Leftarrow & \langle (3.76a) \text{ Weakening, } p \Rightarrow p \vee q \text{ with } q := \circ \Diamond p \rangle \\ & p \quad \blacksquare \end{array}$$

Because $\Diamond p$ equivales $p \vee \circ \Diamond p$, and $p \vee \circ \Diamond p$ follows from p, it follows by mutual transitivity that $\Diamond p$ follows from p.

The following two theorems from LADM provide a further extension to proof steps with implication.

(4.2) **Monotonicity of $\vee$** : $(p \Rightarrow q) \Rightarrow (p \vee r \Rightarrow q \vee r)$
(4.3) **Monotonicity of $\wedge$** : $(p \Rightarrow q) \Rightarrow (p \wedge r \Rightarrow q \wedge r)$

They are required to justify an implication when the antecedent of the implication is a conjunct or disjunct. For example, here is a proof step where the antecedent $p \,\mathcal{U}\, q$ is a disjunct in the expression $\Box p \vee p \,\mathcal{U}\, q$.

$$\begin{array}{rl} & \Box p \vee p \,\mathcal{U}\, q \\ \Rightarrow & \langle (42) \text{ Eventuality and } (4.2) \text{ Monotonicity of } \vee \rangle \\ & \Box p \vee \Diamond q \end{array}$$

Previously proved theorem (42) Eventuality is $p \,\mathcal{U}\, q \Rightarrow \Diamond q$. The application of (4.2) is with the following textual substitution.

$$\begin{array}{rl} & ((p \Rightarrow q) \Rightarrow (p \vee r \Rightarrow q \vee r))[p,q,r := p \,\mathcal{U}\, q, \Diamond q, \Box p] \\ = & \langle \text{Textual substitution} \rangle \\ & (p \,\mathcal{U}\, q \Rightarrow \Diamond q) \Rightarrow (p \,\mathcal{U}\, q \vee \Box p \Rightarrow \Diamond q \vee \Box p) \end{array}$$

In other words, because $p \,\mathcal{U}\, q$ implies $\Diamond q$, $p \,\mathcal{U}\, q \vee \Box p$ implies $\Diamond q \vee \Box p$.

2.1.4 Predicate Calculus

The predicate calculus of the calculational system has a consistent quantification notation that applies to Abelian monoids in both mathematics and logic. Denoting a general Abelian monoid as the infix operator $\star$, the form of a quantification is

$(\star \, dummies \mid range : body)$

All quantifications have explicit scope for the dummy variable denoted by the outer parentheses. Within the parentheses the quantification consists of three parts:

- the infix operator $\star$ and dummy variable(s),
- the range, which is a boolean expression, and
- the body, which is an expression that is type compatible with the operator $\star$.

A vertical bar separates the operator and dummy variable from the range, and a colon separates the range from the body. An abbreviation is to omit the range when it is *true*. For example, $(\forall i |: P)$ is an abbreviation for $(\forall i \mid true : P)$.

For example, the standard mathematical notation for writing the sum of the squares of the first n positive integers is

$$\sum_{i=1}^{n} i^2$$

where $\star$ is Abelian monoid $+$, and Σ is the quantified symbol for addition. The calculational notation for the same expression is

$$(\Sigma i \mid 1 \leq i \leq n : i^2)$$

Similarly, the standard logic notation for writing that there exists a number between 10 and 20 inclusive that divides n is

$$\exists i (10 \leq i \leq 20 \wedge \text{divides}(i,n))$$

where $\star$ is Abelian monoid $\vee$, $\exists$ is the quantified symbol for disjunction, and divides is a predicate that is true when i divides n. The calculational notation for the same expression is

$$(\exists i \mid 10 \leq i \leq 20 : \text{divides}(i,n))$$

Predicate calculus in the calculational system of LADM begins with nine general axioms that apply to all Abelian monoids. For example, here are the first two axioms.

For symmetric and associative binary operator $\star$ with identity u.

(8.13) **Axiom, Empty range:** $(\star x \mid false : P) = u$

(8.14) **Axiom, One-point rule:** Provided $\neg occurs(\text{'}x\text{'}, \text{'}E\text{'})$,
$(\star x \mid x = E : P) = P[x := E]$

It has two axioms for universal quantification.

(9.2) **Axiom, Trading:** $(\forall x \mid R : P) \equiv (\forall x \mid: R \Rightarrow P)$

(9.5) **Axiom, Distributivity of $\vee$ over $\forall$:** Provided $\neg occurs(\text{'}x\text{'}, \text{'}P\text{'})$,
$P \vee (\forall x \mid R : Q) \equiv (\forall x \mid R : P \vee Q)$

And it has one axiom for existential quantification.

(9.17) **Axiom, Generalized De Morgan:** $(\exists x \mid R : P) \equiv \neg(\forall x \mid R : \neg P)$

2.2 Linear Temporal Logic

The operators of propositional calculus, $\neg, =, \wedge, \vee, \Rightarrow, \Leftarrow$, and $\equiv$ are static. That is, they apply at a single point in time. Each operator has a truth table that dictates how to evaluate the truth value of an expression. A state is an assignment of a truth value to each variable in the expression. A given boolean expression may be false in all states, true in some states and false in others, or true in all states, in which case the expression is known as a theorem or validity or tautology.

The operators of temporal logic, $\bigcirc, \Diamond, \square, \mathcal{U}$, and $\mathcal{W}$ are dynamic. That is, they do not apply at a single point in time, but apply over an infinite sequence of states. Each state corresponds to a discrete point in time that represents one point in the execution of a program, possibly having several threads running concurrently but whose instruction executions have been serialized. As one instruction in the program is executed, the state changes, and hence the truth value of an expression may change as well.

2.2.1 Models and Anchored Sequences

A program consists of, among other things, a set of variables and constants. Using state expressions, with operations provided by the programming language, the program changes the values of the variables as it executes. When the value of a variable changes, a property associated with that variable might also change. Properties are described by state formulas (assertions). V is the set of variables that are combined with the operations of the programming language to form state expressions and with the boolean operations of logic to describe properties.

A model σ over V is an infinite sequence of the form

$$\sigma : s_0, s_1, s_2, \ldots$$

where s_0 is the initial state of a computation and each state $s_i, 0 \leq i$ is the state at time i. [21]

For example, suppose x is an integer variable whose value varies at each step of the computation. Then, x and the property $x \geq 10$ might evolve as follows.

σ	s_0	s_1	s_2	s_3	s_4	...
x	8	9	10	11	12	...
$x \geq 10$	F	F	T	T	T	...

The bottom row shows the evaluation of the state formula for each state in the sequence.

An anchored sequence is a pair (σ, j) where j is a natural number (the anchor) that specifies a state in model σ. [24] The anchor point j partitions the states s_i of σ into the past $0 \leq i < j$, present $i = j$, and future $i > j$. The notation

$$(\sigma, j) \models p$$

means that the property p holds at position j in a sequence σ. In this example,

$$(\sigma, 3) \models x \geq 10.$$

The symbol $\models$ means "satisfies", so the above expression is read as "State 3 of sequence σ satisfies $x \geq 10$." Or, using "holds", the same expression is read as, "$x \geq 10$ holds in state 3 of sequence σ."

In the above example, evaluation of the property $x \geq 10$ in the anchored sequence (σ, j) depends only on the value of variable x at the anchor point j. In general, the truth of a temporal assertion at j may depend on future states as well. For example, an informal English assertion is, "p is now, and always will be from this point on, true." The temporal notation for this assertion is $\Box p$. If you assume that x in the above sequence keeps increasing by one, then $\Box p$ holds in state 3 of sequence σ. However, the truth of this assertion depends not only on the fact that p holds at s_3 but that p also holds at $s_4, s_5, \ldots$. A more precise formulation is that p holds at s_3 and that $\Box p$ also holds at each subsequent state. See Schneider [24] for a formal treatment, which depends on the formulation of prefix and suffix anchored sequences.

There is a distinction between the constant *true* and the truth value of an expression T in a given state. The constant *true* is an expression that evaluates to T in every state. Similarly, there is a distinction between the constant *false* and the truth value of an expression F in a given state. The constant *false* is an expression that evaluates to F in every state.

σ	s_0	s_1	s_2	s_3	s_4	...
true	T	T	T	T	T	...
false	F	F	F	F	F	...

The propositional logic system of LADM [13] describes a case analysis metatheorem as follows: If $E[z := \mathit{true}]$ and $E[z := \mathit{false}]$ are theorems, then so is $E[z := p]$. This metatheorem does *not* hold in LTL because the two cases, $z := \mathit{true}$ and $z := \mathit{false}$, account for only two out of an infinite number of possible sequences of T's and F's in σ.

2.2.2 The Temporal Operators

This section uses the anchored sequence (σ, j) and $\models$ to formalize the interpretation of each temporal operator.

The *next* Operator $\bigcirc$

The semantics of the unary prefix operator $\bigcirc$ are

$$(\sigma, j) \models \bigcirc p \quad \text{iff} \quad (\sigma, j+1) \models p$$

That is, $\bigcirc p$ holds at position j iff p holds at position $j+1$.
For example, in the following sequence $\bigcirc\ 10 \leq x < 13$ holds at state s_1 because $10 \leq x < 13$ holds at state s_2.

σ	s_0	s_1	s_2	s_3	s_4	s_5	s_6	...
x	8	9	10	11	12	13	14	...
$10 \leq x < 13$	F	F	T	T	T	F	F	...
$\bigcirc\ 10 \leq x < 13$	F	T	T	T	F	F	F	...

In other words,

$$(\sigma, 1) \models \bigcirc\ 10 \leq x < 13 \quad \text{because} \quad (\sigma, 2) \models 10 \leq x < 13$$

Furthermore, $\bigcirc\ 10 \leq x < 13$ does not hold at state s_4 even though $10 \leq x < 13$ does hold in that state, because $10 \leq x < 13$ does not hold in state s_5.

This definition of $\bigcirc$ assumes an infinite sequence of states. Emerson [10] shows variations of the *next* operator that would apply to a finite sequence of states suitable for modeling a program that terminates.

The *until* Operator $\mathcal{U}$

The semantics of the binary infix operator $\mathcal{U}$ are

$$(\sigma, j) \models p\,\mathcal{U}\,q \quad \text{iff} \quad (\exists k \mid k \geq j : (\sigma, k) \models q \wedge (\forall i \mid j \leq i < k : (\sigma, i) \models p))$$

If $p\,\mathcal{U}\,q$ holds at state s_j, then p holds at state s_j and continues to hold at every state after s_j until q holds at some future state. $p\,\mathcal{U}\,q$ guarantees that q will eventually hold at some future state and that p will continue to hold until then. After the state in which q holds for the first time, there are no restrictions on either p or q.

For example, suppose x and y evolve in the computation as follows.

σ	s_0	s_1	s_2	s_3	s_4	s_5	s_6	s_7	s_8	s_9	...
x	-1	0	1	2	3	4	5	6	7	8	...
y	9	8	7	6	5	4	3	2	1	0	...
$0 < x < y$	F	F	T	T	T	F	F	F	F	F	...
$2 \leq y < 5$	F	F	F	F	F	T	T	T	F	F	...
$(0 < x < y)\,\mathcal{U}\,(2 \leq y < 5)$	F	F	T	T	T	T	T	T	F	F	...

The bottom row shows the evaluation of the expression $p\,\mathcal{U}\,q$ where $p \equiv 0 < x < y$ and $q \equiv 2 \leq y < 5$. In states s_0 and s_1, $p\,\mathcal{U}\,q$ is false because both p and q are false. Starting at state s_2, $p\,\mathcal{U}\,q$ is true because in that state p is true and will remain true until q eventually becomes true in state s_5.

From the semantics of $p\,\mathcal{U}\,q$, if q is true in any state, then $p\,\mathcal{U}\,q$ is true in that state regardless of p. For example, not only is $p\,\mathcal{U}\,q$ true in state s_5, before which p was true in several preceding states, it is also true in states s_6 and s_7, because in those states q is true. This behavior of $p\,\mathcal{U}\,q$ comes from the empty range and one-point rules [13] of the predicate calculus in the case that q holds in state s_j and $k = j$.

Proof:

$\quad\quad (\exists k \mid k \geq j : (\sigma,k) \models q \land (\forall i \mid j \leq i < k : (\sigma,i) \models p))$
$= \quad \langle \text{Case } k = j \rangle$
$\quad\quad (\exists k \mid k = j : (\sigma,k) \models q \land (\forall i \mid j \leq i < j : (\sigma,i) \models p))$
$= \quad \langle j \leq i < j \equiv false \rangle$
$\quad\quad (\exists k \mid k = j : (\sigma,k) \models q \land (\forall i \mid false : (\sigma,i) \models p))$
$= \quad \langle (8.13) \text{ Empty range rule } (\star x \mid false : P) = u \text{ with } true \text{ the identity of } \land \rangle$
$\quad\quad (\exists k \mid k = j : (\sigma,k) \models q \land true)$
$= \quad \langle (3.39) \text{ Identity of } \land, p \land true \equiv p \rangle$
$\quad\quad (\exists k \mid k = j : (\sigma,k) \models q)$
$= \quad \langle (8.14) \text{ One-point rule } (\star x \mid x = E : P) = P[x := E] \rangle$
$\quad\quad ((\sigma,k) \models q)[k := j]$
$= \quad \langle \text{Textual substitution} \rangle$
$\quad\quad (\sigma,j) \models q$
$= \quad \langle \text{Case } q \text{ holds in state } s_j \rangle$
$\quad\quad true \quad \blacksquare$

This result is theorem (20) $p \, \mathcal{U} \, true \equiv true$ listed in Section 3.2. *true* is the right zero of the *until* operator.

The *until* operator $\mathcal{U}$ is not associative as shown by the following sequence.

σ	s_0	s_1	s_2	s_3	s_4	s_5	s_6	s_7	...
p	F	F	T	T	T	T	F	F	...
q	F	T	F	T	F	F	F	F	...
r	F	F	F	T	T	F	T	F	...
$p \, \mathcal{U} \, q$	F	T	T	T	F	F	F	F	...
$q \, \mathcal{U} \, r$	F	F	F	T	T	F	T	F	...
$p \, \mathcal{U} \, (q \, \mathcal{U} \, r)$	F	F	T	T	T	T	T	F	...
$(p \, \mathcal{U} \, q) \, \mathcal{U} \, r$	F	T	T	T	T	F	T	F	...

State s_1 in the last two rows of the above table shows that $(p \, \mathcal{U} \, q) \, \mathcal{U} \, r$ does not imply $p \, \mathcal{U} \, (q \, \mathcal{U} \, r)$, and state s_5 shows that $p \, \mathcal{U} \, (q \, \mathcal{U} \, r)$ does not imply $(p \, \mathcal{U} \, q) \, \mathcal{U} \, r$.

The *eventually* Operator $\diamond$

The semantics of the unary prefix operator $\diamond$ are

$$(\sigma, j) \models \diamond p \quad \text{iff} \quad (\exists k \mid k \geq j : (\sigma,k) \models p)$$

So, $\diamond p$ holds in state s_j if p holds in state s_j or in any other state s_k where $k \geq j$, that is, if p holds in the current state or in any other future state.

For example, suppose x evolves in the computation as follows.

σ	s_0	s_1	s_2	s_3	s_4	s_5	s_6	...
x	1	2	3	4	5	6	7	...
$3 \leq x < 6$	F	F	T	T	T	F	F	...
$\Diamond (3 \leq x < 6)$	T	T	T	T	T	F	F	...

The bottom row shows the evaluation of the expression $\Diamond p$ where $p \equiv 3 \leq x < 6$. In states s_0 and s_1, $\Diamond p$ is true because there is a state, either now or in the future, in which p will hold.

If $\Diamond p$ is ever false in any state s_i in a sequence σ, it must be false in all subsequent states $s_j, j \geq i$. If $\Diamond p$ is ever true in any state s_i in a sequence σ, it must be true in all preceding states $s_j, j \leq i$. For example, suppose p and q evolve in the computation as follows.

σ	s_0	s_1	s_2	s_3	s_4	s_5	s_6	s_7	s_8	s_9	...
p	F	F	T	F	F	T	F	F	F	F	...
q	F	F	T	T	F	F	T	T	F	F	...
$\Diamond p$	T	T	T	T	T	T	F	F	F	F	...
$\Diamond q$	T	T	T	T	T	T	T	T	T	T	...

The bottom two rows show the evaluation of the expressions $\Diamond p$ and $\Diamond q$ assuming that p remains false indefinitely and q continues to switch between true and false indefinitely.

The *eventually* operator is a special case of the *until* operator. Namely, $true \; \mathcal{U} \; q$ is equivalent to $\Diamond q$.

Proof:

$$(\sigma, j) \models true \; \mathcal{U} \; q$$
$= \quad \langle \text{Semantics of } p \; \mathcal{U} \; q \text{ with } p := true \rangle$
$$(\exists k \mid k \geq j : (\sigma, k) \models q \land (\forall i \mid j \leq i < k : (\sigma, i) \models true))$$
$= \quad \langle true \text{ holds in all states} \rangle$
$$(\exists k \mid k \geq j : (\sigma, k) \models q \land (\forall i \mid j \leq i < k : true))$$
$= \quad \langle (9.8) \; (\forall x \mid R : true) \equiv true \rangle$
$$(\exists k \mid k \geq j : (\sigma, k) \models q \land true)$$
$= \quad \langle (3.39) \text{ Identity of } \land, \; p \land true \equiv p \rangle$
$$(\exists k \mid k \geq j : (\sigma, k) \models q)$$
$= \quad \langle \text{Semantics of } \Diamond q \rangle$
$$(\sigma, j) \models \Diamond q \quad \blacksquare$$

This relationship is the basis of the definition of $\Diamond q$ in (38) $\Diamond q \equiv true \; \mathcal{U} \; q$ assumed in Section 3.3.

Pepperdine Papers on LTL

The *always* Operator $\square$

The semantics of the unary prefix operator $\square$ are

$$(\sigma, j) \models \square p \quad \text{iff} \quad (\forall k \mid k \geq j : (\sigma, k) \models p)$$

So, $\square p$ holds in state s_j if p holds in state s_j and in all other states s_k where $k \geq j$, that is, if p holds in the current state and in all other future states.

For example, suppose x evolves in the computation as follows.

σ	s_0	s_1	s_2	s_3	s_4	s_5	s_6	s_7	...
x	1	2	3	4	5	6	7	8	...
$x < 4 \vee x \geq 6$	T	T	T	F	F	T	T	T	...
$\square(x < 4 \vee x \geq 6)$	F	F	F	F	F	T	T	T	...

The bottom row shows the evaluation of the expression $\square p$ where $p \equiv x < 4 \vee x \geq 6$. In states s_3 and s_4, $\square p$ is false because p does not hold in those states. In states s_0, s_1, and s_2, p is true. However, $\square p$ is false in those states because p does no hold in all future states. In states s_5, s_6, s_7, and subsequent states, $\square p$ is true because p holds in in those states and in all future states as well.

If $\square p$ is ever true in any state s_i in a sequence σ, it must be true in all subsequent states s_j, $j \geq i$. If $\square p$ is ever false in any state s_i in a sequence σ, it must be false in all preceding states s_j, $j \leq i$.

For example, suppose p and q evolve in the computation as follows.

σ	s_0	s_1	s_2	s_3	s_4	s_5	s_6	s_7	s_8	s_9	...
p	T	T	F	T	T	F	T	T	T	T	...
q	T	T	F	F	T	T	F	F	T	T	...
$\square p$	F	F	F	F	F	F	T	T	T	T	...
$\square q$	F	F	F	F	F	F	F	F	F	F	...

The bottom two rows show the evaluation of the expressions $\square p$ and $\square q$ assuming that p remains true indefinitely and q continues to switch between true and false indefinitely.

Note that $\square p$ is a universal operator, while $\Diamond p$ is an existential operator. In the same way that $(\forall x \mid R : P)$ is equivalent to $\neg(\exists x \mid R : \neg P)$ through the generalized De Morgan theorem, $\square p$ is equivalent to $\neg \Diamond \neg p$.

Proof:

$\quad (\sigma, j) \models \square p$
$= \quad \langle \text{Semantics of } \square p \rangle$
$\quad (\forall k \mid k \geq j : (\sigma, k) \models p)$
$= \quad \langle \text{(9.18a) Generalized De Morgan } \neg(\exists x \mid R : \neg P) \equiv (\forall x \mid R : P) \rangle$

$$\neg(\exists k \mid k \geq j : \neg((\sigma,k) \models p))$$
= ⟨p does not hold in a state iff $\neg p$ holds in that state⟩
$$\neg(\exists k \mid k \geq j : (\sigma,k) \models \neg p)$$
= ⟨Semantics of $\Diamond p$⟩
$$\neg((\sigma,j) \models \Diamond \neg p)$$
= ⟨p does not hold in a state iff $\neg p$ holds in that state⟩
$$(\sigma,j) \models \neg \Diamond \neg p \quad \blacksquare$$

This relationship is the basis of the definition of $\Box p$ in equation (54) $\Box p \equiv \neg \Diamond \neg p$ assumed in Section 3.4.

The above calculational proof illustrates a common advantage of $\mathcal{E}$ over traditional logic systems. The same equivalence is proved in [3] but only by resorting to proof by mutual implication using proof by contradiction within each case. Gries and Schneider [13] point out that many equivalence proofs are shorter in $\mathcal{E}$, in which equality is central, than in traditional systems, in which implication is central.

The above demonstration that $(\sigma,j) \models \Box p \equiv (\sigma,j) \models \neg \Diamond \neg p$ depends on the rule, "p does not hold in a state iff $\neg p$ holds in that state", written formally as

$$\neg((\sigma,j) \models p) \quad \text{iff} \quad (\sigma,j) \models \neg p$$

The corresponding rules for the binary operators are

$$((\sigma,j) \models p) \wedge ((\sigma,j) \models q) \quad \text{iff} \quad (\sigma,j) \models p \wedge q$$
$$((\sigma,j) \models p) \vee ((\sigma,j) \models q) \quad \text{iff} \quad (\sigma,j) \models p \vee q$$
$$((\sigma,j) \models p) \Rightarrow ((\sigma,j) \models q) \quad \text{iff} \quad (\sigma,j) \models p \Rightarrow q$$
$$((\sigma,j) \models p) \equiv ((\sigma,j) \models q) \quad \text{iff} \quad (\sigma,j) \models p \equiv q$$

The *wait* Operator $\mathcal{W}$

The semantics of the binary infix operator $\mathcal{W}$ in terms of $\mathcal{U}$ and $\Box$ are

$$(\sigma,j) \models p \, \mathcal{W} \, q \quad \text{iff} \quad (\sigma,j) \models p \, \mathcal{U} \, q \vee (\sigma,j) \models \Box p$$

The *wait* operator $\mathcal{W}$ is weaker than the *until* operator $\mathcal{U}$, because $p \, \mathcal{W} \, q$ does not require q to ever be true, while $p \, \mathcal{U} \, q$ does. Furthermore, theorem (174) shows that $p \, \mathcal{U} \, q \Rightarrow p \, \mathcal{W} \, q$. For example, suppose p and q evolve in the computation as follows.

σ	s_0	s_1	s_2	s_3	s_4	s_5	s_6	s_7	s_8	s_9	s_{10}	...
p	F	F	T	T	F	F	F	F	T	T	T	...
q	F	F	F	F	T	T	F	F	F	F	F	...
$\Box p$	F	F	F	F	F	F	F	F	T	T	T	...
$p \, \mathcal{U} \, q$	F	F	T	T	T	T	F	F	F	F	F	...
$p \, \mathcal{W} \, q$	F	F	T	T	T	T	F	F	T	T	T	...

The bottom two rows show the evaluation of the expressions $p \, \mathcal{U} \, q$ and $p \, \mathcal{W} \, q$ assuming that p remains true indefinitely and q remains false indefinitely. From s_0 to s_7, $p \, \mathcal{U} \, q$ and $p \, \mathcal{W} \, q$ hold in the same states. From s_8 on, however, $p \, \mathcal{U} \, q$ does not hold because q never holds thereafter, while $p \, \mathcal{W} \, q$ does hold because p always holds thereafter.

2.2.3 Duality

LADM [13] defines the dual P_D of a boolean expression P to be the expression constructed from P by interchanging occurrences of

> *true* and *false*,
> $\wedge$ and $\vee$,
> $\equiv$ and $\not\equiv$,
> $\Rightarrow$ and $\not\Leftarrow$, and
> $\Leftarrow$ and $\not\Rightarrow$.

This definition of P_D gives rise to the following metatheorem from LADM.

> (2.3) **Metatheorem Duality**
> (a) P is valid iff $\neg P_D$ is valid
> (b) $P \equiv Q$ is valid iff $P_D \equiv Q_D$ is valid

Linear temporal logic extends the definition of P_D for the temporal operators to include interchanging occurrences of

> $\bigcirc$ and $\bigcirc$ (self dual), and
> $\square$ and $\diamond$.

It is not the case that $\mathcal{W}$ is the dual of $\mathcal{U}$. Ben-Ari [3] defines the *release* operator $\mathcal{R}$ as

$$p \mathcal{R} q \equiv \neg(\neg p \mathcal{U} \neg q)$$

to be the dual of the binary operator $\mathcal{U}$. For simplicity, we avoid adding another operator to our system by restricting the LTL binary operators to the more common $\mathcal{U}$ and $\mathcal{W}$.

3 The Calculational Temporal System

This section presents an axiomatic deductive system of temporal logic and proves its theorems with the calculational logic $\mathcal{E}$ of Gries and Schneider's *A Logical Approach to Discrete Math* (LADM). [13] Theorems cited in a proof hint take two forms. A numbered reference enclosed in parentheses *without* a period is a reference to an axiom or a previously-proved theorem in this paper. A numbered reference enclosed in parentheses *with* a period is a reference to an axiom or a theorem from the propositional calculus in LADM. The numbering is consistent with that text with the chapter number followed by the equation number separated by the period. Additional theorems, either not included in LADM or included but not numbered, are indicated by a three-part number with two period separators. The terms "definition" and "axiom" are synonymous. The following exposition includes the theorems from LADM in the proof hints, except that theorems are omitted for (4.2) and (4.3) Monotonicity, as they are described in Section 2.1.

3.1 Next

The following two axioms define the *next* operator $\bigcirc$.

(1) **Axiom, Self-dual:** $\bigcirc \neg p \equiv \neg \bigcirc p$

(2) **Axiom, Distributivity of $\bigcirc$ over $\Rightarrow$:** $\bigcirc (p \Rightarrow q) \equiv \bigcirc p \Rightarrow \bigcirc q$

Self duality states that p not holding in the next state is equivalent to *next* p not holding in the current state. Distributivity states that p implies q in the next state is equivalent to *next* p implies *next* q in the current state. From this axiom, subsequent theorems prove that the *next* operator distributes over all the propositional binary operators.

Linearity follows from self-dual.

(3) **Linearity:** $\bigcirc p \equiv \neg \bigcirc \neg p$

Proof:

$$\begin{aligned}
& \bigcirc p \equiv \neg \bigcirc \neg p \\
= \quad & \langle (3.11)\ \neg p \equiv q \equiv p \equiv \neg q\ \text{with}\ p,q := \bigcirc \neg p, \bigcirc p \rangle \\
& \neg \bigcirc p \equiv \bigcirc \neg p \quad -(1)\ \text{Self-dual} \quad \blacksquare
\end{aligned}$$

The proof that $\bigcirc$ distributes over $\vee$ uses the distributivity of $\bigcirc$ over $\Rightarrow$. The proofs that it also distributes over $\wedge$ and $\equiv$ are similar.

(4) **Distributivity of $\bigcirc$ over $\vee$:** $\bigcirc (p \vee q) \equiv \bigcirc p \vee \bigcirc q$

Proof:

$$\begin{aligned}
& \bigcirc (p \vee q) \\
= \quad & \langle (3.59)\ \text{Implication},\ p \Rightarrow q \equiv \neg p \vee q \rangle \\
& \bigcirc (\neg p \Rightarrow q) \\
= \quad & \langle (2)\ \text{Distributivity of}\ \bigcirc\ \text{over} \Rightarrow \rangle \\
& \bigcirc \neg p \Rightarrow \bigcirc q \\
= \quad & \langle (3.59)\ \text{Implication},\ p \Rightarrow q \equiv \neg p \vee q\ \text{with}\ p,q := \bigcirc \neg p, \bigcirc q \rangle \\
& \neg \bigcirc \neg p \vee \bigcirc q \\
= \quad & \langle (3)\ \text{Linearity} \rangle \\
& \bigcirc p \vee \bigcirc q \quad \blacksquare
\end{aligned}$$

(5) **Distributivity of $\bigcirc$ over $\wedge$:** $\bigcirc (p \wedge q) \equiv \bigcirc p \wedge \bigcirc q$

Proof:

Pepperdine Papers on LTL

$$\circ (p \wedge q)$$
= ⟨(3.12) Double negation, $\neg \neg p \equiv p$, twice⟩
$$\circ (\neg \neg p \wedge \neg \neg q)$$
= ⟨(3.47b) De Morgan, $\neg (p \vee q) \equiv \neg p \wedge \neg q$⟩
$$\circ \neg (\neg p \vee \neg q)$$
= ⟨(1) Self-dual with $p := (\neg p \vee \neg q)$⟩
$$\neg \circ (\neg p \vee \neg q)$$
= ⟨(4) Distributivity of $\circ$ over $\vee$ with $p, q := \neg p, \neg q$⟩
$$\neg (\circ \neg p \vee \circ \neg q)$$
= ⟨(3.47b) De Morgan, $\neg (p \vee q) \equiv \neg p \wedge \neg q$⟩
$$\neg \circ \neg p \wedge \neg \circ \neg q$$
= ⟨(3) Linearity, twice⟩
$$\circ p \wedge \circ q \quad \blacksquare$$

(6) **Distributivity of $\circ$ over $\equiv$:** $\circ (p \equiv q) \equiv \circ p \equiv \circ q$

Proof:

$$\circ (p \equiv q)$$
= ⟨(3.80) Mutual implication, $(p \Rightarrow q) \wedge (q \Rightarrow p) \equiv (p \equiv q)$⟩
$$\circ ((p \Rightarrow q) \wedge (q \Rightarrow p))$$
= ⟨(5) Distributivity of $\circ$ over $\wedge$⟩
$$\circ (p \Rightarrow q) \wedge \circ (p \Rightarrow q)$$
= ⟨(2) Distributivity of $\circ$ over $\Rightarrow$⟩
$$(\circ p \Rightarrow \circ q) \wedge (\circ q \Rightarrow \circ p)$$
= ⟨(3.80) Mutual implication, $(p \Rightarrow q) \wedge (q \Rightarrow p) \equiv (p \equiv q)$⟩
$$\circ p \equiv \circ q \quad \blacksquare$$

Now, *true* holds in the next state, and *false* does not hold in the next state. In the calculational logic of LADM, *true* is theorem (3.4) and is equivalent to all other theorems. Theorem (7) shows that all propositional logic theorems hold at the next state and, by induction, hold in all states. The proof of (7) uses (3.28) Excluded middle. The proof of (8) uses (3.8) Definition of *false*, $false \equiv \neg true$.

(7) **Truth of $\circ$:** $\circ true \equiv true$

Proof:

$$\circ true$$
= ⟨(3.28) Excluded middle, $p \vee \neg p$⟩

$\bigcirc (p \vee \neg p)$
= ⟨(4) Distributivity of $\bigcirc$ over $\vee$⟩
$\bigcirc p \vee \bigcirc \neg p$
= ⟨(1) Self-dual⟩
$\bigcirc p \vee \neg \bigcirc p$
= ⟨(3.28) Excluded middle, $p \vee \neg p$ with $p := \bigcirc p$⟩
true ∎

(8) **Falsehood of $\bigcirc$:** $\bigcirc \textit{false} \equiv \textit{false}$

Proof:

$\bigcirc \textit{false} \equiv \textit{false}$
= ⟨(3.8) Definition of *false*, $\textit{false} \equiv \neg \textit{true}$⟩
$\bigcirc \neg \textit{true} \equiv \neg \textit{true}$
= ⟨(3.11) $\neg p \equiv q \equiv p \equiv \neg q$ with $p, q := \textit{true}, \bigcirc \neg \textit{true}$⟩
$\neg \bigcirc \neg \textit{true} \equiv \textit{true}$
= ⟨(3) Linearity⟩
$\bigcirc \textit{true} \equiv \textit{true}$ —(7) Truth of $\bigcirc$. ∎

3.2 Until

This system defines the *until* operator $\mathcal{U}$ with the following ten axioms. The first axiom, distributivity of $\bigcirc$ over $\mathcal{U}$, implies the distributivity of $\bigcirc$ over $\mathcal{W}$ as Section 3.9 shows. Thus, the *next* operator distributes over all binary operators, both propositional and temporal.

(9) **Axiom, Distributivity of $\bigcirc$ over $\mathcal{U}$:** $\bigcirc (p \mathcal{U} q) \equiv \bigcirc p \mathcal{U} \bigcirc q$

The second axiom, expansion of $\mathcal{U}$, makes the *until* operator different from most propositional binary operators. Its right operand has an existential characteristic and its left operand has a universal characteristic. Expansion states that $p \mathcal{U} q$ is true iff q is true in the current state, or p is true in the current state and $p \mathcal{U} q$ is true in the next state. Thus, q relates to the definition through disjunction, which is existential, while p relates through conjunction, which is universal. Consequently, the *until* operator is neither symmetric (*i.e.* commutative) nor associative.

(10) **Axiom, Expansion of $\mathcal{U}$:** $p \mathcal{U} q \equiv q \vee (p \wedge \bigcirc (p \mathcal{U} q))$

The third axiom states that *false* is the right zero of $\mathcal{U}$ and is not noted in other LTL deductive systems.

(11) **Axiom, Right zero of $\mathcal{U}$:** $p \mathcal{U} \textit{false} \equiv \textit{false}$

Pepperdine Papers on LTL

The next four axioms describe how the *until* operator distributes over conjunction and disjunction. Because $\mathcal{U}$ is not symmetric, this system requires separate axioms for left and right distributivity.

(12) **Axiom, Left distributivity of $\mathcal{U}$ over $\vee$:** $\quad p\,\mathcal{U}\,(q \vee r) \equiv p\,\mathcal{U}\,q \vee p\,\mathcal{U}\,r$

(13) **Axiom, Right distributivity of $\mathcal{U}$ over $\vee$:** $\quad p\,\mathcal{U}\,r \vee q\,\mathcal{U}\,r \Rightarrow (p \vee q)\,\mathcal{U}\,r$

(14) **Axiom, Left distributivity of $\mathcal{U}$ over $\wedge$:** $\quad p\,\mathcal{U}\,(q \wedge r) \Rightarrow p\,\mathcal{U}\,q \wedge p\,\mathcal{U}\,r$

(15) **Axiom, Right distributivity of $\mathcal{U}$ over $\wedge$:** $\quad (p \wedge q)\,\mathcal{U}\,r \equiv p\,\mathcal{U}\,r \wedge q\,\mathcal{U}\,r$

The *until* operator is not associative. The last three axioms describe the ordering property, the right ordering property under disjunction, and the right ordering property under conjunction, of $\mathcal{U}$. These theorems do not appear in other LTL systems. Other systems do, however, list their $\mathcal{W}$ versions, which in this system are (254), (250), and (251).

(16) **Axiom, $\mathcal{U}$ implication ordering:** $\quad p\,\mathcal{U}\,q \wedge \neg q\,\mathcal{U}\,r \Rightarrow p\,\mathcal{U}\,r$

(17) **Axiom, Right $\mathcal{U} \vee$ ordering:** $\quad p\,\mathcal{U}\,(q\,\mathcal{U}\,r) \Rightarrow (p \vee q)\,\mathcal{U}\,r$

(18) **Axiom, Right $\wedge \mathcal{U}$ ordering:** $\quad p\,\mathcal{U}\,(q \wedge r) \Rightarrow (p\,\mathcal{U}\,q)\,\mathcal{U}\,r$

Theorem (19) shows how $\mathcal{U}$ distributes over $\Rightarrow$ and is not listed in other deductive systems.

(19) **Right distributivity of $\mathcal{U}$ over $\Rightarrow$:** $\quad (p \Rightarrow q)\,\mathcal{U}\,r \Rightarrow (p\,\mathcal{U}\,r \Rightarrow q\,\mathcal{U}\,r)$

Proof:

$$(p \Rightarrow q)\,\mathcal{U}\,r \Rightarrow (p\,\mathcal{U}\,r \Rightarrow q\,\mathcal{U}\,r)$$
$$= \quad \langle (3.65) \text{ Shunting}, p \wedge q \Rightarrow r \equiv p \Rightarrow (q \Rightarrow r) \rangle$$
$$(p \Rightarrow q)\,\mathcal{U}\,r \wedge p\,\mathcal{U}\,r \Rightarrow q\,\mathcal{U}\,r$$

And now,

$$(p \Rightarrow q)\,\mathcal{U}\,r \wedge p\,\mathcal{U}\,r$$
$$= \quad \langle (15) \text{ Right distributivity of } \mathcal{U} \text{ over } \wedge \rangle$$
$$(p \wedge (p \Rightarrow q))\,\mathcal{U}\,r$$
$$= \quad \langle (3.66)\ p \wedge (p \Rightarrow q) \equiv p \wedge q \rangle$$
$$(p \wedge q)\,\mathcal{U}\,r$$
$$= \quad \langle (15) \text{ Right distributivity of } \mathcal{U} \text{ over } \wedge \rangle$$
$$p\,\mathcal{U}\,r \wedge q\,\mathcal{U}\,r$$
$$\Rightarrow \quad \langle (3.76b) \text{ Strengthening}, p \wedge q \Rightarrow p \rangle$$
$$q\,\mathcal{U}\,r \quad \blacksquare$$

Theorem (20) shows that *true* is a right zero of $\mathcal{U}$, which is unusual because axiom (11) shows that *false* is also a right zero of $\mathcal{U}$. Theorem (21) shows that *false* is the left identity of $\mathcal{U}$. Proofs of both use (10) Expansion of $\mathcal{U}$. Theorems (11), (20), and (21) cover three of the possibilities of constants *true* and *false* on either side of $\mathcal{U}$. None of these three theorems seem to appear in the temporal logic literature. The fourth possibility with *true* as the left argument is the basis of the definition of the *eventually* operator $\diamond$ in Section 3.3.

(20) **Right zero of** $\mathcal{U}$: $p\,\mathcal{U}\,true \equiv true$

Proof:

$$\begin{aligned}
& p\,\mathcal{U}\,true \\
=\;& \langle (10)\text{ Expansion of }\mathcal{U}\rangle \\
& true \vee (p \wedge \circ(p\,\mathcal{U}\,true)) \\
=\;& \langle (3.29)\text{ Zero of }\vee, p \vee true \equiv true\rangle \\
& true \quad \blacksquare
\end{aligned}$$

(21) **Left identity of** $\mathcal{U}$: $false\,\mathcal{U}\,q \equiv q$

Proof:

$$\begin{aligned}
& false\,\mathcal{U}\,q \\
=\;& \langle (10)\text{ Expansion of }\mathcal{U}\rangle \\
& q \vee (false \wedge \circ(false\,\mathcal{U}\,q)) \\
=\;& \langle (3.40)\text{ Zero of }\wedge, p \wedge false \equiv false\rangle \\
& q \vee false \\
=\;& \langle (3.30)\text{ Identity of }\vee, p \vee false \equiv p\rangle \\
& q \quad \blacksquare
\end{aligned}$$

Theorem (22) shows that the *until* operator is idempotent. Its proof uses (10) Expansion of $\mathcal{U}$ followed by (3.43b) Absorption. Theorem (23) is the *until* version of excluded middle, which is proved from (12) Left distributivity of $\mathcal{U}$ over $\vee$. The proof of (24) illustrates proof by Truth implication. The proof of (25) is similar. The proofs of (26) and (27) use (16) $\mathcal{U}$ implication ordering. The proofs of (28), (29), and (30) require only two steps.

(22) **Idempotency of** $\mathcal{U}$: $p\,\mathcal{U}\,p \equiv p$

Proof:

$$\begin{aligned}
& p\,\mathcal{U}\,p \\
=\;& \langle (10)\text{ Expansion of }\mathcal{U}\rangle \\
& p \vee (p \wedge \circ(p\,\mathcal{U}\,p)) \\
=\;& \langle (3.43b)\text{ Absorption}, p \vee (p \wedge q) \equiv p \text{ with } q := \circ(p\,\mathcal{U}\,p)\rangle \\
& p \quad \blacksquare
\end{aligned}$$

(23) $\mathcal{U}$ **excluded middle:** $p \mathcal{U} q \vee p \mathcal{U} \neg q$

Proof: (Ravi Mohan)

$\quad\quad p \mathcal{U} q \vee p \mathcal{U} \neg q$
$= \quad \langle(12)\text{ Left distributivity of } \mathcal{U} \text{ over } \vee\rangle$
$\quad\quad p \mathcal{U} (q \vee \neg q)$
$= \quad \langle(3.28)\text{ Excluded middle, } p \vee \neg p\rangle$
$\quad\quad p \mathcal{U} \text{ true}$
$= \quad \langle(20)\text{ Right zero of } \mathcal{U}\rangle$
$\quad\quad \text{true} \quad \blacksquare$

(24) $\neg p \mathcal{U} (q \mathcal{U} r) \wedge p \mathcal{U} r \Rightarrow q \mathcal{U} r$

Proof: The proof is by (4.7.1) Truth implication.

$\quad\quad \text{true}$
$\Rightarrow \quad \langle(17)\text{ Right } \mathcal{U} \vee \text{ ordering with } p := \neg p\rangle$
$\quad\quad \neg p \mathcal{U} (q \mathcal{U} r) \Rightarrow (\neg p \vee q) \mathcal{U} r$
$= \quad \langle(3.59)\text{ Implication, } p \Rightarrow q \equiv \neg p \vee q\rangle$
$\quad\quad \neg p \mathcal{U} (q \mathcal{U} r) \Rightarrow (p \Rightarrow q) \mathcal{U} r$
$\Rightarrow \quad \langle(19)\text{ Right distributivity of } \mathcal{U} \text{ over } \Rightarrow \text{ and (3.82a) Transitivity}\rangle$
$\quad\quad \neg p \mathcal{U} (q \mathcal{U} r) \Rightarrow (p \mathcal{U} r \Rightarrow q \mathcal{U} r)$
$= \quad \langle(3.65)\text{ Shunting, } p \wedge q \Rightarrow r \equiv p \Rightarrow (q \Rightarrow r)\rangle$
$\quad\quad \neg p \mathcal{U} (q \mathcal{U} r) \wedge p \mathcal{U} r \Rightarrow q \mathcal{U} r \quad \blacksquare$

(25) $p \mathcal{U} (\neg q \mathcal{U} r) \wedge q \mathcal{U} r \Rightarrow p \mathcal{U} r$

Proof: The proof is by (4.7.1) Truth implication.

$\quad\quad \text{true}$
$\Rightarrow \quad \langle(17)\text{ Right } \mathcal{U} \vee \text{ ordering with } q := \neg q\rangle$
$\quad\quad p \mathcal{U} (\neg q \mathcal{U} r) \Rightarrow (p \vee \neg q) \mathcal{U} r$
$= \quad \langle(3.59)\text{ Implication, } p \Rightarrow q \equiv \neg p \vee q\rangle$
$\quad\quad p \mathcal{U} (\neg q \mathcal{U} r) \Rightarrow (q \Rightarrow p) \mathcal{U} r$
$\Rightarrow \quad \langle(19)\text{ Right distributivity of } \mathcal{U} \text{ over } \Rightarrow \text{ and (3.82a) Transitivity}\rangle$
$\quad\quad p \mathcal{U} (\neg q \mathcal{U} r) \Rightarrow (q \mathcal{U} r \Rightarrow p \mathcal{U} r)$
$= \quad \langle(3.65)\text{ Shunting, } p \wedge q \Rightarrow r \equiv p \Rightarrow (q \Rightarrow r)\rangle$
$\quad\quad p \mathcal{U} (\neg q \mathcal{U} r) \wedge q \mathcal{U} r \Rightarrow p \mathcal{U} r \quad \blacksquare$

(26) $p \mathcal{U} q \wedge \neg q \mathcal{U} p \Rightarrow p$

Proof:

$p \mathcal{U} q \land \neg q \mathcal{U} p$
$\Rightarrow \quad \langle (16) \; \mathcal{U} \text{ implication ordering with } r := p \rangle$
$p \mathcal{U} p$
$= \quad \langle (22) \text{ Idempotency of } \mathcal{U} \rangle$
$p \quad \blacksquare$

(27) $\quad p \land \neg p \mathcal{U} q \Rightarrow q$

Proof: The proof is by (4.7.1) Truth implication.

$true$
$\Rightarrow \quad \langle (16) \; \mathcal{U} \text{ implication ordering with } p,q,r := false, p, q \rangle$
$false \mathcal{U} p \land \neg p \mathcal{U} q \Rightarrow false \mathcal{U} q$
$= \quad \langle (21) \text{ Left identity of } \mathcal{U}, \text{twice} \rangle$
$p \land \neg p \mathcal{U} q \Rightarrow q \quad \blacksquare$

(28) $\quad p \mathcal{U} q \Rightarrow p \lor q$

Proof:

$p \mathcal{U} q$
$= \quad \langle (10) \text{ Expansion of } \mathcal{U} \rangle$
$q \lor (p \land \circ (p \mathcal{U} q))$
$\Rightarrow \quad \langle (3.76d) \; p \lor (q \land r) \Rightarrow p \lor q \text{ with } p,q,r := q, p, \circ (p \mathcal{U} q) \rangle$
$p \lor q \quad \blacksquare$

(29) $\quad \mathcal{U}$ **insertion:** $\quad q \Rightarrow p \mathcal{U} q$

Proof:

$p \mathcal{U} q$
$= \quad \langle (10) \text{ Expansion of } \mathcal{U} \rangle$
$q \lor (p \land \circ (p \mathcal{U} q))$
$\Leftarrow \quad \langle (3.76a) \text{ Weakening, } p \Rightarrow p \lor q \rangle$
$q \quad \blacksquare$

(30) $\quad p \land q \Rightarrow p \mathcal{U} q$

Proof:

$p \land q$
$\Rightarrow \quad \langle (3.76b) \text{ Strengthening, } p \land q \Rightarrow p \rangle$
q
$\Rightarrow \quad \langle (29) \; \mathcal{U} \text{ insertion} \rangle$
$p \mathcal{U} q \quad \blacksquare$

Pepperdine Papers on LTL

This system has the following five absorption properties that do not seem to appear in the temporal logic literature. Most can be proved in one step.

(31) **Absorption:** $p \vee p \, \mathcal{U} \, q \equiv p \vee q$

Proof:

$$p \vee p \, \mathcal{U} \, q$$
$$= \quad \langle (10) \text{ Expansion of } \mathcal{U} \rangle$$
$$p \vee q \vee (p \wedge \circ (p \, \mathcal{U} \, q))$$
$$= \quad \langle (3.43b) \text{ Absorption}, p \vee (p \wedge q) \equiv p \rangle$$
$$p \vee q \quad \blacksquare$$

(32) **Absorption:** $p \, \mathcal{U} \, q \vee q \equiv p \, \mathcal{U} \, q$

Proof:

$$p \, \mathcal{U} \, q \vee q \equiv p \, \mathcal{U} \, q$$
$$= \quad \langle (3.57) \text{ Definition of implication}, p \Rightarrow q \equiv p \vee q \equiv q \rangle$$
$$q \Rightarrow p \, \mathcal{U} \, q \quad -(29) \; \mathcal{U} \text{ insertion} \quad \blacksquare$$

(33) **Absorption:** $p \, \mathcal{U} \, q \wedge q \equiv q$

Proof:

$$p \, \mathcal{U} \, q \wedge q \equiv q$$
$$= \quad \langle (3.60) \text{ Implication}, p \Rightarrow q \equiv p \wedge q \equiv p \rangle$$
$$q \Rightarrow p \, \mathcal{U} \, q \quad -(29) \; \mathcal{U} \text{ insertion} \quad \blacksquare$$

(34) **Absorption:** $p \, \mathcal{U} \, q \vee (p \wedge q) \equiv p \, \mathcal{U} \, q$

Proof:

$$p \, \mathcal{U} \, q \vee (p \wedge q) \equiv p \, \mathcal{U} \, q$$
$$= \quad \langle (3.57) \text{ Definition of implication}, p \Rightarrow q \equiv p \vee q \equiv q \rangle$$
$$p \wedge q \Rightarrow p \, \mathcal{U} \, q \quad -(30). \quad \blacksquare$$

(35) **Absorption:** $p \, \mathcal{U} \, q \wedge (p \vee q) \equiv p \, \mathcal{U} \, q$

Proof: (Ravi Mohan)

$$p \, \mathcal{U} \, q \wedge (p \vee q) \equiv p \, \mathcal{U} \, q$$
$$= \quad \langle (3.60) \text{ Implication}, p \Rightarrow q \equiv p \wedge q \equiv p \rangle$$
$$p \, \mathcal{U} \, q \Rightarrow p \vee q \quad -(28) \quad \blacksquare$$

All systems have the following two absorption theorems. Manna and Pnueli [21] refer to these as idempotence properties. This paper follows Schneider [24], which refers to them as absorption properties. The proof of each uses mutual implication.

(36) **Left absorption of** $\mathcal{U}$: $p \, \mathcal{U} \, (p \, \mathcal{U} \, q) \equiv p \, \mathcal{U} \, q$

Proof: The proof is by (4.7) Mutual implication.
The proof in the first direction follows.

$$\begin{aligned}
& p \, \mathcal{U} \, (p \, \mathcal{U} \, q) \\
\Rightarrow \quad & \langle (17) \text{ Right } \mathcal{U} \vee \text{ ordering} \rangle \\
& (p \vee p) \, \mathcal{U} \, q \\
= \quad & \langle (3.26) \text{ Idempotency of } \vee, p \vee p \equiv p \rangle \\
& p \, \mathcal{U} \, q
\end{aligned}$$

The proof in the second direction follows.

$$\begin{aligned}
& p \, \mathcal{U} \, q \\
\Rightarrow \quad & \langle (29) \, \mathcal{U} \text{ insertion with } q := p \, \mathcal{U} \, q \rangle \\
& p \, \mathcal{U} \, (p \, \mathcal{U} \, q) \quad \blacksquare
\end{aligned}$$

(37) **Right absorption of** $\mathcal{U}$: $(p \, \mathcal{U} \, q) \, \mathcal{U} \, q \equiv p \, \mathcal{U} \, q$

Proof: The proof is by (4.7) Mutual implication.
The proof in the first direction follows.

$$\begin{aligned}
& (p \, \mathcal{U} \, q) \, \mathcal{U} \, q \\
\Rightarrow \quad & \langle (28) \text{ with } p := p \, \mathcal{U} \, q \rangle \\
& p \, \mathcal{U} \, q \vee q \\
= \quad & \langle (32) \text{ Absorption} \rangle \\
& p \, \mathcal{U} \, q
\end{aligned}$$

The proof in the second direction follows.

$$\begin{aligned}
& (p \, \mathcal{U} \, q) \, \mathcal{U} \, q \\
\Leftarrow \quad & \langle (18) \text{ Right } \wedge \, \mathcal{U} \text{ ordering} \rangle \\
& p \, \mathcal{U} \, (q \wedge q) \\
= \quad & \langle (3.38) \text{ Idempotency of } \wedge, p \wedge p \equiv p \rangle \\
& p \, \mathcal{U} \, q \quad \blacksquare
\end{aligned}$$

3.3 Eventually

Eventually $\diamond$ is a special case of $\mathcal{U}$ when the left hand side is *true*. Equation (38) is its only defining axiom.

(38) **Definition of $\diamond$:** $\quad \diamond q \equiv true \,\mathcal{U}\, q$

Theorem (39) shows how the unary operator *eventually* absorbs into the binary operator *until*. Its proof uses (15) Right distributivity of $\mathcal{U}$ over $\wedge$. Theorems (40) and (41) show also that the binary operator *until* absorbs into the unary operator *eventually*. They are proved by mutual implication. Theorem (42) shows that $p \,\mathcal{U}\, q$ guarantees that q will eventually be *true*. Its proof uses (39). Theorems (43) and (44), Truth and Falsehood of $\diamond$, do not appear in other LTL systems. Their proofs are simple.

(39) **Absorption of $\diamond$ into $\mathcal{U}$:** $\quad p\,\mathcal{U}\,q \wedge \diamond q \equiv p\,\mathcal{U}\,q$

Proof:

$\quad p\,\mathcal{U}\,q \wedge \diamond q$
$= \quad \langle (38) \text{ Definition of } \diamond \rangle$
$\quad p\,\mathcal{U}\,q \wedge true\,\mathcal{U}\,q$
$= \quad \langle (15) \text{ Right distributivity of } \mathcal{U} \text{ over } \wedge \rangle$
$\quad (p \wedge true)\,\mathcal{U}\,q$
$= \quad \langle (3.39) \text{ Identity of } \wedge, p \wedge true \equiv p \rangle$
$\quad p\,\mathcal{U}\,q \quad \blacksquare$

(40) **Absorption of $\mathcal{U}$ into $\diamond$:** $\quad p\,\mathcal{U}\,q \vee \diamond q \equiv \diamond q$

Proof: The proof is by (4.7) Mutual implication.
The proof in the first direction follows.

$\quad p\,\mathcal{U}\,\diamond q \vee \diamond q$
$= \quad \langle (38) \text{ Definition of } \diamond \rangle$
$\quad p\,\mathcal{U}\,q \vee true\,\mathcal{U}\,q$
$\Rightarrow \quad \langle (13) \text{ Right distributivity of } \mathcal{U} \text{ over } \vee \rangle$
$\quad (p \vee true)\,\mathcal{U}\,q$
$= \quad \langle (3.29) \text{ Zero of } \vee, p \vee true \equiv true \rangle$
$\quad true\,\mathcal{U}\,q$
$= \quad \langle (38) \text{ Definition of } \diamond \rangle$
$\quad \diamond q$

The proof in the second direction follows.

$\Diamond q$
$\Rightarrow$ ⟨(3.76a) Weakening the consequent, $p \Rightarrow p \vee q$⟩
$p\,\mathcal{U}\,q \vee \Diamond q$ ∎

(41) **Absorption of $\mathcal{U}$ into $\Diamond$:** $p\,\mathcal{U}\,\Diamond q \equiv \Diamond q$

Proof: The proof is by (4.7) Mutual implication.
The proof in the first direction follows.

$p\,\mathcal{U}\,\Diamond q$
$=$ ⟨(38) Definition of $\Diamond$⟩
$p\,\mathcal{U}\,(true\,\mathcal{U}\,q)$
$\Rightarrow$ ⟨(17) Right $\mathcal{U} \vee$ ordering with $q, r := true, q$⟩
$(p \vee true)\,\mathcal{U}\,q$
$=$ ⟨(3.29) Zero of $\vee$, $p \vee true \equiv true$⟩
$true\,\mathcal{U}\,q$
$=$ ⟨(38) Definition of $\Diamond$⟩
$\Diamond q$

The proof in the second direction follows.

$\Diamond q$
$\Rightarrow$ ⟨(29) $\mathcal{U}$ insertion with $q := \Diamond q$⟩
$p\,\mathcal{U}\,\Diamond q$ ∎

(42) **Eventuality:** $p\,\mathcal{U}\,q \Rightarrow \Diamond q$

Proof:

$p\,\mathcal{U}\,q$
$=$ ⟨(39)Absorption of $\Diamond$ into $\mathcal{U}$⟩
$p\,\mathcal{U}\,q \wedge \Diamond q$
$\Rightarrow$ ⟨(3.76b) Strengthening the antecedent, $p \wedge q \Rightarrow p$⟩
$\Diamond q$ ∎

(43) **Truth of $\Diamond$:** $\Diamond true \equiv true$

Proof:

$\Diamond true$
$=$ ⟨(38) Definition of $\Diamond$⟩
$true\,\mathcal{U}\,true$
$=$ ⟨(22) Idempotency of $\mathcal{U}$⟩
$true$ ∎

(44) **Falsehood of $\diamond$:** $\diamond\mathit{false} \equiv \mathit{false}$

Proof:

$\quad\quad \diamond\mathit{false}$
$=\quad \langle\text{(38) Definition of } \diamond \rangle$
$\quad\quad \mathit{true}\,\mathcal{U}\,\mathit{false}$
$=\quad \langle\text{(11) Right zero of } \mathcal{U} \rangle$
$\quad\quad \mathit{false}\quad\blacksquare$

Expansion of $\diamond$, like expansion of $\mathcal{U}$, has two disjuncts. The first describes the current state and the second contains the operation in the next state. The expansion of $\diamond$ follows directly from the expansion of $\mathcal{U}$. The two weakening theorems (46) and (47) follow directly from expansion of $\diamond$.

(45) **Expansion of $\diamond$:** $\diamond p \equiv p \vee \circ \diamond p$

Proof:

$\quad\quad \diamond p$
$=\quad \langle\text{(38) Definition of } \diamond \rangle$
$\quad\quad \mathit{true}\,\mathcal{U}\,p$
$=\quad \langle\text{(10) Expansion of } \mathcal{U} \rangle$
$\quad\quad p \vee (\mathit{true} \wedge \circ (\mathit{true}\,\mathcal{U}\,p))$
$=\quad \langle\text{(38) Definition of } \diamond \rangle$
$\quad\quad p \vee (\mathit{true} \wedge \circ \diamond p)$
$=\quad \langle\text{(3.39) Identity of } \wedge, p \wedge \mathit{true} \equiv p \rangle$
$\quad\quad p \vee \circ \diamond p \quad\blacksquare$

(46) **Weakening of $\diamond$:** $p \Rightarrow \diamond p$

Proof:

$\quad\quad \diamond p$
$=\quad \langle\text{(45) Expansion of } \diamond \rangle$
$\quad\quad p \vee \circ \diamond p$
$\Leftarrow\quad \langle\text{(3.76a) Weakening the consequent, } p \Rightarrow p \vee q \rangle$
$\quad\quad p \quad\blacksquare$

(47) **Weakening of $\diamond$:** $\circ p \Rightarrow \diamond p$

Proof:

$\Diamond p$
$=\quad \langle (45)\text{ Expansion of } \Diamond\rangle$
$p \vee \circ \Diamond p$
$=\quad \langle (45)\text{ Expansion of } \Diamond\rangle$
$p \vee \circ (p \vee \circ \Diamond p)$
$=\quad \langle (4)\text{ Distributivity of } \circ \text{ over } \vee\rangle$
$p \vee \circ p \vee \circ \circ \Diamond p$
$\Leftarrow\quad \langle (3.76a)\text{ Weakening the consequent, } p \Rightarrow p \vee q\rangle$
$\circ p \quad \blacksquare$

The two absorption theorems (48) and (49) do not seem to appear in the temporal logic literature. The following four theorems (50), (51), (52), and (53) are common to all temporal logic systems.

(48) **Absorption of $\vee$ into $\Diamond$:** $\quad p \vee \Diamond p \equiv \Diamond p$

Proof: (Ravi Mohan)

$p \vee \Diamond p \equiv \Diamond p$
$=\quad \langle (3.57)\text{ Definition of Implication, } p \Rightarrow q \equiv p \vee q \equiv q\rangle$
$p \Rightarrow \Diamond p \quad -(46)\text{ Weakening of } \Diamond \quad \blacksquare$

(49) **Absorption of $\Diamond$ into $\wedge$:** $\quad \Diamond p \wedge p \equiv p$

Proof: (Ravi Mohan)

$\Diamond p \wedge p \equiv p$
$=\quad \langle (3.60)\text{ Implication, } p \Rightarrow q \equiv p \wedge q \equiv p\rangle$
$p \Rightarrow \Diamond p \quad -(46)\text{ Weakening of } \Diamond \quad \blacksquare$

(50) **Absorption of $\Diamond$:** $\quad \Diamond \Diamond p \equiv \Diamond p$

Proof:

$\Diamond \Diamond p$
$=\quad \langle (38)\text{ Definition of } \Diamond \text{ with } p := \Diamond p\rangle$
$true\ \mathcal{U}\ \Diamond p$
$=\quad \langle (41)\text{ Absorption of } \mathcal{U} \text{ into } \Diamond \text{ with } p,q := true, p\rangle$
$\Diamond p \quad \blacksquare$

(51) **Exchange of $\circ$ and $\Diamond$:** $\quad \circ \Diamond p \equiv \Diamond \circ p$

Proof:

$\circ \diamond p$
= ⟨(38) Definition of $\diamond$⟩
$\circ (true \, \mathcal{U} \, p)$
= ⟨(9) Distributivity of $\circ$ over $\mathcal{U}$⟩
$\circ \, true \, \mathcal{U} \circ p$
= ⟨(7) Truth of $\circ$⟩
$true \, \mathcal{U} \circ p$
= ⟨(38) Definition of $\diamond$⟩
$\diamond \circ p$ ∎

(52) **Distributivity of $\diamond$ over $\vee$:** $\diamond(p \vee q) \equiv \diamond p \vee \diamond q$

Proof:

$\diamond(p \vee q)$
= ⟨(38) Definition of $\diamond$⟩
$true \, \mathcal{U} \, (p \vee q)$
= ⟨(12) Left distributivity of $\mathcal{U}$ over $\vee$⟩
$true \, \mathcal{U} \, p \vee true \, \mathcal{U} \, q$
= ⟨(38) Definition of $\diamond$, twice⟩
$\diamond p \vee \diamond q$ ∎

(53) **Distributivity of $\diamond$ over $\wedge$:** $\diamond(p \wedge q) \Rightarrow \diamond p \wedge \diamond q$

Proof:

$\diamond(p \wedge q)$
= ⟨(38) Definition of $\diamond$⟩
$true \, \mathcal{U} \, (p \wedge q)$
$\Rightarrow$ ⟨(14) Left distributivity of $\mathcal{U}$ over $\wedge$⟩
$true \, \mathcal{U} \, p \wedge true \, \mathcal{U} \, q$
= ⟨(38) Definition of $\diamond$, twice⟩
$\diamond p \wedge \diamond q$ ∎

3.4 Always

This system defines the *always* operator $\square$ in terms of the *eventually* operator $\diamond$. $\square p$ is true when p is true in the current state and in all future states. The defining equation (54) states that p is always true iff it is not the case that $\neg p$ is eventually true. The induction axiom (55) and the induction theorem (56) do not appear in other LTL systems, either as axioms or theorems.

(54) **Definition of $\square$:** $\square p \equiv \neg \diamond \neg p$

(55) **Axiom, $\mathcal{U}$ induction:** $\square(p \Rightarrow (\circ p \wedge q) \vee r) \Rightarrow (p \Rightarrow \square q \vee q \mathcal{U} r)$

(56) **$\mathcal{U}$ induction:** $\square(p \Rightarrow \circ (p \vee q)) \Rightarrow (p \Rightarrow \square p \vee p \mathcal{U} q)$

Proof: (John Wiegley)   The proof is by (4.7.1) Truth implication.

$\quad\quad true$
$= \quad \langle (55)\ \mathcal{U}\ \text{induction with } p,q,r := p, \circ p, \circ q \rangle$
$\quad\quad \square(p \Rightarrow (\circ p \wedge \circ p) \vee \circ q) \Rightarrow (p \Rightarrow \square \circ p \vee \circ p\, \mathcal{U} \circ q)$
$= \quad \langle (3.38)\ \text{Idempotency of } \wedge,\ p \wedge p \equiv p\ \text{and (4) Distributivity of } \circ \text{ over } \vee \rangle$
$\quad\quad \square(p \Rightarrow \circ(p \vee q)) \Rightarrow (p \Rightarrow \square \circ p \vee \circ p\, \mathcal{U} \circ q)$
$= \quad \langle (3.39)\ \text{Identity of } \wedge,\ p \wedge true \equiv p\ \text{with } p := (p \Rightarrow \square \circ p \vee \circ p\, \mathcal{U} \circ q) \rangle$
$\quad\quad \square(p \Rightarrow \circ(p \vee q)) \Rightarrow (true \wedge (p \Rightarrow \square \circ p \vee \circ p\, \mathcal{U} \circ q))$
$= \quad \langle (3.71)\ \text{Reflexivity of } \Rightarrow,\ p \Rightarrow p \rangle$
$\quad\quad \square(p \Rightarrow \circ(p \vee q)) \Rightarrow ((p \Rightarrow p) \wedge (p \Rightarrow \square \circ p \vee \circ p\, \mathcal{U} \circ q))$
$= \quad \langle (3.63.1)\ \text{Distributivity of } \Rightarrow \text{ over } \wedge,\ p \Rightarrow q \wedge r \equiv (p \Rightarrow q) \wedge (p \Rightarrow r)$
$\quad\quad\quad \text{with } p,q,r := p, p, (\square \circ p \vee \circ p\, \mathcal{U} \circ q) \rangle$
$\quad\quad \square(p \Rightarrow \circ(p \vee q)) \Rightarrow (p \Rightarrow p \wedge (\square \circ p \vee \circ p\, \mathcal{U} \circ q))$
$= \quad \langle (3.46)\ \text{Distributivity of } \wedge \text{ over } \vee,\ p \wedge (q \vee r) \equiv (p \wedge q) \vee (p \wedge r)$
$\quad\quad\quad \text{with } p,q,r := p, \square \circ p, \circ p\, \mathcal{U} \circ q \rangle$
$\quad\quad \square(p \Rightarrow \circ(p \vee q)) \Rightarrow (p \Rightarrow (p \wedge \square \circ p) \vee (p \wedge \circ p\, \mathcal{U} \circ q))$
$= \quad \langle (9)\ \text{Distributivity of } \circ \text{ over } \mathcal{U}\ \text{ and Lemma A:}\ \circ \square p \equiv \square \circ p$
$\quad\quad\quad \text{and Lemma B:}\ \square p \equiv p \wedge \circ \square p \rangle$
$\quad\quad \square(p \Rightarrow \circ(p \vee q)) \Rightarrow (p \Rightarrow \square p \vee (p \wedge \circ(p\, \mathcal{U}\, q)))$
$\Rightarrow \quad \langle (3.76a)\ \text{Weakening},\ p \Rightarrow p \vee q\ \text{with } p := \square p \vee (p \wedge \circ(p\, \mathcal{U}\, q))$
$\quad\quad\quad \text{and (3.82a) Transitivity},\ (p \Rightarrow q) \wedge (q \Rightarrow r) \Rightarrow (p \Rightarrow r)$
$\quad\quad\quad \text{with } p,q,r := p, \square p \vee (p \wedge \circ(p\, \mathcal{U}\, q)), \square p \vee (p \wedge \circ(p\, \mathcal{U}\, q)) \vee q \rangle$
$\quad\quad \square(p \Rightarrow \circ(p \vee q)) \Rightarrow (p \Rightarrow \square p \vee (p \wedge \circ(p\, \mathcal{U}\, q)) \vee q)$
$= \quad \langle (3.24)\ \text{Symmetry of } \vee,\ p \vee q \equiv q \vee p\ \text{and (10) Expansion of } \mathcal{U} \rangle$
$\quad\quad \square(p \Rightarrow \circ(p \vee q)) \Rightarrow (p \Rightarrow \square p \vee p\, \mathcal{U}\, q) \quad \blacksquare$

Lemma A: $\circ \square p \equiv \square \circ p$

Proof:

$\quad\quad \circ \square p$
$= \quad \langle (54)\ \text{Definition of } \square \rangle$

$\quad\quad\quad \bigcirc \neg \Diamond \neg p$
$= \quad \langle (1) \text{ Self-dual} \rangle$
$\quad\quad\quad \neg \bigcirc \Diamond \neg p$
$= \quad \langle (51) \text{ Exchange of } \bigcirc \text{ and } \Diamond \text{ with } p := \neg p \rangle$
$\quad\quad\quad \neg \Diamond \bigcirc \neg p$
$= \quad \langle (1) \text{ Self-dual} \rangle$
$\quad\quad\quad \neg \Diamond \neg \bigcirc p$
$= \quad \langle (54) \text{ Definition of } \Box \rangle$
$\quad\quad\quad \Box \bigcirc p \quad \blacksquare$

Lemma B: $\Box p \equiv p \wedge \bigcirc \Box p$
Proof:

$\quad\quad\quad \Box p$
$= \quad \langle (54) \text{ Definition of } \Box \rangle$
$\quad\quad\quad \neg \Diamond \neg p$
$= \quad \langle (45) \text{ Expansion of } \Diamond \text{ with } p := \neg p \rangle$
$\quad\quad\quad \neg(\neg p \vee \bigcirc \Diamond \neg p)$
$= \quad \langle (3.47b) \text{ De Morgan}, \neg(p \vee q) \equiv \neg p \wedge \neg q \rangle$
$\quad\quad\quad \neg \neg p \wedge \neg \bigcirc \Diamond \neg p$
$= \quad \langle (3.12) \text{ Double negation}, \neg \neg p \equiv p \rangle$
$\quad\quad\quad p \wedge \neg \bigcirc \Diamond \neg p$
$= \quad \langle (1) \text{ Self dual} \rangle$
$\quad\quad\quad p \wedge \bigcirc \neg \Diamond \neg p$
$= \quad \langle (54) \text{ Definition of } \Box \rangle$
$\quad\quad\quad p \wedge \bigcirc \Box p \quad \blacksquare$

Theorem (57) $\Box$ induction is common to many systems. It follows from (56) with $q := false$. The negation of the dual of theorem (58) is equivalent to theorem (57). While the proof of (58) $\Diamond$ induction is by (2.3a) Metatheorem Duality, it is also possible to prove it without appeal to any metatheorems as the alternate proof shows.

(57) $\quad \Box$ **induction:** $\quad \Box(p \Rightarrow \bigcirc p) \Rightarrow (p \Rightarrow \Box p)$

Proof:

$\quad\quad\quad true$
$= \quad \langle (56) \text{ with } q := false \rangle$
$\quad\quad\quad \Box(p \Rightarrow \bigcirc(p \vee false)) \Rightarrow (p \Rightarrow \Box p \vee p \, \mathcal{U} \, false)$
$= \quad \langle (11) \text{ Right zero of } \mathcal{U} \rangle$
$\quad\quad\quad \Box(p \Rightarrow \bigcirc(p \vee false)) \Rightarrow (p \Rightarrow \Box p \vee false)$
$= \quad \langle (3.30) \text{ Identity of } \vee, \ p \vee false \equiv p, \text{twice} \rangle$
$\quad\quad\quad \Box(p \Rightarrow \bigcirc p) \Rightarrow (p \Rightarrow \Box p) \quad \blacksquare$

(58) $\diamond$ **induction:** $\square(\circ p \Rightarrow p) \Rightarrow (\diamond p \Rightarrow p)$

Proof: The proof is by (2.3a) Metatheorem Duality. The negation of the dual of (58) is

$$\neg(\diamond(\circ p \not\equiv p) \not\equiv (\square p \not\equiv p))$$
$= \quad \langle\text{Definition of} / \text{ and (3.58) Consequence}, p \Leftarrow q \equiv q \Rightarrow p\rangle$
$$\neg\neg((\square p \not\equiv p) \Rightarrow \diamond(\circ p \not\equiv p))$$
$= \quad \langle(3.12) \text{ Double negation}, \neg\neg p \equiv p\rangle$
$$(\square p \not\equiv p) \Rightarrow \diamond(\circ p \not\equiv p)$$
$= \quad \langle\text{Definition of} / \text{ and (3.58) Consequence}, p \Leftarrow q \equiv q \Rightarrow p, \text{twice}\rangle$
$$\neg(p \Rightarrow \square p) \Rightarrow \diamond\neg(p \Rightarrow \circ p)$$
$= \quad \langle(3.61) \text{ Contrapositive}, p \Rightarrow q \equiv \neg q \Rightarrow \neg p\rangle$
$$\neg\diamond\neg(p \Rightarrow \circ p) \Rightarrow (p \Rightarrow \square p)$$
$= \quad \langle(54) \text{ Definition of } \square \rangle$
$$\square(p \Rightarrow \circ p) \Rightarrow (p \Rightarrow \square p) \quad -(57) \square \text{ induction} \quad \blacksquare$$

Alternate Proof: (John Wiegley)

$\quad true$
$= \quad \langle(56)\ \mathcal{U} \text{ induction with } p, q := \neg p, false\rangle$
$\quad \square(\neg p \Rightarrow \circ(\neg p \vee false)) \Rightarrow (\neg p \Rightarrow \square\neg p \vee \neg p\ \mathcal{U}\ false)$
$= \quad \langle(3.30) \text{ Identity of } \vee, p \vee false \equiv p, \text{twice and (11) Right zero of } \mathcal{U} \rangle$
$\quad \square(\neg p \Rightarrow \circ \neg p) \Rightarrow (\neg p \Rightarrow \square \neg p)$
$= \quad \langle(3.61) \text{ Contrapositive}, p \Rightarrow q \equiv \neg q \Rightarrow \neg p, \text{twice}\rangle$
$\quad \square(\neg\circ\neg p \Rightarrow p) \Rightarrow (\neg\square\neg p \Rightarrow p)$
$= \quad \langle(3) \text{ Linearity and (59) } \diamond p \equiv \neg\square\neg p\rangle$
$\quad \square(\circ p \Rightarrow p) \Rightarrow (\diamond p \Rightarrow p) \quad \blacksquare$

Theorem (59) expresses $\diamond p$ in terms of $\square p$ and is the dual of the defining equation (54).

(59) $\diamond p \equiv \neg\square\neg p$

Proof:

$\quad \neg\square\neg p$
$= \quad \langle(54) \text{ Definition of } \square \rangle$
$\quad \neg\neg\diamond\neg\neg p$
$= \quad \langle(3.12) \text{ Double negation}, \neg\neg p \equiv p, \text{twice}\rangle$
$\quad \diamond p \quad \blacksquare$

Pepperdine Papers on LTL

Whereas the *next* operator $\circ$ is its own dual, the *eventually* operator $\diamond$ and the *always* operator $\square$ are mutually dual, as are $\diamond\square$ and $\square\diamond$. Each of the following four theorems can be proved directly without invoking (2.3) Metatheorem Duality. However, with P and Q defined as the expressions $P : \neg\square p$ and $Q : \diamond\neg p$, the dual expressions are $P_D : \neg\diamond p$ and $Q_D : \square\neg p$. Because theorem (60) is the expression $P \equiv Q$ and theorem (61) is the expression $P_D \equiv Q_D$, the validity of (61) can be asserted by invoking (2.3b) Metatheorem Duality with theorem (60). Similarly, the validity of (63) can be asserted by invoking duality with theorem (62).

(60) **Dual of $\square$:** $\neg\square p \equiv \diamond\neg p$

Proof:

$\qquad \neg\square p \equiv \diamond\neg p$
$= \quad \langle (3.11)\ \neg p \equiv q \equiv p \equiv \neg q\ \text{with}\ p,q := \square p, \diamond\neg p \rangle$
$\qquad \square p \equiv \neg\diamond\neg p \quad -(54)\ \text{Definition of}\ \square \quad \blacksquare$

(61) **Dual of $\diamond$:** $\neg\diamond p \equiv \square\neg p$

Proof:

$\qquad \square\neg p$
$= \quad \langle (54)\ \text{Definition of}\ \square \rangle$
$\qquad \neg\diamond\neg\neg p$
$= \quad \langle (3.12)\ \text{Double negation},\ \neg\neg p \equiv p \rangle$
$\qquad \neg\diamond p \quad \blacksquare$

(62) **Dual of $\diamond\square$:** $\neg\diamond\square p \equiv \square\diamond\neg p$

Proof:

$\qquad \neg\diamond\square p$
$= \quad \langle (61)\ \text{Dual of}\ \diamond\ \text{with}\ p := \square p \rangle$
$\qquad \square\neg\square p$
$= \quad \langle (60)\ \text{Dual of}\ \square \rangle$
$\qquad \square\diamond\neg p \quad \blacksquare$

(63) **Dual of $\square\diamond$:** $\neg\square\diamond p \equiv \diamond\square\neg p$

Proof:

$\qquad \neg\square\diamond p$
$= \quad \langle (60)\ \text{Dual of}\ \square\ \text{with}\ p := \diamond p \rangle$
$\qquad \diamond\neg\diamond p$
$= \quad \langle (61)\ \text{Dual of}\ \diamond \rangle$
$\qquad \diamond\square\neg p \quad \blacksquare$

Theorems (64) and (65), Truth and Falsehood of $\square$, do not appear in other LTL systems.

(64) **Truth of $\square$:** $\square\,true \equiv true$

Proof:

$\quad\square\,true$
$= \quad \langle(54)\text{ Definition of }\square\rangle$
$\quad\neg\Diamond\neg true$
$= \quad \langle(3.8)\text{ Definition of }false, false \equiv \neg true\rangle$
$\quad\neg\Diamond\,false$
$= \quad \langle(44)\text{ Falsehood of }\Diamond\rangle$
$\quad\neg false$
$= \quad \langle(3.13)\text{ Negation of }false, \neg false \equiv true\rangle$
$\quad true$ ∎

(65) **Falsehood of $\square$:** $\square\,false \equiv false$

Proof:

$\quad\square\,false \equiv false$
$= \quad \langle(3.8)\text{ Definition of }false, false \equiv \neg true, \text{twice}\rangle$
$\quad\square\neg true \equiv \neg true$
$= \quad \langle(3.11)\ \neg p \equiv q \equiv p \equiv \neg q\rangle$
$\quad\neg\square\neg\,true \equiv true$
$= \quad \langle(59)\ \Diamond p \equiv \neg\square\neg p\rangle$
$\quad\Diamond\,true \equiv true \quad -(43)\text{ Truth of }\Diamond$ ∎

While the expansions of $\mathcal{U}$ and $\Diamond$ have two disjuncts, the expansion of $\square$ has two conjuncts. As usual, the first describes the current state and the second contains the operation in the next state. Theorem (66) is the dual of (45) which can be used in its direct proof.

(66) **Expansion of $\square$:** $\square p \equiv p \wedge \bigcirc \square p$

Proof:

$\quad\square p$
$= \quad \langle(54)\text{ Definition of }\square\rangle$
$\quad\neg\Diamond\neg p$
$= \quad \langle(45)\text{ Expansion of }\Diamond \text{ with } p := \neg p\rangle$
$\quad\neg(\neg p \vee \bigcirc\Diamond\neg p)$
$= \quad \langle(3.47\text{b})\text{ De Morgan}, \neg(p \vee q) \equiv \neg p \wedge \neg q\rangle$

Pepperdine Papers on LTL

$$\neg\neg p \land \neg \bigcirc \Diamond \neg p$$
$= \quad \langle (3.12) \text{ Double negation, } \neg\neg p \equiv p \rangle$
$$p \land \neg \bigcirc \Diamond \neg p$$
$= \quad \langle (1) \text{ Self dual} \rangle$
$$p \land \bigcirc \neg \Diamond \neg p$$
$= \quad \langle (54) \text{ Definition of } \Box \rangle$
$$p \land \bigcirc \Box p \quad \blacksquare$$

(67) **Expansion of $\Box$:** $\Box p \equiv p \land \bigcirc p \land \bigcirc \Box p$

Proof:

$$p \land \bigcirc p \land \bigcirc \Box p$$
$= \quad \langle (5) \text{ Distributivity of } \bigcirc \text{ over } \land \rangle$
$$p \land \bigcirc (p \land \Box p)$$
$= \quad \langle (66) \text{ Expansion of } \Box \rangle$
$$p \land \bigcirc (p \land p \land \bigcirc \Box p)$$
$= \quad \langle (3.38) \text{ Idempotency of } \land, p \land p \equiv p \rangle$
$$p \land \bigcirc (p \land \bigcirc \Box p)$$
$= \quad \langle (66) \text{ Expansion of } \Box \rangle$
$$p \land \bigcirc \Box p$$
$= \quad \langle (66) \text{ Expansion of } \Box \rangle$
$$\Box p \quad \blacksquare$$

Theorem (68) absorption of $\land$ into $\Box$, is the dual of (48) the absorption of $\lor$ into $\Diamond$, while (69) absorption of $\Box$ into $\lor$ is the dual of (49) the absorption of $\Diamond$ into $\land$. As with (48) and (49), neither seem to appear in the temporal logic literature.

Conjunction $\land$ and the *always* operator $\Box$ are both universal, while disjunction $\lor$ and the *eventually* operator $\Diamond$ are both existential. When the left side of the equivalence contains both existential (or both universal) operators as in (48) and (68), the right side retains the same type of unary operator. When existential and universal operators are mixed on the left side, as in (49) and (69), the equivalence is just a statement about p at the current time.

(68) **Absorption of $\land$ into $\Box$:** $p \land \Box p \equiv \Box p$

Proof:

$$p \land \Box p$$
$= \quad \langle (66) \text{ Expansion of } \Box \rangle$
$$p \land p \land \bigcirc \Box p$$
$= \quad \langle (3.38) \text{ Idempotency of } \land, p \land p \equiv p \rangle$
$$p \land \bigcirc \Box p$$
$= \quad \langle (66) \text{ Expansion of } \Box \rangle$
$$\Box p \quad \blacksquare$$

(69) **Absorption of $\Box$ into $\lor$:** $\Box p \lor p \equiv p$

Proof:

$\quad \Box p \lor p$
$= \quad \langle$(66) Expansion of $\Box\rangle$
$\quad (p \land \circ \Box p) \lor p$
$= \quad \langle$(3.43b) Absorption, $p \lor (p \land q) \equiv p\rangle$
$\quad p \quad \blacksquare$

The absorption of $\Diamond$ into $\Box$ and of $\Box$ into $\Diamond$ do not appear in the LTL systems we survey. Their proofs are straightforward applications of the previous absorption theorems.

(70) **Absorption of $\Diamond$ into $\Box$:** $\Diamond p \land \Box p \equiv \Box p$

Proof:

$\quad \Diamond p \land \Box p$
$= \quad \langle$(68) Absorption of $\land$ into $\Box\rangle$
$\quad \Diamond p \land p \land \Box p$
$= \quad \langle$(49) Absorption of $\Diamond$ into $\land\rangle$
$\quad p \land \Box p$
$= \quad \langle$(68) Absorption of $\land$ into $\Box\rangle$
$\quad \Box p \quad \blacksquare$

(71) **Absorption of $\Box$ into $\Diamond$:** $\Box p \lor \Diamond p \equiv \Diamond p$

Proof:

$\quad \Box p \lor \Diamond p$
$= \quad \langle$(48) Absorption of $\lor$ into $\Diamond\rangle$
$\quad \Box p \lor p \lor \Diamond p$
$= \quad \langle$(69) Absorption of $\Box$ into $\lor\rangle$
$\quad p \lor \Diamond p$
$= \quad \langle$(48) Absorption of $\lor$ into $\Diamond\rangle$
$\quad \Diamond p \quad \blacksquare$

Theorem (72) absorption of $\Box$ is the dual of (50) the absorption of $\Diamond$, and (73) the exchange of $\circ$ and $\Box$ is the dual of (51) the exchange of $\circ$ and $\Diamond$.

(72) **Absorption of $\Box$:** $\Box\Box p \equiv \Box p$

Proof: (Kyle Sundman)

$\square\square p$

= ⟨(54) Definition of $\square$ with $p := \square p$⟩

$\neg\Diamond\neg\square p$

= ⟨(60) Dual of $\square$⟩

$\neg\Diamond\Diamond\neg p$

= ⟨(50) Absorption of $\Diamond$⟩

$\neg\Diamond\neg p$

= ⟨(54) Definition of $\square$⟩

$\square p$ ■

(73) **Exchange of $\circ$ and $\square$:** $\circ\square p \equiv \square\circ p$

Proof:

$\circ\square p$

= ⟨(54) Definition of $\square$⟩

$\circ\neg\Diamond\neg p$

= ⟨(1) Self-dual⟩

$\neg\circ\Diamond\neg p$

= ⟨(51) Exchange of $\circ$ and $\Diamond$ with $p := \neg p$⟩

$\neg\Diamond\circ\neg p$

= ⟨(1) Self-dual⟩

$\neg\Diamond\neg\circ p$

= ⟨(54) Definition of $\square$⟩

$\square\circ p$ ■

Theorem (74) does not appear in other LTL systems. Theorem (75) states that if p holds and eventually $\neg p$ holds, there must eventually be a state where p holds in that state and it does not hold in the next state. Theorem (75) is the contrapositive of, and therefore equivalent to, (57) $\square$ induction.

(74) $p \Rightarrow \square p \equiv p \Rightarrow \circ\square p$

Proof: (Ravi Mohan)

$p \Rightarrow \square p$

= ⟨(66) Expansion of $\square$⟩

$p \Rightarrow p \wedge \circ\square p$

= ⟨(3.63.1) Distributivity of $\Rightarrow$ over $\wedge$, $p \Rightarrow q \wedge r \equiv (p \Rightarrow q) \wedge (p \Rightarrow r)$⟩

$(p \Rightarrow p) \wedge (p \Rightarrow \circ\square p)$

= ⟨(3.71) Reflexivity of $\Rightarrow$, $p \Rightarrow p$⟩

$true \wedge (p \Rightarrow \circ\square p)$

= ⟨(3.39) Identity of $\wedge$, $p \wedge true \equiv p$⟩

$p \Rightarrow \circ\square p$ ■

(75) $\quad p \wedge \Diamond \neg p \Rightarrow \Diamond (p \wedge \circ \neg p)$

Proof: By (3.61) Contrapositive, prove $\neg \Diamond (p \wedge \circ \neg p) \Rightarrow \neg (p \wedge \Diamond \neg p)$

$\qquad \neg \Diamond (p \wedge \circ \neg p)$
$= \quad \langle (61) \text{ Dual of } \Diamond \rangle$
$\qquad \Box \neg (p \wedge \circ \neg p)$
$= \quad \langle (3.47a) \text{ De Morgan}, \neg (p \wedge q) \equiv \neg p \vee \neg q \rangle$
$\qquad \Box (\neg p \vee \neg \circ \neg p)$
$= \quad \langle (3) \text{ Linearity} \rangle$
$\qquad \Box (\neg p \vee \circ p)$
$= \quad \langle (3.59) \text{ Implication}, p \Rightarrow q \equiv \neg p \vee q \rangle$
$\qquad \Box (p \Rightarrow \circ p)$
$\Rightarrow \quad \langle (57) \Box \text{ induction} \rangle$
$\qquad p \Rightarrow \Box p$
$= \quad \langle (3.59) \text{ Implication}, p \Rightarrow q \equiv \neg p \vee q \rangle$
$\qquad \neg p \vee \Box p$
$= \quad \langle (54) \text{ Definition of } \Box \rangle$
$\qquad \neg p \vee \neg \Diamond \neg p$
$= \quad \langle (3.47a) \text{ De Morgan}, \neg (p \wedge q) \equiv \neg p \vee \neg q \rangle$
$\qquad \neg (p \wedge \Diamond \neg p)$ ∎

The following four strengthening theorems for $\Box$ are common and contrast with the weakening theorems (46) and (47) for $\Diamond$. Theorem (80) is listed in one other system. Theorem (81) is unique to this one.

(76) **Strengthening of $\Box$:** $\quad \Box p \Rightarrow p$

Proof:

$\qquad \Box p$
$= \quad \langle (66) \text{ Expansion of } \Box \rangle$
$\qquad p \wedge \circ \Box p$
$\Rightarrow \quad \langle (3.76b) \text{ Strengthening}, p \wedge q \Rightarrow p \rangle$
$\qquad p$ ∎

(77) **Strengthening of $\Box$:** $\quad \Box p \Rightarrow \Diamond p$

Proof:

$\qquad \Box p$
$\Rightarrow \quad \langle (76) \text{ Strengthening of } \Box \rangle$
$\qquad p$
$\Rightarrow \quad \langle (46) \text{ Weakening of } \Diamond \rangle$
$\qquad \Diamond p$ ∎

(78) **Strengthening of $\square$:** $\square p \Rightarrow \bigcirc p$

Proof:

$\quad\quad \square p$
$= \quad \langle(67)\text{ Expansion of }\square\rangle$
$\quad\quad p \wedge \bigcirc p \wedge \bigcirc \square p$
$\Rightarrow \quad \langle(3.76\text{b})\text{ Strengthening, }p \wedge q \Rightarrow p\rangle$
$\quad\quad \bigcirc p \quad\blacksquare$

(79) **Strengthening of $\square$:** $\square p \Rightarrow \bigcirc \square p$

Proof:

$\quad\quad \square p$
$= \quad \langle(66)\text{ Expansion of }\square\rangle$
$\quad\quad p \wedge \bigcirc \square p$
$\Rightarrow \quad \langle(3.76\text{b})\text{ Strengthening, }p \wedge q \Rightarrow p\rangle$
$\quad\quad \bigcirc \square p \quad\blacksquare$

(80) **$\bigcirc$ generalization:** $\square p \Rightarrow \square \bigcirc p$

Proof:

$\quad\quad \square p \Rightarrow \square \bigcirc p$
$= \quad \langle(73)\text{ Exchange of }\bigcirc\text{ and }\square\rangle$
$\quad\quad \square p \Rightarrow \bigcirc \square p \quad -(79)\text{ Strengthening of }\square \quad\blacksquare$

(81) $\square p \Rightarrow \neg(q\,\mathcal{U}\,\neg p)$

Proof:

$\quad\quad \square p \Rightarrow \neg(q\,\mathcal{U}\,\neg p)$
$= \quad \langle(3.61)\text{ Contrapositive, }p \Rightarrow q \equiv \neg q \Rightarrow \neg p\rangle$
$\quad\quad q\,\mathcal{U}\,\neg p \Rightarrow \neg \square p$
$= \quad \langle(60)\text{ Dual of }\square\rangle$
$\quad\quad q\,\mathcal{U}\,\neg p \Rightarrow \Diamond \neg p \quad -(42)\text{ Eventuality with }p,q := q,\neg p \quad\blacksquare$

3.5 Temporal Deduction

The following deduction proof technique is a metatheorem in LADM.

(4.4) **Deduction (assume conjuncts of antecedent):**
To prove $P_1 \wedge P_2 \Rightarrow Q$, assume P_1 and P_2, and prove Q.
You cannot use textual substitution in P_1 or P_2.

Corresponding to the deduction metatheorem of the propositional calculus is the following temporal deduction metatheorem.

(82) **Temporal deduction:**

To prove $\Box P_1 \wedge \Box P_2 \Rightarrow Q$, assume P_1 and P_2, and prove Q.
You cannot use textual substitution in P_1 or P_2.

Temporal deduction is Theorem (2.1.6) of Kröger and Merz [20], who also give the justification. Note that if you assume P in a step of an LTL proof of Q, you have *not* proved that $P \Rightarrow Q$, but rather that $\Box P \Rightarrow Q$.

The temporal deduction metatheorem says $\Box P_1 \wedge \Box P_2 \Rightarrow Q$ is a theorem if you can assume P_1 and P_2, and prove Q. You cannot use textual substitution in P_1 or P_2.

But, if you can assume P_1 and P_2, and prove Q, then $P_1 \wedge P_2 \Rightarrow Q$ is a theorem. This will be taken as an assumption in the following proof of (82).

Proof: By (4.9) Proof by contradiction. To prove $\Box P_1 \wedge \Box P_2 \Rightarrow Q$ show that

$\neg(\Box P_1 \wedge \Box P_2 \Rightarrow Q) \Rightarrow \mathit{false}$

$\quad \neg(\Box P_1 \wedge \Box P_2 \Rightarrow Q)$
$= \quad \langle\text{(68) Absorption of } \wedge \text{ into } \Box, \text{ twice}\rangle$
$\quad \neg((P_1 \wedge \Box P_1) \wedge (P_2 \wedge \Box P_2) \Rightarrow Q)$
$= \quad \langle\text{(3.36) Symmetry of } \wedge, \, p \wedge q \equiv q \wedge p\rangle$
$\quad \neg((P_1 \wedge P_2) \wedge (\Box P_1 \wedge \Box P_2) \Rightarrow Q)$
$= \quad \langle\text{(3.78.1) } (p \Rightarrow r) \vee (q \Rightarrow r) \equiv p \wedge q \Rightarrow r$
$\quad\quad \text{with } p,q,r := (P_1 \wedge P_2), (\Box P_1 \wedge \Box P_2), Q\rangle$
$\quad \neg((P_1 \wedge P_2 \Rightarrow Q) \vee (\Box P_1 \wedge \Box P_2 \Rightarrow Q))$
$= \quad \langle\text{Assume: } P_1 \wedge P_2 \Rightarrow Q \text{ is a theorem}\rangle$
$\quad \neg(\mathit{true} \vee (\Box P_1 \wedge \Box P_2 \Rightarrow Q))$
$= \quad \langle\text{(3.29) Zero of } \vee, \, p \vee \mathit{true} \equiv \mathit{true}\rangle$
$\quad \neg \mathit{true}$
$= \quad \langle\text{(3.8) Definition of } \mathit{false}, \, \mathit{false} \equiv \neg \mathit{true}\rangle$
$\quad \mathit{false}$ ∎

3.6 Always, Continued

The following two theorems, (83) and (84), do not appear in other LTL systems. However, they are required for the proof of later theorems that are included in other systems. In particular, the proofs of (85) and (86) depend on (83) Distributivity of $\wedge$ over $\mathcal{U}$. The proof of (83) illustrates temporal deduction in a calculational proof. The proof of (84) is similar.

(83) **Distributivity of $\wedge$ over $\mathcal{U}$:** $\quad \Box p \wedge q \, \mathcal{U} \, r \Rightarrow (p \wedge q) \, \mathcal{U} \, (p \wedge r)$

Proof: The proof is by (82) Temporal deduction.

$$\begin{aligned}
& \Box p \wedge q \, \mathcal{U} \, r \Rightarrow (p \wedge q) \, \mathcal{U} \, (p \wedge r) \\
={} & \langle (3.65) \text{ Shunting}, p \wedge q \Rightarrow r \equiv p \Rightarrow (q \Rightarrow r) \rangle \\
& \Box p \Rightarrow (q \, \mathcal{U} \, r \Rightarrow (p \wedge q) \, \mathcal{U} \, (p \wedge r))
\end{aligned}$$

And now,

$$\begin{aligned}
& q \, \mathcal{U} \, r \Rightarrow (p \wedge q) \, \mathcal{U} \, (p \wedge r) \\
={} & \langle \text{Assume antecedent } p \rangle \\
& q \, \mathcal{U} \, r \Rightarrow (true \wedge q) \, \mathcal{U} \, (true \wedge r) \\
={} & \langle (3.39) \text{ Identity of } \wedge, p \wedge true \equiv p, \text{ twice} \rangle \\
& q \, \mathcal{U} \, r \Rightarrow q \, \mathcal{U} \, r \quad -(3.71) \text{ Reflexivity of } \Rightarrow, p \Rightarrow p \quad \blacksquare
\end{aligned}$$

(84) **$\mathcal{U}$ implication:** $\quad \Box p \wedge \Diamond q \Rightarrow p \, \mathcal{U} \, q$

Proof: The proof is by (82) Temporal deduction.

$$\begin{aligned}
& \Box p \wedge \Diamond q \Rightarrow p \, \mathcal{U} \, q \\
={} & \langle (3.65) \text{ Shunting}, p \wedge q \Rightarrow r \equiv p \Rightarrow (q \Rightarrow r) \rangle \\
& \Box p \Rightarrow (\Diamond q \Rightarrow p \, \mathcal{U} \, q)
\end{aligned}$$

And now,

$$\begin{aligned}
& \Diamond q \Rightarrow p \, \mathcal{U} \, q \\
={} & \langle \text{Assume antecedent } p \rangle \\
& \Diamond q \Rightarrow true \, \mathcal{U} \, q \\
={} & \langle (38) \text{ Definition of } \Diamond \rangle \\
& \Diamond q \Rightarrow \Diamond q \quad -(3.71) \text{ Reflexivity of } \Rightarrow, p \Rightarrow p \quad \blacksquare
\end{aligned}$$

(85) **Right monotonicity of $\mathcal{U}$:** $\quad \Box(p \Rightarrow q) \Rightarrow (r \, \mathcal{U} \, p \Rightarrow r \, \mathcal{U} \, q)$

Proof:

$$\Box(p \Rightarrow q) \Rightarrow (r \,\mathcal{U}\, p \Rightarrow r \,\mathcal{U}\, q)$$
$= \quad \langle(3.65)\text{ Shunting}, p \wedge q \Rightarrow r \equiv p \Rightarrow (q \Rightarrow r)\rangle$
$$\Box(p \Rightarrow q) \wedge r \,\mathcal{U}\, p \Rightarrow r \,\mathcal{U}\, q$$

And now,

$$\Box(p \Rightarrow q) \wedge r \,\mathcal{U}\, p$$
$\Rightarrow \quad \langle(83)\text{ Distributivity of } \wedge \text{ over } \mathcal{U}\rangle$
$$((p \Rightarrow q) \wedge r) \,\mathcal{U}\, ((p \Rightarrow q) \wedge p)$$
$= \quad \langle(3.66)\; p \wedge (p \Rightarrow q) \equiv p \wedge q\rangle$
$$((p \Rightarrow q) \wedge r) \,\mathcal{U}\, (p \wedge q)$$
$= \quad \langle(15)\text{ Right Distributivity of } \mathcal{U} \text{ over } \wedge\rangle$
$$(p \Rightarrow q) \,\mathcal{U}\, (p \wedge q) \wedge r \,\mathcal{U}\, (p \wedge q)$$
$\Rightarrow \quad \langle(14)\text{ Left Distributivity of } \mathcal{U} \text{ over } \wedge \text{ and } (4.3)\text{ Monotonicity of } \wedge\rangle$
$$(p \Rightarrow q) \,\mathcal{U}\, (p \wedge q) \wedge r \,\mathcal{U}\, p \wedge r \,\mathcal{U}\, q$$
$\Rightarrow \quad \langle(3.76\text{b})\text{ Strengthening}, p \wedge q \Rightarrow p\rangle$
$$r \,\mathcal{U}\, q \quad \blacksquare$$

(86) **Left monotonicity of** $\mathcal{U}$: $\quad \Box(p \Rightarrow q) \Rightarrow (p \,\mathcal{U}\, r \Rightarrow q \,\mathcal{U}\, r)$

Proof:

$$\Box(p \Rightarrow q) \Rightarrow (p \,\mathcal{U}\, r \Rightarrow q \,\mathcal{U}\, r)$$
$= \quad \langle(3.65)\text{ Shunting}, p \wedge q \Rightarrow r \equiv p \Rightarrow (q \Rightarrow r)\rangle$
$$\Box(p \Rightarrow q) \wedge p \,\mathcal{U}\, r \Rightarrow q \,\mathcal{U}\, r$$

And now,

$$\Box(p \Rightarrow q) \wedge p \,\mathcal{U}\, r$$
$\Rightarrow \quad \langle(83)\text{ Distributivity of } \wedge \text{ over } \mathcal{U}\rangle$
$$((p \Rightarrow q) \wedge p) \,\mathcal{U}\, ((p \Rightarrow q) \wedge r)$$
$= \quad \langle(3.66)\; p \wedge (p \Rightarrow q) \equiv p \wedge q\rangle$
$$(p \wedge q) \,\mathcal{U}\, ((p \Rightarrow q) \wedge r)$$
$= \quad \langle(15)\text{ Right Distributivity of } \mathcal{U} \text{ over } \wedge\rangle$
$$p \,\mathcal{U}\, ((p \Rightarrow q) \wedge r) \wedge q \,\mathcal{U}\, ((p \Rightarrow q) \wedge r)$$
$\Rightarrow \quad \langle(14)\text{ Left Distributivity of } \mathcal{U} \text{ over } \wedge \text{ and } (4.3)\text{ Monotonicity of } \wedge\rangle$
$$p \,\mathcal{U}\, ((p \Rightarrow q) \wedge r) \wedge q \,\mathcal{U}\, (p \Rightarrow q) \wedge q \,\mathcal{U}\, r$$
$\Rightarrow \quad \langle(3.76\text{b})\text{ Strengthening}, p \wedge q \Rightarrow p\rangle$
$$q \,\mathcal{U}\, r \quad \blacksquare$$

Pepperdine Papers on LTL

Theorem (87) states that if it is always the case that p is false then it is not the case that p is always true, but not the converse. Suppose, for example, that p continually oscillates between true and false over time. Then, the consequent of (87) is true, but the antecedent is false. Theorem (88) shows how $\Diamond$ distributes over $\wedge$.

(87) **Distributivity of $\neg$ over $\Box$:** $\quad \Box \neg p \Rightarrow \neg \Box p$

Proof:

$$\Box \neg p$$
$$\Rightarrow \quad \langle (77) \text{ Strengthening of } \Box \rangle$$
$$\Diamond \neg p$$
$$= \quad \langle (60) \text{ Dual of } \Box \rangle$$
$$\neg \Box p \quad \blacksquare$$

(88) **Distributivity of $\Diamond$ over $\wedge$:** $\quad \Box p \wedge \Diamond q \Rightarrow \Diamond (p \wedge q)$

Proof: (Ravi Mohan)

$$\Box p \wedge \Diamond q$$
$$= \quad \langle (38) \text{ Definition of } \Diamond \rangle$$
$$\Box p \wedge true \, \mathcal{U} \, q$$
$$\Rightarrow \quad \langle (83) \text{ Distributivity of } \wedge \text{ over } \mathcal{U} \text{ with } q, r := true, q \rangle$$
$$(p \wedge true) \, \mathcal{U} \, (p \wedge q)$$
$$\Rightarrow \quad \langle (42) \text{ Eventuality with } p, q := p \wedge true, p \wedge q \rangle$$
$$\Diamond (p \wedge q) \quad \blacksquare$$

Theorems (89), (90), and (91) are the linear temporal versions of the excluded middle axiom of propositional logic, (3.28) $p \vee \neg p$. Theorems (92), (93), and (94) are the linear temporal versions of the contradiction theorem of propositional logic, (3.42) $p \wedge \neg p \equiv false$. Theorems (95), (96), (97), and (98) are variations. These theorems are obvious, and their proofs are simple, which is perhaps why they do not appear in other LTL systems.

(89) **$\Diamond$ excluded middle:** $\quad \Diamond p \vee \Box \neg p$

Proof:

$$\Diamond p \vee \Box \neg p$$
$$= \quad \langle (61) \text{ Dual of } \Diamond \rangle$$
$$\Diamond p \vee \neg \Diamond p \quad -(3.28) \text{ Excluded middle, } p \vee \neg p \text{ with } p := \Diamond p \quad \blacksquare$$

(90) **$\Box$ excluded middle:** $\quad \Box p \vee \Diamond \neg p$

Proof: (Ray McIntyre)

$\square p \vee \Diamond \neg p$
$= \quad \langle (60) \text{ Dual of } \square \rangle$
$\square p \vee \neg \square p \quad -(3.28) \text{ Excluded middle, } p \vee \neg p \text{ with } p := \square p \quad \blacksquare$

(91) **Temporal excluded middle:** $\Diamond p \vee \Diamond \neg p$

Proof:

$\Diamond p \vee \Diamond \neg p$
$= \quad \langle (60) \text{ Dual of } \square \rangle$
$\Diamond p \vee \neg \square p$
$= \quad \langle (3.59) \text{ Implication, } p \Rightarrow q \equiv \neg p \vee q \rangle$
$\square p \Rightarrow \Diamond p \quad -(77) \text{ Strengthening of } \square \quad \blacksquare$

(92) $\Diamond$ **contradiction:** $\Diamond p \wedge \square \neg p \equiv \textit{false}$

Proof: (Ray McIntyre)

$\Diamond p \wedge \square \neg p \equiv \textit{false}$
$= \quad \langle (61) \text{ Dual of } \Diamond \rangle$
$\Diamond p \wedge \neg \Diamond p \equiv \textit{false} \quad -(3.42) \text{ Contradiction, } p \wedge \neg p \equiv \textit{false} \text{ with } p := \Diamond p \quad \blacksquare$

(93) $\square$ **contradiction:** $\square p \wedge \Diamond \neg p \equiv \textit{false}$

Proof: (Ray McIntyre)

$\square p \wedge \Diamond \neg p \equiv \textit{false}$
$= \quad \langle (60) \text{ Dual of } \square \rangle$
$\square p \wedge \neg \square p \equiv \textit{false} \quad -(3.42) \text{ Contradiction, } p \wedge \neg p \equiv \textit{false} \text{ with } p := \square p \quad \blacksquare$

(94) **Temporal contradiction:** $\square p \wedge \square \neg p \equiv \textit{false}$

Proof:

$\square p \wedge \square \neg p \equiv \textit{false}$
$= \quad \langle (3.15) \neg p \equiv p \equiv \textit{false} \rangle$
$\neg(\square p \wedge \square \neg p)$
$= \quad \langle (3.47a) \text{ De Morgan, } \neg(p \wedge q) \equiv \neg p \vee \neg q \rangle$
$\neg \square p \vee \neg \square \neg p$
$= \quad \langle (60) \text{ Dual of } \square \text{ and } (59) \rangle$
$\Diamond \neg p \vee \Diamond p \quad -(91) \text{ Temporal excluded middle} \quad \blacksquare$

(95) $\square \Diamond$ **excluded middle:** $\square \Diamond p \vee \Diamond \square \neg p$

Proof:

$\quad\quad \Box\Diamond p \lor \Diamond\Box\neg p$
$=\quad \langle(63)\text{ Dual of }\Box\Diamond\rangle$
$\quad\quad \Box\Diamond p \lor \neg\Box\Diamond p \quad -(3.28)\text{ Excluded middle, } p \lor \neg p \text{ with } p := \Box\Diamond p \quad \blacksquare$

(96) $\quad \Diamond\Box$ **excluded middle:** $\quad \Diamond\Box p \lor \Box\Diamond\neg p$

Proof:

$\quad\quad \Diamond\Box p \lor \Box\Diamond\neg p$
$=\quad \langle(62)\text{ Dual of }\Diamond\Box\rangle$
$\quad\quad \Diamond\Box p \lor \neg\Diamond\Box p \quad -(3.28)\text{ Excluded middle, } p \lor \neg p \text{ with } p := \Diamond\Box p \quad \blacksquare$

(97) $\quad \Box\Diamond$ **contradiction:** $\quad \Box\Diamond p \land \Diamond\Box\neg p \equiv \textit{false}$

Proof:

$\quad\quad \Box\Diamond p \land \Diamond\Box\neg p \equiv \textit{false}$
$=\quad \langle(63)\text{ Dual of }\Box\Diamond\rangle$
$\quad\quad \Box\Diamond p \land \neg\Box\Diamond p \equiv \textit{false} \quad -(3.42)\text{ Contradiction, } p \land \neg p \equiv \textit{false} \text{ with } p := \Box\Diamond p \quad \blacksquare$

(98) $\quad \Diamond\Box$ **contradiction:** $\quad \Diamond\Box p \land \Box\Diamond\neg p \equiv \textit{false}$

Proof:

$\quad\quad \Diamond\Box p \land \Box\Diamond\neg p \equiv \textit{false}$
$=\quad \langle(62)\text{ Dual of }\Diamond\Box\rangle$
$\quad\quad \Diamond\Box p \land \neg\Diamond\Box p \equiv \textit{false} \quad -(3.42)\text{ Contradiction, } p \land \neg p \equiv \textit{false} \text{ with } p := \Diamond\Box p \quad \blacksquare$

Theorem (99) shows that $\Box$, a universal operator, distributes over conjunction. Because disjunction is existential, (100) shows that $\Box$ distributes over conjunction in only one direction. Theorems (101), (102), and (103) reflect the concept of logical equivalence $\cong$ in [20]. Theorems (104) and (105) show how $\Diamond$ distributes over $\Rightarrow$.

(99) **Distributivity of $\Box$ over $\land$:** $\quad \Box(p \land q) \equiv \Box p \land \Box q$

Proof:

$\quad\quad \Box(p \land q)$
$=\quad \langle(54)\text{ Definition of }\Box\rangle$
$\quad\quad \neg\Diamond\neg(p \land q)$
$=\quad \langle(3.47\text{a})\text{ De Morgan, }\neg(p \land q) \equiv \neg p \lor \neg q\rangle$
$\quad\quad \neg\Diamond(\neg p \lor \neg q)$
$=\quad \langle(52)\text{ Distributivity of }\Diamond\text{ over }\lor\rangle$
$\quad\quad \neg(\Diamond\neg p \lor \Diamond\neg q)$
$=\quad \langle(3.47\text{b})\text{ De Morgan, }\neg(p \lor q) \equiv \neg p \land \neg q\rangle$
$\quad\quad \neg\Diamond\neg p \land \neg\Diamond\neg q$
$=\quad \langle(54)\text{ Definition of }\Box\text{, twice}\rangle$
$\quad\quad \Box p \land \Box q \quad \blacksquare$

(100) **Distributivity of $\Box$ over $\lor$:** $\Box p \lor \Box q \Rightarrow \Box (p \lor q)$

Proof:

$\quad\quad \Box p \lor \Box q \Rightarrow \Box (p \lor q)$
$= \quad \langle (3.60)\text{ Implication}, p \Rightarrow q \equiv p \land q \equiv p \rangle$
$\quad\quad (\Box p \lor \Box q) \land \Box (p \lor q) \equiv \Box p \lor \Box q$
$= \quad \langle (3.46)\text{ Distributivity of } \land \text{ over } \lor, p \land (q \lor r) \equiv (p \land q) \lor (p \land r) \rangle$
$\quad\quad (\Box p \land \Box (p \lor q)) \lor (\Box q \land \Box (p \lor q)) \equiv \Box p \lor \Box q$
$= \quad \langle (99)\text{ Distributivity of } \Box \text{ over } \land, \text{twice} \rangle$
$\quad\quad \Box (p \land (p \lor q)) \lor \Box (q \land (p \lor q)) \equiv \Box p \lor \Box q$
$= \quad \langle (3.43a)\text{ Absorption}, p \land (p \lor q) \equiv p, \text{twice} \rangle$
$\quad\quad \Box p \lor \Box q \equiv \Box p \lor \Box q \quad -(3.5)\text{ Reflexivity of } \equiv, p \equiv p \text{ with } p := \Box p \lor \Box q \quad \blacksquare$

(101) **Logical equivalence law of $\circ$:** $\Box (p \equiv q) \Rightarrow (\circ p \equiv \circ q)$

Proof:

$\quad\quad \Box (p \equiv q)$
$\Rightarrow \quad \langle (78)\text{ Strengthening of } \Box \text{ with } p := (p \equiv q) \rangle$
$\quad\quad \circ (p \equiv q)$
$= \quad \langle (6)\text{ Distributivity of } \circ \text{ over } \equiv \rangle$
$\quad\quad \circ p \equiv \circ q \quad \blacksquare$

(102) **Logical equivalence law of $\Diamond$:** $\Box (p \equiv q) \Rightarrow (\Diamond p \equiv \Diamond q)$

Proof: The proof is by (82) Temporal deduction.

$\quad\quad \Diamond p \equiv \Diamond q$
$= \quad \langle \text{Assume antecedent } p \equiv q, \text{twice} \rangle$
$\quad\quad \Diamond (p \land (p \equiv q)) \equiv \Diamond (q \land (p \equiv q))$
$= \quad \langle (3.50)\ p \land (q \equiv p) \equiv p \land q, \text{twice} \rangle$
$\quad\quad \Diamond (p \land q) \equiv \Diamond (p \land q) \quad -(3.5)\text{ Reflexivity of } \equiv, p \equiv p \text{ with } p := \Diamond (p \land q). \quad \blacksquare$

(103) **Logical equivalence law of $\Box$:** $\Box (p \equiv q) \Rightarrow (\Box p \equiv \Box q)$

Proof:

$\quad\quad \Box (p \equiv q) \Rightarrow (\Box p \equiv \Box q)$
$= \quad \langle (3.62)\ p \Rightarrow (q \equiv r) \equiv p \land q \equiv p \land r \rangle$
$\quad\quad \Box (p \equiv q) \land \Box p \equiv \Box (p \equiv q) \land \Box q$
$= \quad \langle (99)\text{ Distributivity of } \Box \text{ over } \land, \text{twice} \rangle$
$\quad\quad \Box ((p \equiv q) \land p) \equiv \Box ((p \equiv q) \land q)$
$= \quad \langle (3.50)\ p \land (q \equiv p) \equiv p \land q, \text{twice} \rangle$
$\quad\quad \Box (p \land q) \equiv \Box (p \land q) \quad -(3.5)\text{ Reflexivity of } \equiv, p \equiv p \text{ with } p := \Box (p \land q) \quad \blacksquare$

(104) **Distributivity of $\Diamond$ over $\Rightarrow$:** $\Diamond(p \Rightarrow q) \equiv (\Box p \Rightarrow \Diamond q)$

Proof:

$\quad\quad \Diamond(p \Rightarrow q)$
$= \quad \langle(3.59) \text{ Implication}, p \Rightarrow q \equiv \neg p \vee q\rangle$
$\quad\quad \Diamond(\neg p \vee q)$
$= \quad \langle(52) \text{ Distributivity of } \Diamond \text{ over } \vee\rangle$
$\quad\quad \Diamond \neg p \vee \Diamond q$
$= \quad \langle(60) \text{ Dual of } \Box\rangle$
$\quad\quad \neg \Box p \vee \Diamond q$
$= \quad \langle(3.59) \text{ Implication}, p \Rightarrow q \equiv \neg p \vee q\rangle$
$\quad\quad \Box p \Rightarrow \Diamond q \quad \blacksquare$

(105) **Distributivity of $\Diamond$ over $\Rightarrow$:** $(\Diamond p \Rightarrow \Diamond q) \Rightarrow \Diamond(p \Rightarrow q)$

Proof:

$\quad\quad (\Diamond p \Rightarrow \Diamond q) \Rightarrow \Diamond(p \Rightarrow q)$
$= \quad \langle(104) \text{ Distributivity of } \Diamond \text{ over } \Rightarrow\rangle$
$\quad\quad (\Diamond p \Rightarrow \Diamond q) \Rightarrow (\Box p \Rightarrow \Diamond q)$
$= \quad \langle(3.65) \text{ Shunting}, p \wedge q \Rightarrow r \equiv p \Rightarrow (q \Rightarrow r)\rangle$
$\quad\quad (\Diamond p \Rightarrow \Diamond q) \wedge \Box p \Rightarrow \Diamond q$

And now,

$\quad\quad (\Diamond p \Rightarrow \Diamond q) \wedge \Box p$
$\Rightarrow \quad \langle(77) \text{ Strengthening of } \Box \text{ and } (4.3) \text{ Monotonicity of } \wedge\rangle$
$\quad\quad (\Diamond p \Rightarrow \Diamond q) \wedge \Diamond p$
$\Rightarrow \quad \langle(3.77) \text{ Modus ponens}, p \wedge (p \Rightarrow q) \Rightarrow q\rangle$
$\quad\quad \Diamond q \quad \blacksquare$

The next three frame laws (106), (107), and (108) state that if $\Box p$ holds then p may be "added" by conjunction under each temporal operator [20]. For completeness, we show that they may be added under disjunction, implication, and equivalence as well. Theorems (148) to (150) extend the frame laws to $\mathcal{U}$ and theorems (210) to (212) extend them to $\mathcal{W}$.

(106) $\wedge$ **frame law of $\circ$:** $\Box p \Rightarrow (\circ q \Rightarrow \circ(p \wedge q))$

Proof:

$\quad\quad \Box p \Rightarrow (\circ q \Rightarrow \circ(p \wedge q))$
$= \quad \langle(3.65) \text{ Shunting}, p \wedge q \Rightarrow r \equiv p \Rightarrow (q \Rightarrow r)\rangle$
$\quad\quad \Box p \wedge \circ q \Rightarrow \circ(p \wedge q)$

And now,

$$\square p \wedge \circ q$$
$\Rightarrow$ $\langle$(78) Strengthening of $\square$ and (4.3) Monotonicity of $\wedge\rangle$
$$\circ p \wedge \circ q$$
$=$ $\langle$(5) Distributivity of $\circ$ over $\wedge\rangle$
$$\circ(p \wedge q) \quad \blacksquare$$

(107) $\wedge$ **frame law of** $\diamond$: $\square p \Rightarrow (\diamond q \Rightarrow \diamond(p \wedge q))$

Proof:

$$\square p \Rightarrow (\diamond q \Rightarrow \diamond(p \wedge q))$$
$=$ $\langle$(3.65) Shunting, $p \wedge q \Rightarrow r \equiv p \Rightarrow (q \Rightarrow r)\rangle$
$$\square p \wedge \diamond q \Rightarrow \diamond(p \wedge q) \quad -\text{(88) Distributivity of } \diamond \text{ over } \wedge \quad \blacksquare$$

(108) $\wedge$ **frame law of** $\square$: $\square p \Rightarrow (\square q \Rightarrow \square(p \wedge q))$

Proof:

$$\square p \Rightarrow (\square q \Rightarrow \square(p \wedge q))$$
$=$ $\langle$(3.65) Shunting, $p \wedge q \Rightarrow r \equiv p \Rightarrow (q \Rightarrow r)\rangle$
$$\square p \wedge \square q \Rightarrow \square(p \wedge q)$$
$=$ $\langle$(99) Distributivity of $\square$ over $\wedge\rangle$
$$\square(p \wedge q) \Rightarrow \square(p \wedge q) \quad -\text{(3.71) Reflexivity of } \Rightarrow, p \Rightarrow p \quad \blacksquare$$

(109) $\vee$ **frame law of** $\circ$: $\square p \Rightarrow (\circ q \Rightarrow \circ(p \vee q))$

Proof:

$$\square p \Rightarrow (\circ q \Rightarrow \circ(p \vee q))$$
$=$ $\langle$(3.65) Shunting, $p \wedge q \Rightarrow r \equiv p \Rightarrow (q \Rightarrow r)\rangle$
$$\square p \wedge \circ q \Rightarrow \circ(p \vee q)$$

And now,

$$\square p \wedge \circ q$$
$\Rightarrow$ $\langle$(78) Strengthening of $\square$ and (4.3) Monotonicity of $\wedge\rangle$
$$\circ p \wedge \circ q$$
$\Rightarrow$ $\langle$(3.76c) (Weakening/strengthening), $p \wedge q \Rightarrow p \vee q\rangle$
$$\circ p \vee \circ q$$
$=$ $\langle$(4) Distributivity of $\circ$ over $\vee\rangle$
$$\circ(p \vee q) \quad \blacksquare$$

Pepperdine Papers on LTL 61

(110) ∨ **frame law of** ◇: $\square p \Rightarrow (\Diamond q \Rightarrow \Diamond(p \vee q))$

Proof:

$\qquad \square p \Rightarrow (\Diamond q \Rightarrow \Diamond (p \vee q))$
$= \quad \langle(3.65) \text{ Shunting}, p \wedge q \Rightarrow r \equiv p \Rightarrow (q \Rightarrow r)\rangle$
$\qquad \square p \wedge \Diamond q \Rightarrow \Diamond (p \vee q)$

And now,

$\qquad \square p \wedge \Diamond q$
$\Rightarrow \quad \langle(77) \text{ Strengthening of } \square \text{ and } (4.3) \text{ Monotonicity of } \wedge\rangle$
$\qquad \Diamond p \wedge \Diamond q$
$\Rightarrow \quad \langle(3.76c) \text{ (Weakening/strengthening)}, p \wedge q \Rightarrow p \vee q\rangle$
$\qquad \Diamond p \vee \Diamond q$
$= \quad \langle(52) \text{ Distributivity of } \Diamond \text{ over } \vee\rangle$
$\qquad \Diamond (p \vee q)$ ∎

(111) ∨ **frame law of** □: $\square p \Rightarrow (\square q \Rightarrow \square(p \vee q))$

Proof:

$\qquad \square p \Rightarrow (\square q \Rightarrow \square (p \vee q))$
$= \quad \langle(3.65) \text{ Shunting}, p \wedge q \Rightarrow r \equiv p \Rightarrow (q \Rightarrow r)\rangle$
$\qquad \square p \wedge \square q \Rightarrow \square (p \vee q)$

And now,

$\qquad \square p \wedge \square q$
$\Rightarrow \quad \langle(3.76c) \text{ (Weakening/strengthening)}, p \wedge q \Rightarrow p \vee q\rangle$
$\qquad \square p \vee \square q$
$\Rightarrow \quad \langle(100) \text{ Distributivity of } \square \text{ over } \vee\rangle$
$\qquad \square (p \vee q)$ ∎

(112) ⇒ **frame law of** ○: $\square p \Rightarrow (\circ q \Rightarrow \circ(p \Rightarrow q))$

Proof:

$\qquad \square p \Rightarrow (\circ q \Rightarrow \circ (p \Rightarrow q))$
$= \quad \langle(3.65) \text{ Shunting}, p \wedge q \Rightarrow r \equiv p \Rightarrow (q \Rightarrow r)\rangle$
$\qquad \square p \wedge \circ q \Rightarrow \circ (p \Rightarrow q)$
$= \quad \langle(2) \text{ Axiom, Distributivity of } \circ \text{ over } \Rightarrow\rangle$
$\qquad \square p \wedge \circ q \Rightarrow (\circ p \Rightarrow \circ q)$
$= \quad \langle(3.65) \text{ Shunting}, p \wedge q \Rightarrow r \equiv p \Rightarrow (q \Rightarrow r)\rangle$
$\qquad \square p \wedge \circ q \wedge \circ p \Rightarrow \circ q$
$\qquad \quad -(3.76b) \text{ Strengthening}, p \wedge q \Rightarrow p \text{ with } p,q := \circ q, \square p \wedge \circ p$ ∎

(113) $\Rightarrow$ **frame law of** $\Diamond$: $\Box p \Rightarrow (\Diamond q \Rightarrow \Diamond (p \Rightarrow q))$

Proof:

$\quad \Box p \Rightarrow (\Diamond q \Rightarrow \Diamond (p \Rightarrow q))$
$= \quad \langle (3.65) \text{ Shunting}, p \wedge q \Rightarrow r \equiv p \Rightarrow (q \Rightarrow r) \rangle$
$\quad \Box p \wedge \Diamond q \Rightarrow \Diamond (p \Rightarrow q)$
$= \quad \langle (104) \text{ Distributivity of } \Diamond \text{ over } \Rightarrow \rangle$
$\quad \Box p \wedge \Diamond q \Rightarrow (\Box p \Rightarrow \Diamond q)$
$= \quad \langle (3.65) \text{ Shunting}, p \wedge q \Rightarrow r \equiv p \Rightarrow (q \Rightarrow r) \rangle$
$\quad \Box p \wedge \Diamond q \wedge \Box p \Rightarrow \Diamond q$
$\qquad -(3.76\text{b}) \text{ Strengthening}, p \wedge q \Rightarrow p \text{ with } p, q := \Diamond q, \Box p \wedge \Box p \quad \blacksquare$

(114) $\Rightarrow$ **frame law of** $\Box$: $\Box p \Rightarrow (\Box q \Rightarrow \Box (p \Rightarrow q))$

Proof:

$\quad \Box p \Rightarrow (\Box q \Rightarrow \Box (p \Rightarrow q))$
$= \quad \langle (3.65) \text{ Shunting}, p \wedge q \Rightarrow r \equiv p \Rightarrow (q \Rightarrow r) \rangle$
$\quad \Box p \wedge \Box q \Rightarrow \Box (p \Rightarrow q)$
$= \quad \langle (99) \text{ Distributivity of } \Box \text{ over } \wedge \rangle$
$\quad \Box (p \wedge q) \Rightarrow \Box (p \Rightarrow q)$
$= \quad \langle (3.66) \ p \wedge (p \Rightarrow q) \equiv p \wedge q \rangle$
$\quad \Box (p \wedge (p \Rightarrow q)) \Rightarrow \Box (p \Rightarrow q)$
$= \quad \langle (99) \text{ Distributivity of } \Box \text{ over } \wedge \rangle$
$\quad \Box p \wedge \Box (p \Rightarrow q) \Rightarrow \Box (p \Rightarrow q)$
$\qquad -(3.76\text{b}) \text{ Strengthening}, p \wedge q \Rightarrow p \text{ with } p, q := \Box (p \Rightarrow q), \Box p \quad \blacksquare$

(115) $\equiv$ **frame law of** $\circ$: $\Box p \Rightarrow (\circ q \Rightarrow \circ (p \equiv q))$

Proof: The proof is by (82) Temporal deduction.

$\quad \circ q \Rightarrow \circ (p \equiv q)$
$= \quad \langle \text{Assume antecedent } p \rangle$
$\quad \circ q \Rightarrow \circ (\textit{true} \equiv q)$
$= \quad \langle (3.3) \text{ Identity of } \equiv, \textit{true} \equiv p \equiv p \rangle$
$\quad \circ q \Rightarrow \circ q \quad -(3.71) \text{ Reflexivity of } \Rightarrow, p \Rightarrow p \quad \blacksquare$

(116) $\equiv$ **frame law of** $\Diamond$: $\Box p \Rightarrow (\Diamond q \Rightarrow \Diamond (p \equiv q))$

Proof:

$$\square p \Rightarrow (\lozenge q \Rightarrow \lozenge (p \equiv q))$$
$= \quad \langle (3.70)\ p \vee q \Rightarrow p \wedge q \equiv p \equiv q \rangle$
$$\square p \Rightarrow (\lozenge q \Rightarrow \lozenge (p \vee q \Rightarrow p \wedge q))$$
$= \quad \langle (104)\ \text{Distributivity of } \lozenge \text{ over } \Rightarrow \rangle$
$$\square p \Rightarrow (\lozenge q \Rightarrow (\square (p \vee q) \Rightarrow \lozenge (p \wedge q)))$$
$= \quad \langle (3.65)\ \text{Shunting},\ p \wedge q \Rightarrow r \equiv p \Rightarrow (q \Rightarrow r),\ \text{twice} \rangle$
$$\square p \wedge \lozenge q \wedge \square (p \vee q) \Rightarrow \lozenge (p \wedge q)$$
$= \quad \langle (99)\ \text{Distributivity of } \square \text{ over } \wedge \rangle$
$$\square (p \wedge (p \vee q)) \wedge \lozenge q \Rightarrow \lozenge (p \wedge q)$$
$= \quad \langle (3.43a)\ \text{Absorption},\ p \wedge (p \vee q) \equiv p \rangle$
$$\square p \wedge \lozenge q \Rightarrow \lozenge (p \wedge q) \quad -(88)\ \text{Distributivity of } \lozenge \text{ over } \wedge \quad \blacksquare$$

(117) $\equiv$ **frame law of** $\square$: $\quad \square p \Rightarrow (\square q \Rightarrow \square (p \equiv q))$

Proof: The proof is by (82) Temporal deduction.

$$\square q \Rightarrow \square (p \equiv q)$$
$= \quad \langle \text{Assume antecedent } p \rangle$
$$\square q \Rightarrow \square (\mathit{true} \equiv q)$$
$= \quad \langle (3.3)\ \text{Identity of } \equiv,\ \mathit{true} \equiv p \equiv p \rangle$
$$\square q \Rightarrow \square q \quad -(3.71)\ \text{Reflexivity of } \Rightarrow,\ p \Rightarrow p \quad \blacksquare$$

Theorems (118), (119), and (120) show that all unary temporal operators are monotonic. Theorem (120) can also be considered distributivity of $\square$ over $\Rightarrow$. Proofs of the consequence rules (121), (122), and (123) use the monotonicity theorems as shown in the proof of (121). Proofs of the catenation rules (124), (125), and (126) are similar.

(118) **Monotonicity of** $\circ$: $\quad \square (p \Rightarrow q) \Rightarrow (\circ p \Rightarrow \circ q)$

Proof:

$$\square (p \Rightarrow q)$$
$\Rightarrow \quad \langle (78)\ \text{Strengthening of } \square \rangle$
$$\circ (p \Rightarrow q)$$
$= \quad \langle (2)\ \text{Distributivity of } \circ \text{ over } \Rightarrow \rangle$
$$\circ p \Rightarrow \circ q \quad \blacksquare$$

(119) **Monotonicity of** $\lozenge$: $\quad \square (p \Rightarrow q) \Rightarrow (\lozenge p \Rightarrow \lozenge q)$

Proof:

$\square(p \Rightarrow q) \Rightarrow (\Diamond p \Rightarrow \Diamond q)$
$= \quad \langle(3.59)\text{ Implication, } p \Rightarrow q \equiv \neg p \vee q, \text{ thrice}\rangle$
$\neg\square(\neg p \vee q) \vee \neg\Diamond p \vee \Diamond q$
$= \quad \langle(60)\text{ Dual of } \square\rangle$
$\Diamond\neg(\neg p \vee q) \vee \neg\Diamond p \vee \Diamond q$
$= \quad \langle(3.47b)\text{ De Morgan, } \neg(p \vee q) \equiv \neg p \wedge \neg q\rangle$
$\Diamond(p \wedge \neg q) \vee \neg\Diamond p \vee \Diamond q$
$= \quad \langle(52)\text{ Distributivity of } \Diamond \text{ over } \vee\rangle$
$\Diamond((p \wedge \neg q) \vee q) \vee \neg\Diamond p$
$= \quad \langle(3.44b)\text{ Absorption, } p \vee (\neg p \wedge q) \equiv p \vee q\rangle$
$\Diamond(p \vee q) \vee \neg\Diamond p$
$= \quad \langle(52)\text{ Distributivity of } \Diamond \text{ over } \vee\rangle$
$\Diamond p \vee \Diamond q \vee \neg\Diamond p$
$= \quad \langle(3.28)\text{ Excluded middle, } p \vee \neg p \text{ with } p := \Diamond p\rangle$
$\Diamond q \vee \mathit{true}$
$= \quad \langle(3.29)\text{ Zero of } \vee, p \vee \mathit{true} \equiv \mathit{true}\rangle$
true ∎

(120) Monotonicity of $\square$: $\square(p \Rightarrow q) \Rightarrow (\square p \Rightarrow \square q)$

Proof:

$\square(p \Rightarrow q)$
$= \quad \langle(3.60)\text{ Implication, } p \Rightarrow q \equiv p \wedge q \equiv p\rangle$
$\square(p \wedge q \equiv p)$
$\Rightarrow \quad \langle(103)\text{ Logical equivalence law of } \square\rangle$
$\square(p \wedge q) \equiv \square p$
$= \quad \langle(99)\text{ Distributivity of } \square \text{ over } \wedge\rangle$
$\square p \wedge \square q \equiv \square p$
$= \quad \langle(3.60)\text{ Implication, } p \Rightarrow q \equiv p \wedge q \equiv p\rangle$
$\square p \Rightarrow \square q$ ∎

(121) Consequence rule of $\circ$: $\square((p \Rightarrow q) \wedge (q \Rightarrow \circ r) \wedge (r \Rightarrow s)) \Rightarrow (p \Rightarrow \circ s)$

Proof:

$\square((p \Rightarrow q) \wedge (q \Rightarrow \circ r) \wedge (r \Rightarrow s))$
$= \quad \langle(99)\text{ Distributivity of } \square \text{ over } \wedge\rangle$
$\square(p \Rightarrow q) \wedge \square(q \Rightarrow \circ r) \wedge \square(r \Rightarrow s)$
$\Rightarrow \quad \langle(76)\text{ Strengthening of } \square \text{ and } (4.3)\text{ Monotonicity of } \wedge, \text{ twice}\rangle$

$$(p \Rightarrow q) \wedge (q \Rightarrow \circ r) \wedge \Box (r \Rightarrow s)$$
$\Rightarrow \quad \langle(3.82a) \text{ Transitivity}, (p \Rightarrow q) \wedge (q \Rightarrow r) \Rightarrow (p \Rightarrow r) \text{ and } (4.3) \text{ Monotonicity of } \wedge\rangle$
$$(p \Rightarrow \circ r) \wedge \Box (r \Rightarrow s)$$
$\Rightarrow \quad \langle(118) \text{ Monotonicity of } \circ \text{ and } (4.3) \text{ Monotonicity of } \wedge\rangle$
$$(p \Rightarrow \circ r) \wedge (\circ r \Rightarrow \circ s)$$
$\Rightarrow \quad \langle(3.82a) \text{ Transitivity}, (p \Rightarrow q) \wedge (q \Rightarrow r) \Rightarrow (p \Rightarrow r)\rangle$
$$p \Rightarrow \circ s \quad \blacksquare$$

(122) **Consequence rule of** $\Diamond$: $\quad \Box((p \Rightarrow q) \wedge (q \Rightarrow \Diamond r) \wedge (r \Rightarrow s)) \Rightarrow (p \Rightarrow \Diamond s)$

Proof:

$$\Box((p \Rightarrow q) \wedge (q \Rightarrow \Diamond r) \wedge (r \Rightarrow s))$$
$= \quad \langle(99) \text{ Distributivity of } \Box \text{ over } \wedge\rangle$
$$\Box(p \Rightarrow q) \wedge \Box(q \Rightarrow \Diamond r) \wedge \Box(r \Rightarrow s)$$
$\Rightarrow \quad \langle(76) \text{ Strengthening of } \Box \text{ and } (4.3) \text{ Monotonicity of } \wedge, \text{ twice}\rangle$
$$(p \Rightarrow q) \wedge (q \Rightarrow \Diamond r) \wedge \Box(r \Rightarrow s)$$
$\Rightarrow \quad \langle(3.82a) \text{ Transitivity}, (p \Rightarrow q) \wedge (q \Rightarrow r) \Rightarrow (p \Rightarrow r) \text{ and } (4.3) \text{ Monotonicity of } \wedge\rangle$
$$(p \Rightarrow \Diamond r) \wedge \Box(r \Rightarrow s)$$
$\Rightarrow \quad \langle(119) \text{ Monotonicity of } \Diamond \text{ and } (4.3) \text{ Monotonicity of } \wedge\rangle$
$$(p \Rightarrow \Diamond r) \wedge (\Diamond r \Rightarrow \Diamond s)$$
$\Rightarrow \quad \langle(3.82a) \text{ Transitivity}, (p \Rightarrow q) \wedge (q \Rightarrow r) \Rightarrow (p \Rightarrow r)\rangle$
$$p \Rightarrow \Diamond s \quad \blacksquare$$

(123) **Consequence rule of** $\Box$: $\quad \Box((p \Rightarrow q) \wedge (q \Rightarrow \Box r) \wedge (r \Rightarrow s)) \Rightarrow (p \Rightarrow \Box s)$

Proof:

$$\Box((p \Rightarrow q) \wedge (q \Rightarrow \Box r) \wedge (r \Rightarrow s))$$
$= \quad \langle(99) \text{ Distributivity of } \Box \text{ over } \wedge\rangle$
$$\Box(p \Rightarrow q) \wedge \Box(q \Rightarrow \Box r) \wedge \Box(r \Rightarrow s)$$
$\Rightarrow \quad \langle(76) \text{ Strengthening of } \Box \text{ and } (4.3) \text{ Monotonicity of } \wedge, \text{ twice}\rangle$
$$(p \Rightarrow q) \wedge (q \Rightarrow \Box r) \wedge \Box(r \Rightarrow s)$$
$\Rightarrow \quad \langle(3.82a) \text{ Transitivity}, (p \Rightarrow q) \wedge (q \Rightarrow r) \Rightarrow (p \Rightarrow r) \text{ and } (4.3) \text{ Monotonicity of } \wedge\rangle$
$$(p \Rightarrow \Box r) \wedge \Box(r \Rightarrow s)$$
$\Rightarrow \quad \langle(120) \text{ Monotonicity of } \Box \text{ and } (4.3) \text{ Monotonicity of } \wedge\rangle$
$$(p \Rightarrow \Box r) \wedge (\Box r \Rightarrow \Box s)$$
$\Rightarrow \quad \langle(3.82a) \text{ Transitivity}, (p \Rightarrow q) \wedge (q \Rightarrow r) \Rightarrow (p \Rightarrow r)\rangle$
$$p \Rightarrow \Box s \quad \blacksquare$$

(124) **Catenation rule of $\Diamond$:** $\Box((p \Rightarrow \Diamond q) \wedge (q \Rightarrow \Diamond r)) \Rightarrow (p \Rightarrow \Diamond r)$

Proof:

$\quad\quad\quad \Box((p \Rightarrow \Diamond q) \wedge (q \Rightarrow \Diamond r))$
$= \quad \langle(99) \text{ Distributivity of } \Box \text{ over } \wedge\rangle$
$\quad\quad\quad \Box(p \Rightarrow \Diamond q) \wedge \Box(q \Rightarrow \Diamond r)$
$\Rightarrow \quad \langle(76) \text{ Strengthening of } \Box \text{ and (4.3) Monotonicity of } \wedge\rangle$
$\quad\quad\quad (p \Rightarrow \Diamond q) \wedge \Box(q \Rightarrow \Diamond r)$
$\Rightarrow \quad \langle(119) \text{ Monotonicity of } \Diamond \text{ and (4.3) Monotonicity of } \wedge\rangle$
$\quad\quad\quad (p \Rightarrow \Diamond q) \wedge (\Diamond q \Rightarrow \Diamond \Diamond r)$
$= \quad \langle(50) \text{ Absorption of } \Diamond\rangle$
$\quad\quad\quad (p \Rightarrow \Diamond q) \wedge (\Diamond q \Rightarrow \Diamond r)$
$\Rightarrow \quad \langle(3.82a) \text{ Transitivity, } (p \Rightarrow q) \wedge (q \Rightarrow r) \Rightarrow (p \Rightarrow r)\rangle$
$\quad\quad\quad p \Rightarrow \Diamond r \quad \blacksquare$

(125) **Catenation rule of $\Box$:** $\Box((p \Rightarrow \Box q) \wedge (q \Rightarrow \Box r)) \Rightarrow (p \Rightarrow \Box r)$

Proof:

$\quad\quad\quad \Box((p \Rightarrow \Box q) \wedge (q \Rightarrow \Box r))$
$\Rightarrow \quad \langle(76) \text{ Strengthening of } \Box\rangle$
$\quad\quad\quad (p \Rightarrow \Box q) \wedge (q \Rightarrow \Box r)$
$= \quad \langle(76) \text{ Strengthening of } \Box \text{ and (3.39) Identity of } \wedge, p \wedge true \equiv p\rangle$
$\quad\quad\quad (p \Rightarrow \Box q) \wedge (\Box q \Rightarrow q) \wedge (q \Rightarrow \Box r)$
$\Rightarrow \quad \langle(3.82a) \text{ Transitivity, } (p \Rightarrow q) \wedge (q \Rightarrow r) \Rightarrow (p \Rightarrow r) \text{ and (4.3) Monotonicity of } \wedge\rangle$
$\quad\quad\quad (p \Rightarrow q) \wedge (q \Rightarrow \Box r)$
$\Rightarrow \quad \langle(3.82a) \text{ Transitivity, } (p \Rightarrow q) \wedge (q \Rightarrow r) \Rightarrow (p \Rightarrow r)\rangle$
$\quad\quad\quad p \Rightarrow \Box r \quad \blacksquare$

(126) **Catenation rule of $\mathcal{U}$:** $\Box((p \Rightarrow q\,\mathcal{U}\,r) \wedge (r \Rightarrow q\,\mathcal{U}\,s)) \Rightarrow (p \Rightarrow q\,\mathcal{U}\,s)$

Proof:

$\quad\quad\quad \Box((p \Rightarrow q\,\mathcal{U}\,r) \wedge (r \Rightarrow q\,\mathcal{U}\,s)) \Rightarrow (p \Rightarrow q\,\mathcal{U}\,s)$
$= \quad \langle(3.65) \text{ Shunting, } p \wedge q \Rightarrow r \equiv p \Rightarrow (q \Rightarrow r)\rangle$
$\quad\quad\quad \Box((p \Rightarrow q\,\mathcal{U}\,r) \wedge (r \Rightarrow q\,\mathcal{U}\,s)) \wedge p \Rightarrow q\,\mathcal{U}\,s$

And now,

Pepperdine Papers on LTL

$$\square((p \Rightarrow q\,\mathcal{U}\,r) \wedge (r \Rightarrow q\,\mathcal{U}\,s)) \wedge p$$
$= \quad \langle(99) \text{ Distributivity of } \square \text{ over } \wedge\rangle$
$$\square(p \Rightarrow q\,\mathcal{U}\,r) \wedge \square(r \Rightarrow q\,\mathcal{U}\,s) \wedge p$$
$\Rightarrow \quad \langle(76) \text{ Strengthening of } \square \text{ and (4.3) Monotonicity of } \wedge\rangle$
$$(p \Rightarrow q\,\mathcal{U}\,r) \wedge \square(r \Rightarrow q\,\mathcal{U}\,s) \wedge p$$
$\Rightarrow \quad \langle(3.77) \text{ Modus ponens, } p \wedge (p \Rightarrow q) \Rightarrow q \text{ and (4.3) Monotonicity of } \wedge\rangle$
$$q\,\mathcal{U}\,r \wedge \square(r \Rightarrow q\,\mathcal{U}\,s)$$
$\Rightarrow \quad \langle(85) \text{ Right monotonicity of } \mathcal{U} \text{ with } p,q,r := r, q\,\mathcal{U}\,s, q$
$\quad\quad \text{ and (4.3) Monotonicity of } \wedge\rangle$
$$q\,\mathcal{U}\,r \wedge (q\,\mathcal{U}\,r \Rightarrow q\,\mathcal{U}\,(q\,\mathcal{U}\,s))$$
$\Rightarrow \quad \langle(3.77) \text{ Modus ponens, } p \wedge (p \Rightarrow q) \Rightarrow q\rangle$
$$q\,\mathcal{U}\,(q\,\mathcal{U}\,s)$$
$= \quad \langle(36) \text{ Left absorption of } \mathcal{U}\rangle$
$$q\,\mathcal{U}\,s \quad \blacksquare$$

Most of the remaining theorems in this section are included in a single source in our survey.

(127) $\mathcal{U}$ **strengthening rule:** $\square((p \Rightarrow r) \wedge (q \Rightarrow s)) \Rightarrow (p\,\mathcal{U}\,q \Rightarrow r\,\mathcal{U}\,s)$

Proof:

$$\square((p \Rightarrow r) \wedge (q \Rightarrow s)) \Rightarrow (p\,\mathcal{U}\,q \Rightarrow r\,\mathcal{U}\,s)$$
$= \quad \langle(3.65) \text{ Shunting, } p \wedge q \Rightarrow r \equiv p \Rightarrow (q \Rightarrow r)\rangle$
$$\square((p \Rightarrow r) \wedge (q \Rightarrow s)) \wedge p\,\mathcal{U}\,q \Rightarrow r\,\mathcal{U}\,s$$

And now,

$$\square((p \Rightarrow r) \wedge (q \Rightarrow s)) \wedge p\,\mathcal{U}\,q$$
$= \quad \langle(99) \text{ Distributivity of } \square \text{ over } \wedge\rangle$
$$\square(p \Rightarrow r) \wedge \square(q \Rightarrow s) \wedge p\,\mathcal{U}\,q$$
$\Rightarrow \quad \langle(85) \text{ Right monotonicity of } \mathcal{U} \text{ with } p,q,r := q,s,p$
$\quad\quad \text{ and (4.3) Monotonicity of } \wedge\rangle$
$$\square(p \Rightarrow r) \wedge (p\,\mathcal{U}\,q \Rightarrow p\,\mathcal{U}\,s) \wedge p\,\mathcal{U}\,q$$
$\Rightarrow \quad \langle(3.77) \text{ Modus ponens, } p \wedge (p \Rightarrow q) \Rightarrow q \text{ and (4.3) Monotonicity of } \wedge\rangle$
$$\square(p \Rightarrow r) \wedge p\,\mathcal{U}\,s$$
$\Rightarrow \quad \langle(86) \text{ Left monotonicity of } \mathcal{U} \text{ with } q,r := r,s$
$\quad\quad \text{ and (4.3) Monotonicity of } \wedge\rangle$
$$(p\,\mathcal{U}\,s \Rightarrow r\,\mathcal{U}\,s) \wedge p\,\mathcal{U}\,s$$
$\Rightarrow \quad \langle(3.77) \text{ Modus ponens, } p \wedge (p \Rightarrow q) \Rightarrow q\rangle$
$$r\,\mathcal{U}\,s \quad \blacksquare$$

(128) **Induction rule $\Diamond$:** $\Box(p \lor \bigcirc q \Rightarrow q) \Rightarrow (\Diamond p \Rightarrow q)$

Proof:

$\quad\quad \Box(p \lor \bigcirc q \Rightarrow q)$
$= \quad \langle(3.78)\ (p \Rightarrow r) \land (q \Rightarrow r) \equiv p \lor q \Rightarrow r\rangle$
$\quad\quad \Box((p \Rightarrow q) \land (\bigcirc q \Rightarrow q))$
$= \quad \langle(99)\ \text{Distributivity of } \Box \text{ over } \land\rangle$
$\quad\quad \Box(p \Rightarrow q) \land \Box(\bigcirc q \Rightarrow q)$
$\Rightarrow \quad \langle(58)\ \Diamond \text{ induction and (4.3) Monotonicity of } \land\rangle$
$\quad\quad \Box(p \Rightarrow q) \land (\Diamond q \Rightarrow q)$
$\Rightarrow \quad \langle(119)\ \text{Monotonicity of } \Diamond \text{ and (4.3) Monotonicity of } \land\rangle$
$\quad\quad (\Diamond p \Rightarrow \Diamond q) \land (\Diamond q \Rightarrow q)$
$\Rightarrow \quad \langle(3.82a)\ \text{Transitivity}, (p \Rightarrow q) \land (q \Rightarrow r) \Rightarrow (p \Rightarrow r)\rangle$
$\quad\quad \Diamond p \Rightarrow q \quad \blacksquare$

(129) **Induction rule $\Box$:** $\Box(p \Rightarrow \bigcirc p \land q) \Rightarrow (p \Rightarrow \Box q)$

Proof:

$\quad\quad \textit{true}$
$= \quad \langle(55)\ \mathcal{U} \text{ induction with } r := \textit{false}\rangle$
$\quad\quad \Box(p \Rightarrow (\bigcirc p \land q) \lor \textit{false}) \Rightarrow (p \Rightarrow \Box q \lor q\,\mathcal{U}\,\textit{false})$
$= \quad \langle(11)\ \text{Right zero of } \mathcal{U}\rangle$
$\quad\quad \Box(p \Rightarrow (\bigcirc p \land q) \lor \textit{false}) \Rightarrow (p \Rightarrow \Box q \lor \textit{false})$
$= \quad \langle(3.30)\ \text{Identity of } \lor, p \lor \textit{false} \equiv p, \text{twice}\rangle$
$\quad\quad \Box(p \Rightarrow \bigcirc p \land q) \Rightarrow (p \Rightarrow \Box q) \quad \blacksquare$

(130) **Induction rule $\mathcal{U}$:** $\Box(p \Rightarrow \neg q \land \bigcirc p) \Rightarrow (p \Rightarrow \neg(r\,\mathcal{U}\,q))$

Proof:

$\quad\quad \Box(p \Rightarrow \neg q \land \bigcirc p)$
$\Rightarrow \quad \langle(129)\ \text{Induction rule } \Box\rangle$
$\quad\quad p \Rightarrow \Box \neg q$
$\Rightarrow \quad \langle(81)\ \text{with } p, q := \neg q, r,\ (3.12)\ \text{Double negation}, \neg\neg p \equiv p \text{ and}$
$\quad\quad\quad (3.82a)\ \text{Transitivity}, (p \Rightarrow q) \land (q \Rightarrow r) \Rightarrow (p \Rightarrow r)\rangle$
$\quad\quad p \Rightarrow \neg(r\,\mathcal{U}\,q) \quad \blacksquare$

(131) **$\Diamond$ confluence:** $\Box((p \Rightarrow \Diamond(q \lor r)) \land (q \Rightarrow \Diamond t) \land (r \Rightarrow \Diamond t)) \Rightarrow (p \Rightarrow \Diamond t)$

Proof:

Pepperdine Papers on LTL

$\square((p \Rightarrow \Diamond(q \vee r)) \wedge (q \Rightarrow \Diamond t) \wedge (r \Rightarrow \Diamond t)) \Rightarrow (p \Rightarrow \Diamond t)$
= ⟨(3.65) Shunting, $p \wedge q \Rightarrow r \equiv p \Rightarrow (q \Rightarrow r)$⟩
$\square((p \Rightarrow \Diamond(q \vee r)) \wedge (q \Rightarrow \Diamond t) \wedge (r \Rightarrow \Diamond t)) \wedge p \Rightarrow \Diamond t$

And now,

$\square((p \Rightarrow \Diamond(q \vee r)) \wedge (q \Rightarrow \Diamond t) \wedge (r \Rightarrow \Diamond t)) \wedge p$
= ⟨(99) Distributivity of $\square$ over $\wedge$⟩
$\square(p \Rightarrow \Diamond(q \vee r)) \wedge \square(q \Rightarrow \Diamond t) \wedge \square(r \Rightarrow \Diamond t) \wedge p$
⇒ ⟨(76) Strengthening of $\square$ and (4.3) Monotonicity of $\wedge$⟩
$(p \Rightarrow \Diamond(q \vee r)) \wedge \square(q \Rightarrow \Diamond t) \wedge \square(r \Rightarrow \Diamond t) \wedge p$
⇒ ⟨(3.77) Modus ponens, $p \wedge (p \Rightarrow q) \Rightarrow q$ and (4.3) Monotonicity of $\wedge$⟩
$\Diamond(q \vee r) \wedge \square(q \Rightarrow \Diamond t) \wedge \square(r \Rightarrow \Diamond t)$
= ⟨(52) Distributivity of $\Diamond$ over $\vee$⟩
$(\Diamond q \vee \Diamond r) \wedge \square(q \Rightarrow \Diamond t) \wedge \square(r \Rightarrow \Diamond t)$
⇒ ⟨(119) Monotonicity of $\Diamond$ and (4.3) Monotonicity of $\wedge$, twice⟩
$(\Diamond q \vee \Diamond r) \wedge (\Diamond q \Rightarrow \Diamond \Diamond t) \wedge (\Diamond r \Rightarrow \Diamond \Diamond t)$
= ⟨(50) Absorption of $\Diamond$, twice⟩
$(\Diamond q \vee \Diamond r) \wedge (\Diamond q \Rightarrow \Diamond t) \wedge (\Diamond r \Rightarrow \Diamond t)$
⇒ ⟨(3.76.3) $(p \vee q) \wedge (q \Rightarrow r) \Rightarrow p \vee r$ and (4.3) Monotonicity of $\wedge$⟩
$(\Diamond r \vee \Diamond t) \wedge (\Diamond r \Rightarrow \Diamond t)$
⇒ ⟨(3.76.3) $(p \vee q) \wedge (q \Rightarrow r) \Rightarrow p \vee r$⟩
$\Diamond t \vee \Diamond t$
= ⟨(3.26) Idempotency of $\vee$, $p \vee p \equiv p$⟩
$\Diamond t$ ∎

(132) **Temporal generalization law:** $\square(\square p \Rightarrow q) \Rightarrow (\square p \Rightarrow \square q)$

Proof:

$\square(\square p \Rightarrow q)$
⇒ ⟨(120) Monotonicity of $\square$ with $p := \square p$⟩
$\square\square p \Rightarrow \square q$
= ⟨(72) Absorption of $\square$⟩
$\square p \Rightarrow \square q$ ∎

(133) **Temporal particularization law:** $\square(p \Rightarrow \Diamond q) \Rightarrow (\Diamond p \Rightarrow \Diamond q)$

Proof:

$$
\begin{aligned}
& \square(p \Rightarrow \Diamond q) \\
\Rightarrow\ & \langle(119)\ \text{Monotonicity of}\ \Diamond\ \text{with}\ q := \Diamond q\rangle \\
& \Diamond p \Rightarrow \Diamond \Diamond q \\
=\ & \langle(50)\ \text{Absorption of}\ \Diamond\rangle \\
& \Diamond p \Rightarrow \Diamond q \quad \blacksquare
\end{aligned}
$$

(134) $\square(p \Rightarrow \circ q) \Rightarrow (p \Rightarrow \Diamond q)$

Proof:

$$
\begin{aligned}
& \square(p \Rightarrow \circ q) \\
\Rightarrow\ & \langle(76)\ \text{Strengthening of}\ \square\rangle \\
& p \Rightarrow \circ q \\
\Rightarrow\ & \langle(47)\ \text{Weakening of}\ \Diamond\ \text{and}\ (3.82\text{a})\ \text{Transitivity}\rangle \\
& p \Rightarrow \Diamond q \quad \blacksquare
\end{aligned}
$$

(135) $\square(p \Rightarrow \circ \neg p) \Rightarrow (p \Rightarrow \neg \square p)$

Proof:

$$
\begin{aligned}
& \square(p \Rightarrow \circ \neg p) \\
\Rightarrow\ & \langle(134)\ \text{with}\ q := \neg p\rangle \\
& p \Rightarrow \Diamond \neg p \\
=\ & \langle(60)\ \text{Dual of}\ \square\rangle \\
& p \Rightarrow \neg \square p \quad \blacksquare
\end{aligned}
$$

Because the implication relation is reflexive, antisymmetric, and transitive, it defines a partially ordered set on linear temporal logic expressions. Figure 3 is a collection of seven Hasse diagrams showing some implication relations. Each number in parentheses is a linear temporal logic theorem. A number that labels an edge in a Hasse diagram is an implication theorem, and a number that labels a box is an equivalence theorem. For example, edge (87) represents the theorem that $\square \neg p$ implies $\neg \square p$, and box (61) represents the theorem that $\neg \Diamond p$ is equivalent to $\square \neg p$.

The collection of theorems in this paper omit some implication theorems that are trivially derived by mutual transitivity. For example, one such theorem is that $\square \neg p$ implies $\Diamond \neg p$, which follows from the theorems that $\square \neg p$ implies $\neg \square p$ and that $\neg \square p$ is equivalent to $\Diamond \neg p$. The edge labeled by (87) thus represents four implication theorems, one for each combination of the two antecedents $\square \neg p$ and $\neg \Diamond p$ and the two consequents $\Diamond \neg p$ and $\neg \square p$. Likewise, the edge labeled by (105) represents two implication theorems.

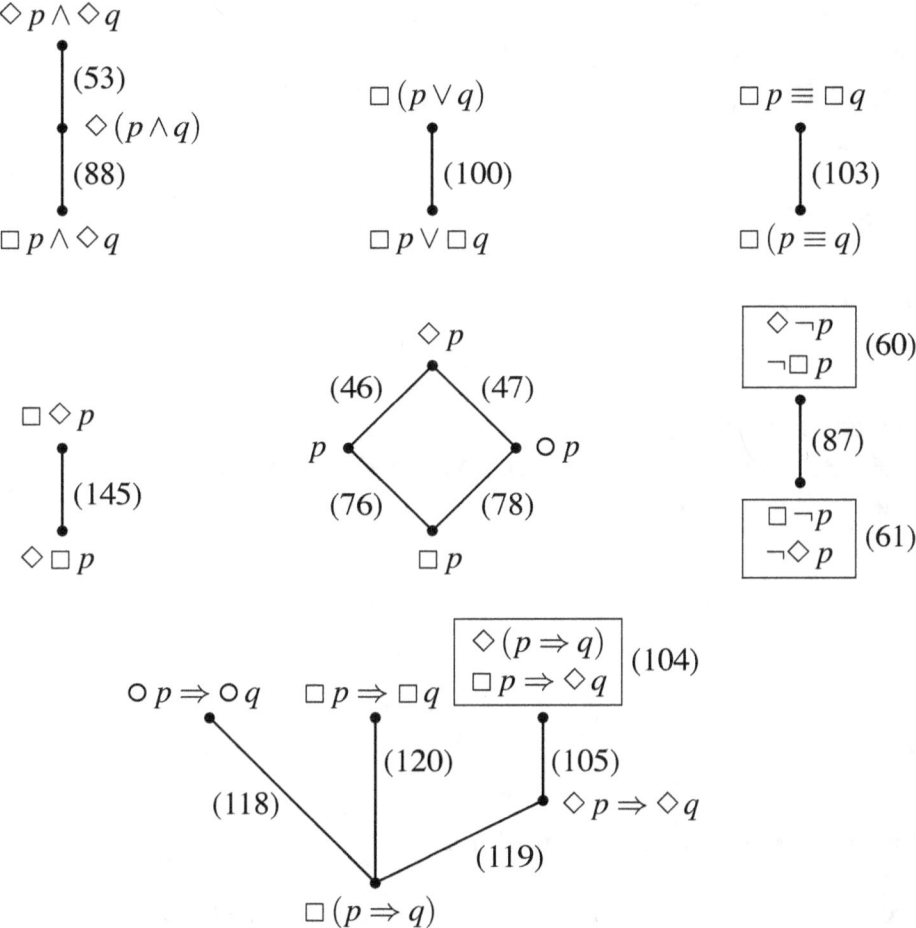

Figure 3: Seven Hasse diagrams showing some implication relations of linear temporal logic.

3.7 Proof Metatheorems

In the calculational system $\mathcal{E}$, Gries and Schneider (LADM) prove the metatheorem (9.16), which states that P is a theorem iff $(\forall x \,|\colon P)$ is a theorem. [13] Theorems in $\mathcal{E}$ are thus said to be "implicitly universally quantified." A similar concept applies to temporal logic theorems in $\mathcal{E}$ except that the implicit application is in the temporal dimension. The following metatheorem shows that theorems "implicitly always hold." Case 1 in the proof below is known as the temporal generalization rule [24], and Case 2 is known as the specialization rule [21].

(136) **Metatheorem:** P is a theorem iff $\Box P$ is a theorem.

Proof: The proof is by (4.7) Mutual implication. The proof of each case is by (4.4) Deduction.

Case 1. If P is a theorem then $\Box P$ is a theorem.
Suppose P is a theorem. Because all theorems are equivalent to each other, and (3.4) *true* is a theorem, P is equivalent to *true*. Then, $\Box P$ can be proved to be a theorem as follows.

$$
\begin{array}{rl}
& \Box P \\
= & \langle P \text{ is a theorem} \rangle \\
& \Box \mathit{true} \\
= & \langle (64) \text{ Truth of } \Box \rangle \\
& \mathit{true}
\end{array}
$$

Case 2. If $\Box P$ is a theorem then P is a theorem.
Suppose $\Box P$ is a theorem. Then, $\Box P$ is equivalent to *true*. P can be proved to be a theorem by (4.7.1) Truth implication as follows.

$$
\begin{array}{rl}
& P \\
\Leftarrow & \langle (76) \text{ Strengthening of } \Box \rangle \\
& \Box P \\
= & \langle \Box P \text{ is a theorem} \rangle \\
& \mathit{true} \quad \blacksquare
\end{array}
$$

Proofs of the following three metatheorems are similar.

(137) **Metatheorem** $\circ$: If $P \Rightarrow Q$ is a theorem then $\circ P \Rightarrow \circ Q$ is a theorem.

Proof: The proof is by (4.4) Deduction. Suppose $P \Rightarrow Q$ is a theorem. Then, by (136) Metatheorem, $\Box(P \Rightarrow Q)$ is a theorem. Because all theorems are equivalent to each other, and (3.4) *true* is a theorem, $\Box(P \Rightarrow Q)$ is equivalent to *true*. Then, $\circ P \Rightarrow \circ Q$ can be proved to be a theorem by (4.7.1) Truth implication as follows.

$$
\begin{array}{rl}
& \circ P \Rightarrow \circ Q \\
\Leftarrow & \langle (118) \text{ Monotonicity of } \circ \rangle \\
& \Box(P \Rightarrow Q) \\
= & \langle \Box(P \Rightarrow Q) \text{ is a theorem} \rangle \\
& \mathit{true} \quad \blacksquare
\end{array}
$$

(138) **Metatheorem** $\diamond$: If $P \Rightarrow Q$ is a theorem then $\diamond P \Rightarrow \diamond Q$ is a theorem.

Proof: The proof is by (4.4) Deduction. Suppose $P \Rightarrow Q$ is a theorem. Then, by (136) Metatheorem, $\Box(P \Rightarrow Q)$ is a theorem. Because all theorems are equivalent to each other, and (3.4) *true* is a theorem, $\Box(P \Rightarrow Q)$ is equivalent to *true*. Then, $\diamond P \Rightarrow \diamond Q$ can be proved to be a theorem by (4.7.1) Truth implication as follows.

Pepperdine Papers on LTL

$$\diamond P \Rightarrow \diamond Q$$
$$\Leftarrow \quad \langle (119) \text{ Monotonicity of } \diamond \rangle$$
$$\square(P \Rightarrow Q)$$
$$= \quad \langle \square(P \Rightarrow Q) \text{ is a theorem}\rangle$$
$$true \quad \blacksquare$$

(139) **Metatheorem $\square$:** If $P \Rightarrow Q$ is a theorem then $\square P \Rightarrow \square Q$ is a theorem.

Proof: The proof is by (4.4) Deduction. Suppose $P \Rightarrow Q$ is a theorem. Then, by (136) Metatheorem, $\square(P \Rightarrow Q)$ is a theorem. Because all theorems are equivalent to each other, and (3.4) *true* is a theorem, $\square(P \Rightarrow Q)$ is equivalent to *true*. Then, $\square P \Rightarrow \square Q$ can be proved to be a theorem by (4.7.1) Truth implication as follows.

$$\square P \Rightarrow \square Q$$
$$\Leftarrow \quad \langle (120) \text{ Monotonicity of } \square \rangle$$
$$\square(P \Rightarrow Q)$$
$$= \quad \langle \square(P \Rightarrow Q) \text{ is a theorem}\rangle$$
$$true \quad \blacksquare$$

The proof of (142) in Section 3.8 illustrates the use of (136) Metatheorem. The proof of (166) illustrates the use of (138) Metatheorem $\diamond$ and (139) Metatheorem $\square$.

3.8 Always, Continued

Theorems (140) and (141) do not seem to appear in the LTL literature. However, they play a key role here in the proofs of several later theorems that do exist in the literature. The proof of (140) is based on (129) Induction rule $\square$ with $p, q := p \, \mathcal{U} \, \square q, p \, \mathcal{U} \, q$. It establishes the truth of the antecedent with the help of two lemmas, each of which is proved with metatheorem (136). The strengthening and ordering theorems are also unique to this system.

(140) $\mathcal{U}\square$ **implication:** $p \, \mathcal{U} \, \square q \Rightarrow \square(p \, \mathcal{U} \, q)$

Proof: Theorem (129) Induction rule $\square$ with $p, q := p \, \mathcal{U} \, \square q, p \, \mathcal{U} \, q$ is

$$\square(p \, \mathcal{U} \, \square q \Rightarrow (p \, \mathcal{U} \, q \wedge \circ (p \, \mathcal{U} \, \square q))) \Rightarrow (p \, \mathcal{U} \, \square q \Rightarrow \square(p \, \mathcal{U} \, q))$$

Because the consequent is the theorem to be proved, it suffices to prove the truth of the antecedent.

$$\square(p \, \mathcal{U} \, \square q \Rightarrow (p \, \mathcal{U} \, q \wedge \circ (p \, \mathcal{U} \, \square q)))$$
$$= \quad \langle (3.63.1) \text{ Distributivity of } \Rightarrow \text{ over } \wedge, p \Rightarrow q \wedge r \equiv (p \Rightarrow q) \wedge (p \Rightarrow r)\rangle$$
$$\square((p \, \mathcal{U} \, \square q \Rightarrow p \, \mathcal{U} \, q) \wedge (p \, \mathcal{U} \, \square q \Rightarrow \circ(p \, \mathcal{U} \, \square q)))$$

$$\begin{aligned}
&= \quad \langle(99)\text{ Distributivity of }\Box\text{ over }\wedge\rangle \\
&\quad \Box(p\,\mathcal{U}\,\Box q \Rightarrow p\,\mathcal{U}\,q) \wedge \Box(p\,\mathcal{U}\,\Box q \Rightarrow \circ(p\,\mathcal{U}\,\Box q)) \\
&= \quad \langle\text{Lemma A: }\Box(p\,\mathcal{U}\,\Box q \Rightarrow p\,\mathcal{U}\,q)\rangle \\
&\quad true \wedge \Box(p\,\mathcal{U}\,\Box q \Rightarrow \circ(p\,\mathcal{U}\,\Box q)) \\
&= \quad \langle\text{Lemma B: }\Box(p\,\mathcal{U}\,\Box q \Rightarrow \circ(p\,\mathcal{U}\,\Box q))\rangle \\
&\quad true \wedge true \\
&= \quad \langle(3.38)\text{ Idempotency of }\wedge,\, p\wedge p \equiv p\rangle \\
&\quad true \quad \blacksquare
\end{aligned}$$

Lemma A: $\Box(p\,\mathcal{U}\,\Box q \Rightarrow p\,\mathcal{U}\,q)$

Proof: The proof is by (4.7.1) Truth implication.

$$\begin{aligned}
&\quad true \\
&= \quad \langle(76)\text{ Strengthening of }\Box\text{ with }p := q\rangle \\
&\quad \Box q \Rightarrow q \\
&= \quad \langle(136)\text{ Metatheorem with the above theorem}\rangle \\
&\quad \Box(\Box q \Rightarrow q) \\
&\Rightarrow \quad \langle(85)\text{ Right monotonicity of }\mathcal{U}\text{ with }p,q,r := \Box q, q, p\rangle \\
&\quad p\,\mathcal{U}\,\Box q \Rightarrow p\,\mathcal{U}\,q \\
&= \quad \langle(136)\text{ Metatheorem with the above theorem}\rangle \\
&\quad \Box(p\,\mathcal{U}\,\Box q \Rightarrow p\,\mathcal{U}\,q) \quad \blacksquare
\end{aligned}$$

Lemma B: $\Box(p\,\mathcal{U}\,\Box q \Rightarrow \circ(p\,\mathcal{U}\,\Box q))$

Proof: By (136) Metatheorem, Lemma B is valid iff

$$p\,\mathcal{U}\,\Box q \Rightarrow \circ(p\,\mathcal{U}\,\Box q)$$

is valid.

$$\begin{aligned}
&\quad p\,\mathcal{U}\,\Box q \\
&= \quad \langle(10)\text{ Expansion of }\mathcal{U}\rangle \\
&\quad \Box q \vee (p \wedge \circ(p\,\mathcal{U}\,\Box q)) \\
&= \quad \langle(3.45)\text{ Distributivity of }\vee\text{ over }\wedge,\, p\vee(q\wedge r) \equiv (p\vee q)\wedge(p\vee r)\rangle \\
&\quad (\Box q \vee p) \wedge (\Box q \vee \circ(p\,\mathcal{U}\,\Box q)) \\
&\Rightarrow \quad \langle(3.76b)\text{ Strengthening, }p\wedge q \Rightarrow p\rangle \\
&\quad \Box q \vee \circ(p\,\mathcal{U}\,\Box q) \\
&= \quad \langle(66)\text{ Expansion of }\Box\rangle \\
&\quad (q \wedge \circ\Box q) \vee \circ(p\,\mathcal{U}\,\Box q) \\
&= \quad \langle(3.45)\text{ Distributivity of }\vee\text{ over }\wedge,\, p\vee(q\wedge r) \equiv (p\vee q)\wedge(p\vee r)\rangle
\end{aligned}$$

$\quad (q \vee \circ (p\,\mathcal{U}\,\Box q)) \wedge (\circ \Box q \vee \circ (p\,\mathcal{U}\,\Box q))$
$\Rightarrow \quad \langle\text{(3.76b) Strengthening, } p \wedge q \Rightarrow p\rangle$
$\quad \circ \Box q \vee \circ (p\,\mathcal{U}\,\Box q)$
$= \quad \langle\text{(9) Distributivity of } \circ \text{ over } \mathcal{U}\rangle$
$\quad \circ \Box q \vee \circ p\,\mathcal{U} \circ \Box q$
$= \quad \langle\text{(32) Absorption with } p,q := \circ p, \circ \Box q\rangle$
$\quad \circ p\,\mathcal{U} \circ \Box q$
$= \quad \langle\text{(9) Distributivity of } \circ \text{ over } \mathcal{U}\rangle$
$\quad \circ (p\,\mathcal{U}\,\Box q)$ ∎

(141) **Absorption of $\mathcal{U}$ into $\Box$:** $\quad p\,\mathcal{U}\,\Box p \equiv \Box p$

Proof: The proof is by (4.7) Mutual implication. The proof in the first direction follows.

$\quad p\,\mathcal{U}\,\Box p$
$\Rightarrow \quad \langle\text{(140) } \mathcal{U}\,\Box \text{ implication with } q := p\rangle$
$\quad \Box (p\,\mathcal{U}\,p)$
$= \quad \langle\text{(22) Idempotency of } \mathcal{U}\rangle$
$\quad \Box p$

The proof in the second direction follows.

$\quad \Box p$
$\Rightarrow \quad \langle\text{(29) } \mathcal{U} \text{ insertion with } q := \Box p\rangle$
$\quad p\,\mathcal{U}\,\Box p$ ∎

(142) **Right $\wedge\,\mathcal{U}$ strengthening:** $\quad p\,\mathcal{U}\,(q \wedge r) \Rightarrow p\,\mathcal{U}\,(q\,\mathcal{U}\,r)$

Proof: The proof is by (4.7.1) Truth implication.

$\quad true$
$= \quad \langle\text{(136) Metatheorem and (30) } p \wedge q \Rightarrow p\,\mathcal{U}\,q \text{ with } p,q := q,r\rangle$
$\quad \Box(q \wedge r \Rightarrow q\,\mathcal{U}\,r)$
$\Rightarrow \quad \langle\text{(85) Right monotonicity of } \mathcal{U} \text{ with } p,q,r := q \wedge r, q\,\mathcal{U}\,r, p\rangle$
$\quad p\,\mathcal{U}\,(q \wedge r) \Rightarrow p\,\mathcal{U}\,(q\,\mathcal{U}\,r)$ ∎

(143) **Left $\wedge\,\mathcal{U}$ strengthening:** $\quad (p \wedge q)\,\mathcal{U}\,r \Rightarrow (p\,\mathcal{U}\,q)\,\mathcal{U}\,r$

Proof: The proof is by (4.7.1) Truth implication.

$$
\begin{aligned}
& \quad \textit{true} \\
&= \quad \langle (136) \text{ Metatheorem and } (30)\ p \wedge q \Rightarrow p\,\mathcal{U}\,q \rangle \\
& \quad \Box(p \wedge q \Rightarrow p\,\mathcal{U}\,q) \\
&\Rightarrow \quad \langle (86) \text{ Left monotonicity of } \mathcal{U} \text{ with } p,q := p \wedge q, p\,\mathcal{U}\,q \rangle \\
& \quad (p \wedge q)\,\mathcal{U}\,r \Rightarrow (p\,\mathcal{U}\,q)\,\mathcal{U}\,r \quad \blacksquare
\end{aligned}
$$

(144) **Left $\wedge\,\mathcal{U}$ ordering:** $\quad (p \wedge q)\,\mathcal{U}\,r \Rightarrow p\,\mathcal{U}\,(q\,\mathcal{U}\,r)$

Proof: Theorem (127) $\mathcal{U}$ strengthening rule with $p,q,r,s := p \wedge q, r, p, q\,\mathcal{U}\,r$ is

$$\Box((p \wedge q \Rightarrow p) \wedge (r \Rightarrow q\,\mathcal{U}\,r)) \Rightarrow (p \wedge q)\,\mathcal{U}\,r \Rightarrow p\,\mathcal{U}\,(q\,\mathcal{U}\,r)$$

Because the consequent is the theorem to be proved, it suffices to prove the truth of the antecedent.

$$
\begin{aligned}
& \quad \Box((p \wedge q \Rightarrow p) \wedge (r \Rightarrow q\,\mathcal{U}\,r)) \\
&= \quad \langle (3.76\text{b}) \text{ Strengthening, } p \wedge q \Rightarrow p \text{ and } (29)\ \mathcal{U} \text{ insertion} \rangle \\
& \quad \Box(\textit{true} \wedge \textit{true}) \\
&= \quad \langle (3.38) \text{ Idempotency of } \wedge,\ p \wedge p \equiv p \rangle \\
& \quad \Box\,\textit{true} \\
&= \quad \langle (64) \text{ Truth of } \Box \rangle \\
& \quad \textit{true} \quad \blacksquare
\end{aligned}
$$

The $\Diamond\Box$ implication theorem states that $\Diamond\Box p$ ensures that p will always eventually hold, but not the converse. Suppose, for example, that p continually oscillates between true and false over time. Then, the consequent of (145) is true, but the antecedent is false. Its proof uses (140). Theorem (146) is a second version of (95) $\Box\Diamond$ excluded middle, and is perhaps less intuitive. Its proof uses (145).

(145) $\Diamond\Box$ **implication:** $\quad \Diamond\Box p \Rightarrow \Box\Diamond p$

Proof:

$$
\begin{aligned}
& \quad \Diamond\Box p \\
&= \quad \langle (38) \text{ Definition of } \Diamond \rangle \\
& \quad \textit{true}\,\mathcal{U}\,\Box p \\
&\Rightarrow \quad \langle (140)\ \mathcal{U}\,\Box \text{ implication} \rangle \\
& \quad \Box(\textit{true}\,\mathcal{U}\,p) \\
&= \quad \langle (38) \text{ Definition of } \Diamond \rangle \\
& \quad \Box\Diamond p \quad \blacksquare
\end{aligned}
$$

Pepperdine Papers on LTL 77

(146) $\square\diamond$ **excluded middle:** $\square\diamond p \vee \square\diamond \neg p$

Proof:

$\quad\quad \square\diamond p \vee \square\diamond \neg p$
$= \quad \langle(62) \text{ Dual of } \diamond\square\rangle$
$\quad\quad \square\diamond p \vee \neg\diamond\square p$
$= \quad \langle(3.59) \text{ Implication, } p \Rightarrow q \equiv \neg p \vee q\rangle$
$\quad\quad \diamond\square p \Rightarrow \square\diamond p \quad -(145) \diamond\square \text{ implication} \quad \blacksquare$

(147) $\diamond\square$ **contradiction:** $\diamond\square p \wedge \diamond\square \neg p \equiv \text{false}$

Proof:

$\quad\quad \diamond\square p \wedge \diamond\square \neg p \equiv \text{false}$
$= \quad \langle(3.15) \neg p \equiv p \equiv \text{false}\rangle$
$\quad\quad \neg(\diamond\square p \wedge \diamond\square \neg p)$
$= \quad \langle(3.47a) \text{ De Morgan, } \neg(p \wedge q) \equiv \neg p \vee \neg q\rangle$
$\quad\quad \neg\diamond\square p \vee \neg\diamond\square \neg p$
$= \quad \langle(62) \text{ Dual of } \diamond\square \text{ and } (63) \text{ Dual of } \square\diamond\rangle$
$\quad\quad \square\diamond \neg p \vee \square\diamond \neg\neg p$
$= \quad \langle(3.12) \text{ Double negation, } \neg\neg p \equiv p\rangle$
$\quad\quad \square\diamond \neg p \vee \square\diamond p \quad -(146) \square\diamond \text{ excluded middle} \quad \blacksquare$

(148) $\mathcal{U}$ **frame law of** $\circ$: $\square p \Rightarrow (\circ q \Rightarrow \circ(p \mathcal{U} q))$

Proof:

$\quad\quad \square p \Rightarrow (\circ q \Rightarrow \circ(p \mathcal{U} q))$
$= \quad \langle(2) \text{ Distributivity of } \circ \text{ over } \Rightarrow\rangle$
$\quad\quad \square p \Rightarrow \circ(q \Rightarrow p \mathcal{U} q)$
$= \quad \langle(29) \mathcal{U} \text{ insertion}\rangle$
$\quad\quad \square p \Rightarrow \circ \text{ true}$
$= \quad \langle(7) \text{ Truth of } \circ\rangle$
$\quad\quad \square p \Rightarrow \text{true} \quad -(3.72) \text{ Right zero of } \Rightarrow, p \Rightarrow \text{true} \equiv \text{true} \quad \blacksquare$

(149) $\mathcal{U}$ **frame law of** $\diamond$: $\square p \Rightarrow (\diamond q \Rightarrow \diamond(p \mathcal{U} q))$

Proof:

$\quad\quad \square p \Rightarrow (\diamond q \Rightarrow \diamond(p \mathcal{U} q))$
$= \quad \langle(3.65) \text{ Shunting, } p \wedge q \Rightarrow r \equiv p \Rightarrow (q \Rightarrow r)\rangle$
$\quad\quad \square p \wedge \diamond q \Rightarrow \diamond(p \mathcal{U} q)$

And now,

$$\Box p \land \Diamond q$$
$\Rightarrow \quad \langle (84)\ \mathcal{U}\ \text{implication} \rangle$
$$p\ \mathcal{U}\ q$$
$\Rightarrow \quad \langle (46)\ \text{Weakening of}\ \Diamond,\ p \Rightarrow \Diamond p\ \text{with}\ p := p\ \mathcal{U}\ q \rangle$
$$\Diamond (p\ \mathcal{U}\ q)\ \blacksquare$$

(150) **$\mathcal{U}$ frame law of $\Box$:** $\quad \Box p \Rightarrow (\Box q \Rightarrow \Box (p\ \mathcal{U}\ q))$

Proof:

$$\Box p \Rightarrow (\Box q \Rightarrow \Box (p\ \mathcal{U}\ q))$$
$= \quad \langle (3.65)\ \text{Shunting},\ p \land q \Rightarrow r \equiv p \Rightarrow (q \Rightarrow r) \rangle$
$$\Box p \land \Box q \Rightarrow \Box (p\ \mathcal{U}\ q)$$

And now,

$$\Box p \land \Box q$$
$\Rightarrow \quad \langle (3.76b)\ \text{Strengthening},\ p \land q \Rightarrow p \rangle$
$$\Box q$$
$\Rightarrow \quad \langle (29)\ \mathcal{U}\ \text{insertion with}\ q := \Box q \rangle$
$$p\ \mathcal{U}\ \Box q$$
$\Rightarrow \quad \langle (140)\ \mathcal{U}\ \Box\ \text{implication} \rangle$
$$\Box (p\ \mathcal{U}\ q)\ \blacksquare$$

The absorption theorems, (151) and (152), together with absorption theorems (50) and (72), allow any arbitrary string of $\Diamond$ and $\Box$ operators of any arbitrary length to be collapsed into one of four expressions: $\Diamond p$, $\Box p$, $\Box \Diamond p$, or $\Diamond \Box p$. These theorems are common to all systems. The remaining absorption theorems (153) to (156) are simple extensions mentioned in a single source in our survey.

(151) **Absorption of $\Diamond$ into $\Box \Diamond$:** $\quad \Diamond \Box \Diamond p \equiv \Box \Diamond p$

Proof: The proof is by (4.7) Mutual implication.
The proof in the first direction follows.

$$\Diamond \Box \Diamond p$$
$\Rightarrow \quad \langle (145)\ \Diamond \Box\ \text{implication} \rangle$
$$\Box \Diamond \Diamond p$$
$= \quad \langle (50)\ \text{Absorption of}\ \Diamond \rangle$
$$\Box \Diamond p$$

The proof in the second direction follows.

$$\square\lozenge p$$
$$\Rightarrow \quad \langle (46) \text{ Weakening of } \lozenge \rangle$$
$$\lozenge\square\lozenge p \quad \blacksquare$$

(152) **Absorption of $\square$ into $\lozenge\square$:** $\quad \square\lozenge\square p \equiv \lozenge\square p$

Proof:

$$\square\lozenge\square p$$
$$= \quad \langle (3.12) \text{ Double negation, } \neg\neg p \equiv p \rangle$$
$$\square\lozenge\square\neg\neg p$$
$$= \quad \langle (63) \text{ Dual of } \square\lozenge \text{ and } (61) \text{ Dual of } \lozenge \rangle$$
$$\neg\lozenge\square\lozenge\neg p$$
$$= \quad \langle (151) \text{ Absorption of } \lozenge \text{ into } \square\lozenge \rangle$$
$$\neg\square\lozenge\neg p$$
$$= \quad \langle (62) \text{ Dual of } \lozenge\square \rangle$$
$$\neg\neg\lozenge\square p$$
$$= \quad \langle (3.12) \text{ Double negation, } \neg\neg p \equiv p \rangle$$
$$\lozenge\square p \quad \blacksquare$$

(153) **Absorption of $\square\lozenge$:** $\quad \square\lozenge\square\lozenge p \equiv \square\lozenge p$

Proof:

$$\square\lozenge\square\lozenge p$$
$$= \quad \langle (152) \text{ Absorption of } \square \text{ into } \lozenge\square \text{ with } p := \lozenge p \rangle$$
$$\lozenge\square\lozenge p$$
$$= \quad \langle (151) \text{ Absorption of } \lozenge \text{ into } \square\lozenge \rangle$$
$$\square\lozenge p \quad \blacksquare$$

(154) **Absorption of $\lozenge\square$:** $\quad \lozenge\square\lozenge\square p \equiv \lozenge\square p$

Proof:

$$\lozenge\square\lozenge\square p$$
$$= \quad \langle (151) \text{ Absorption of } \lozenge \text{ into } \square\lozenge \text{ with } p := \square p \rangle$$
$$\square\lozenge\square p$$
$$= \quad \langle (152) \text{ Absorption of } \square \text{ into } \lozenge\square \rangle$$
$$\lozenge\square p \quad \blacksquare$$

(155) **Absorption of $\circ$ into $\square\diamond$:** $\circ\square\diamond p \equiv \square\diamond p$

Proof: The proof is by (4.7) Mutual implication.
The proof in the first direction follows.

$\quad\quad \circ\square\diamond p$
$\Rightarrow \quad \langle(47)$ Weakening of $\diamond$ with $p := \square\diamond p\rangle$
$\quad\quad \diamond\square\diamond p$
$= \quad \langle(151)$ Absorption of $\diamond$ into $\square\diamond\rangle$
$\quad\quad \square\diamond p$

The proof in the second direction follows.

$\quad\quad \square\diamond p$
$\Rightarrow \quad \langle(79)$ Strengthening of $\square\rangle$
$\quad\quad \circ\square\diamond p$ ∎

(156) **Absorption of $\circ$ into $\diamond\square$:** $\circ\diamond\square p \equiv \diamond\square p$

Proof: The proof is by (4.7) Mutual implication.
The proof in the first direction follows.

$\quad\quad \circ\diamond\square p$
$\Rightarrow \quad \langle(47)$ Weakening of $\diamond$ with $p := \diamond\square p\rangle$
$\quad\quad \diamond\diamond\square p$
$= \quad \langle(50)$ Absorption of $\diamond\rangle$
$\quad\quad \diamond\square p$

The proof in the second direction follows.

$\quad\quad \diamond\square p$
$= \quad \langle(152)$ Absorption of $\square$ into $\diamond\square\rangle$
$\quad\quad \square\diamond\square p$
$\Rightarrow \quad \langle(78)$ Strengthening of $\square$ with $p := \diamond\square p\rangle$
$\quad\quad \circ\diamond\square p$ ∎

The proof of monotonicity theorem (157) uses (139) Metatheorem $\square$ and (119) Monotonicity of $\diamond$. The proof of monotonicity theorem (158) does the same with (120) Monotonicity of $\square$.

(157) **Monotonicity of $\square\diamond$:** $\square(p \Rightarrow q) \Rightarrow (\square\diamond p \Rightarrow \square\diamond q)$

Proof: The proof is by (4.7.1) Truth implication.

$\qquad$ *true*
$=\quad \langle$(139) Metatheorem $\Box$ and (119) Monotonicity of $\Diamond\rangle$
$\qquad \Box\Box(p \Rightarrow q) \Rightarrow \Box(\Diamond p \Rightarrow \Diamond q)$
$=\quad \langle$(72) Absorption of $\Box\rangle$
$\qquad \Box(p \Rightarrow q) \Rightarrow \Box(\Diamond p \Rightarrow \Diamond q)$
$=\quad \langle$(3.39) Identity of $\wedge$, $p \wedge true \equiv p$ and (120) Monotonicity of $\Box$
$\qquad$ with $p, q := \Diamond p, \Diamond q\rangle$
$\qquad (\Box(p \Rightarrow q) \Rightarrow \Box(\Diamond p \Rightarrow \Diamond q)) \wedge (\Box(\Diamond p \Rightarrow \Diamond q) \Rightarrow (\Box\Diamond p \Rightarrow \Box\Diamond q))$
$\Rightarrow\quad \langle$(3.82a) Transitivity, $(p \Rightarrow q) \wedge (q \Rightarrow r) \Rightarrow (p \Rightarrow r)\rangle$
$\qquad \Box(p \Rightarrow q) \Rightarrow (\Box\Diamond p \Rightarrow \Box\Diamond q)$ ∎

(158) **Monotonicity of $\Diamond\Box$:** $\quad \Box(p \Rightarrow q) \Rightarrow (\Diamond\Box p \Rightarrow \Diamond\Box q)$

Proof: The proof is by (4.7.1) Truth implication.

$\qquad$ *true*
$=\quad \langle$(139) Metatheorem $\Box$ and (120) Monotonicity of $\Box\rangle$
$\qquad \Box\Box(p \Rightarrow q) \Rightarrow \Box(\Box p \Rightarrow \Box q)$
$=\quad \langle$(72) Absorption of $\Box\rangle$
$\qquad \Box(p \Rightarrow q) \Rightarrow \Box(\Box p \Rightarrow \Box q)$
$=\quad \langle$(3.39) Identity of $\wedge$, $p \wedge true \equiv p$ and (119) Monotonicity of $\Diamond$
$\qquad$ with $p, q := \Box p, \Box q\rangle$
$\qquad (\Box(p \Rightarrow q) \Rightarrow \Box(\Box p \Rightarrow \Box q)) \wedge (\Box(\Box p \Rightarrow \Box q) \Rightarrow (\Diamond\Box p \Rightarrow \Diamond\Box q))$
$\Rightarrow\quad \langle$(3.82a) Transitivity, $(p \Rightarrow q) \wedge (q \Rightarrow r) \Rightarrow (p \Rightarrow r)\rangle$
$\qquad \Box(p \Rightarrow q) \Rightarrow (\Diamond\Box p \Rightarrow \Diamond\Box q)$ ∎

The next group of four distributivity theorems show how $\Box\Diamond$ and $\Diamond\Box$ distribute over conjunction and disjunction. Theorem (159) shows that $\Box\Diamond$ distributes over conjunction only in one direction. Similarly, Theorem (160) shows that $\Diamond\Box$ distributes over disjunction only in one direction. However, Theorems (161) and (162) show that $\Box\Diamond$ distributes over disjunction and $\Diamond\Box$ distributes over conjunction in both directions.

(159) **Distributivity of $\Box\Diamond$ over $\wedge$:** $\quad \Box\Diamond(p \wedge q) \Rightarrow \Box\Diamond p \wedge \Box\Diamond q$

Proof:

$\qquad \Box\Diamond(p \wedge q)$
$\Rightarrow\quad \langle$(139) Metatheorem $\Box$ and (53) Distributivity of $\Diamond$ over $\wedge$,
$\qquad \Box\Diamond(p \wedge q) \Rightarrow \Box(\Diamond p \wedge \Diamond q)\rangle$
$\qquad \Box(\Diamond p \wedge \Diamond q)$
$=\quad \langle$(99) Distributivity of $\Box$ over $\wedge\rangle$
$\qquad \Box\Diamond p \wedge \Box\Diamond q$ ∎

(160) **Distributivity of $\Diamond\Box$ over $\lor$:** $\Diamond\Box p \lor \Diamond\Box q \Rightarrow \Diamond\Box(p \lor q)$

Proof:

$\quad\quad \Diamond\Box p \lor \Diamond\Box q$
$= \quad \langle(52) \text{ Distributivity of } \Diamond \text{ over } \lor\rangle$
$\quad\quad \Diamond(\Box p \lor \Box q)$
$\Rightarrow \quad \langle(138) \text{ Metatheorem } \Diamond \text{ and } (100) \text{ Distributivity of } \Box \text{ over } \lor,$
$\quad\quad\quad \Diamond(\Box p \lor \Box q) \Rightarrow \Diamond\Box(p \lor q)\rangle$
$\quad\quad \Diamond\Box(p \lor q) \quad \blacksquare$

(161) **Distributivity of $\Box\Diamond$ over $\lor$:** $\Box\Diamond(p \lor q) \equiv \Box\Diamond p \lor \Box\Diamond q$

Proof: The proof is by (4.7) Mutual implication.
The proof in the first direction follows.

$\quad\quad \Box\Diamond p \lor \Box\Diamond q$
$\Rightarrow \quad \langle(100) \text{ Distributivity of } \Box \text{ over } \lor \text{ with } p,q := \Diamond p, \Diamond q\rangle$
$\quad\quad \Box(\Diamond p \lor \Diamond q)$
$= \quad \langle(52) \text{ Distributivity of } \Diamond \text{ over } \lor\rangle$
$\quad\quad \Box\Diamond(p \lor q)$

The proof in the second direction is by (4.7.1) Truth implication.

$\quad\quad \textit{true}$
$= \quad \langle(88) \text{ with } p,q := \Diamond(p \lor q), \Box\neg p\rangle$
$\quad\quad \Box\Diamond(p \lor q) \land \Diamond\Box\neg p \Rightarrow \Diamond(\Diamond(p \lor q) \land \Box\neg p)$
$\Rightarrow \quad \langle\text{Lemma A: } \Diamond(\Diamond(p \lor q) \land \Box\neg p) \Rightarrow \Diamond q \text{ and } (3.82a) \text{ Transitivity}\rangle$
$\quad\quad \Box\Diamond(p \lor q) \land \Diamond\Box\neg p \Rightarrow \Diamond q$
$= \quad \langle(139) \text{ Metatheorem } \Box \text{ with the above theorem}\rangle$
$\quad\quad \Box(\Box\Diamond(p \lor q) \land \Diamond\Box\neg p) \Rightarrow \Box\Diamond q$
$= \quad \langle(99) \text{ Distributivity of } \Box \text{ over } \land\rangle$
$\quad\quad \Box\Box\Diamond(p \lor q) \land \Box\Diamond\Box\neg p \Rightarrow \Box\Diamond q$
$= \quad \langle(72) \text{ Absorption of } \Box \text{ and } (152) \text{ Absorption of } \Box \text{ into } \Diamond\Box\rangle$
$\quad\quad \Box\Diamond(p \lor q) \land \Diamond\Box\neg p \Rightarrow \Box\Diamond q$
$= \quad \langle(3.65) \text{ Shunting}, p \land q \Rightarrow r \equiv p \Rightarrow (q \Rightarrow r)\rangle$
$\quad\quad \Box\Diamond(p \lor q) \Rightarrow (\Diamond\Box\neg p \Rightarrow \Box\Diamond q)$
$= \quad \langle(63) \text{ Dual of } \Box\Diamond\rangle$
$\quad\quad \Box\Diamond(p \lor q) \Rightarrow (\neg\Box\Diamond p \Rightarrow \Box\Diamond q)$
$= \quad \langle(3.59) \text{ Implication}, p \Rightarrow q \equiv \neg p \lor q\rangle$
$\quad\quad \Box\Diamond(p \lor q) \Rightarrow \Box\Diamond p \lor \Box\Diamond q \quad \blacksquare$

Pepperdine Papers on LTL 83

Lemma A: $\Diamond(\Diamond(p \vee q) \wedge \Box \neg p) \Rightarrow \Diamond q$
Proof: The proof is by (4.7.1) Truth implication.

$\quad\quad$ *true*
$=\quad \langle$(88) Distributivity of $\Diamond$ over $\wedge$ with $p, q := \neg p, p \vee q\rangle$
$\quad\quad \Box \neg p \wedge \Diamond(p \vee q) \Rightarrow \Diamond(\neg p \wedge (p \vee q))$
$=\quad \langle$(3.44a) Absorption, $p \wedge (\neg p \vee q) \equiv p \wedge q\rangle$
$\quad\quad \Box \neg p \wedge \Diamond(p \vee q) \Rightarrow \Diamond(\neg p \wedge q)$
$=\quad \langle$(138) Metatheorem $\Diamond$ with the above theorem$\rangle$
$\quad\quad \Diamond(\Box \neg p \wedge \Diamond(p \vee q)) \Rightarrow \Diamond\Diamond(\neg p \wedge q)$
$=\quad \langle$(50) Absorption of $\Diamond$ and (3.36) Symmetry of $\wedge$, $p \wedge q \equiv q \wedge p\rangle$
$\quad\quad \Diamond(\Diamond(p \vee q) \wedge \Box \neg p) \Rightarrow \Diamond(\neg p \wedge q)$
$\Rightarrow\quad \langle$(53) Distributivity of $\Diamond$ over $\wedge$ and (3.82a) Transitivity$\rangle$
$\quad\quad \Diamond(\Diamond(p \vee q) \wedge \Box \neg p) \Rightarrow \Diamond \neg p \wedge \Diamond q$
$\Rightarrow\quad \langle$(3.76b) Strengthening, $p \wedge q \Rightarrow p$ and (3.82a) Transitivity$\rangle$
$\quad\quad \Diamond(\Diamond(p \vee q) \wedge \Box \neg p) \Rightarrow \Diamond q$ ∎

(162) **Distributivity of $\Diamond\Box$ over $\wedge$:** $\quad \Diamond\Box(p \wedge q) \equiv \Diamond\Box p \wedge \Diamond\Box q$

Proof:

$\quad\quad \Diamond\Box(p \wedge q)$
$=\quad \langle$(3.12) Double negation, $\neg\neg p \equiv p$, twice$\rangle$
$\quad\quad \Diamond\Box(\neg\neg p \wedge \neg\neg q)$
$=\quad \langle$(3.47b) De Morgan, $\neg(p \vee q) \equiv \neg p \wedge \neg q\rangle$
$\quad\quad \Diamond\Box\neg(\neg p \vee \neg q)$
$=\quad \langle$(63) Dual of $\Box\Diamond\rangle$
$\quad\quad \neg\Box\Diamond(\neg p \vee \neg q)$
$=\quad \langle$(161) Distributivity of $\Box\Diamond$ over $\vee\rangle$
$\quad\quad \neg(\Box\Diamond\neg p \vee \Box\Diamond\neg q)$
$=\quad \langle$(62) Dual of $\Diamond\Box\rangle$
$\quad\quad \neg(\neg\Diamond\Box p \vee \neg\Diamond\Box q)$
$=\quad \langle$(3.47a) De Morgan, $\neg(p \wedge q) \equiv \neg p \vee \neg q\rangle$
$\quad\quad \neg\neg(\Diamond\Box p \wedge \Diamond\Box q)$
$=\quad \langle$(3.12) Double negation, $\neg\neg p \equiv p\rangle$
$\quad\quad \Diamond\Box p \wedge \Diamond\Box q$ ∎

Theorem (163) is Problem 4.2 in Manna and Pnueli. [21] Theorems (164) and (165) are Exercise 14.6 and 14.7 respectively in Ben-Ari. [3]

(163) **Eventual latching:** $\quad \Diamond\Box(p \Rightarrow \Box q) \equiv \Diamond\Box\neg p \vee \Diamond\Box q$

Proof: The proof is by (4.7) Mutual implication.
The proof in the first direction follows.

$\Diamond\Box(p \Rightarrow \Box q)$
$\Rightarrow$ ⟨Lemma A: $\Diamond\Box(p \Rightarrow \Box q) \Rightarrow \Diamond(\Box\Diamond p \Rightarrow \Diamond\Box q)$⟩
$\Diamond(\Box\Diamond p \Rightarrow \Diamond\Box q)$
$=$ ⟨(104) Distributivity of $\Diamond$ over $\Rightarrow$⟩
$\Box\Box\Diamond p \Rightarrow \Diamond\Diamond\Box q$
$=$ ⟨(72) Absorption of $\Box$ and (50) Absorption of $\Diamond$⟩
$\Box\Diamond p \Rightarrow \Diamond\Box q$
$=$ ⟨(3.59) Implication, $p \Rightarrow q \equiv \neg p \vee q$⟩
$\neg\Box\Diamond p \vee \Diamond\Box q$
$=$ ⟨(63) Dual of $\Box\Diamond$⟩
$\Diamond\Box\neg p \vee \Diamond\Box q$

The proof in the second direction follows.

$\Diamond\Box(p \Rightarrow \Box q)$
$=$ ⟨(3.59) Implication, $p \Rightarrow q \equiv \neg p \vee q$⟩
$\Diamond\Box(\neg p \vee \Box q)$
$\Leftarrow$ ⟨(160) Distributivity of $\Diamond\Box$ over $\vee$⟩
$\Diamond\Box\neg p \vee \Diamond\Box\Box q$
$=$ ⟨(72) Absorption of $\Box$⟩
$\Diamond\Box\neg p \vee \Diamond\Box q$ ∎

Lemma A: $\Diamond\Box(p \Rightarrow \Box q) \Rightarrow \Diamond(\Box\Diamond p \Rightarrow \Diamond\Box q)$

Proof: The proof is by (4.7.1) Truth implication.

true
$=$ ⟨(119) Monotonicity of $\Diamond$ with $q := \Box q$⟩
$\Box(p \Rightarrow \Box q) \Rightarrow (\Diamond p \Rightarrow \Diamond\Box q)$
$=$ ⟨(136) Metatheorem with the above theorem⟩
$\Box(\Box(p \Rightarrow \Box q) \Rightarrow (\Diamond p \Rightarrow \Diamond\Box q))$
$\Rightarrow$ ⟨(157) Monotonicity of $\Box\Diamond$⟩
$\Box\Diamond\Box(p \Rightarrow \Box q) \Rightarrow \Box\Diamond(\Diamond p \Rightarrow \Diamond\Box q)$
$=$ ⟨(152) Absorption of $\Box$ into $\Diamond\Box$⟩
$\Diamond\Box(p \Rightarrow \Box q) \Rightarrow \Box\Diamond(\Diamond p \Rightarrow \Diamond\Box q)$
$\Rightarrow$ ⟨(76) Strengthening of $\Box$ and (3.82a) Transitivity⟩
$\Diamond\Box(p \Rightarrow \Box q) \Rightarrow \Diamond(\Diamond p \Rightarrow \Diamond\Box q)$
$=$ ⟨(104) Distributivity of $\Diamond$ over $\Rightarrow$⟩
$\Diamond\Box(p \Rightarrow \Box q) \Rightarrow (\Box\Diamond p \Rightarrow \Diamond\Diamond\Box q)$

$$\begin{aligned}
&= \quad \langle (50) \text{ Absorption of } \Diamond \rangle \\
&\quad \Diamond\Box(p \Rightarrow \Box q) \Rightarrow (\Box\Diamond p \Rightarrow \Diamond\Box q) \\
&\Rightarrow \quad \langle (46) \text{ Weakening of } \Diamond \text{ and } (3.82a) \text{ Transitivity} \rangle \\
&\quad \Diamond\Box(p \Rightarrow \Box q) \Rightarrow \Diamond(\Box\Diamond p \Rightarrow \Diamond\Box q) \quad \blacksquare
\end{aligned}$$

(164) $\Box(\Box\Diamond p \Rightarrow \Diamond q) \equiv \Diamond\Box\neg p \lor \Box\Diamond q$

Proof: The proof is by (4.7) Mutual implication.
The proof in the first direction follows.

$$\begin{aligned}
&\quad \Box(\Box\Diamond p \Rightarrow \Diamond q) \\
&\Rightarrow \quad \langle (120) \text{ Monotonicity of } \Box \text{ with } p,q := \Box\Diamond p, \Diamond q \rangle \\
&\quad \Box\Box\Diamond p \Rightarrow \Box\Diamond q \\
&= \quad \langle (72) \text{ Absorption of } \Box \rangle \\
&\quad \Box\Diamond p \Rightarrow \Box\Diamond q \\
&= \quad \langle (3.59) \text{ Implication, } p \Rightarrow q \equiv \neg p \lor q \rangle \\
&\quad \neg\Box\Diamond p \lor \Box\Diamond q \\
&= \quad \langle (63) \text{ Dual of } \Box\Diamond \rangle \\
&\quad \Diamond\Box\neg p \lor \Box\Diamond q
\end{aligned}$$

The proof in the second direction follows.

$$\begin{aligned}
&\quad \Box(\Box\Diamond p \Rightarrow \Diamond q) \\
&= \quad \langle (3.59) \text{ Implication, } p \Rightarrow q \equiv \neg p \lor q \rangle \\
&\quad \Box(\neg\Box\Diamond p \lor \Diamond q) \\
&\Leftarrow \quad \langle (100) \text{ Distributivity of } \Box \text{ over } \lor \rangle \\
&\quad \Box\neg\Box\Diamond p \lor \Box\Diamond q \\
&= \quad \langle (63) \text{ Dual of } \Box\Diamond \rangle \\
&\quad \Box\Diamond\Box\neg p \lor \Box\Diamond q \\
&= \quad \langle (152) \text{ Absorption of } \Box \text{ into } \Diamond\Box \rangle \\
&\quad \Diamond\Box\neg p \lor \Box\Diamond q \quad \blacksquare
\end{aligned}$$

(165) $\Box((p \lor \Box q) \land (\Box p \lor q)) \equiv \Box p \lor \Box q$

Proof: The proof is by (4.7) Mutual implication and is based on the following lemmas, where R is defined as the expression

$$R : (p \lor \Box q) \land (\Box p \lor q)$$

With this definition, the theorem to be proved is

$$\Box R \equiv \Box p \lor \Box q$$

Lemma A: $R \equiv \Box p \vee \Box q \vee (p \wedge q)$

Proof:

$\quad R$
$= \quad \langle \text{Definition of } R \rangle$
$\quad (p \vee \Box q) \wedge (\Box p \vee q)$
$= \quad \langle (3.46) \text{ Distributivity of } \wedge \text{ over } \vee, p \wedge (q \vee r) \equiv (p \wedge q) \vee (p \wedge r) \rangle$
$\quad (p \wedge \Box p) \vee (p \wedge q) \vee (\Box p \wedge \Box q) \vee (q \wedge \Box q)$
$= \quad \langle (68) \text{ Absorption of } \wedge \text{ into } \Box, \text{twice} \rangle$
$\quad \Box p \vee (p \wedge q) \vee (\Box p \wedge \Box q) \vee \Box q$
$= \quad \langle (3.43\text{b}) \text{ Absorption}, p \vee (p \wedge q) \equiv p \rangle$
$\quad \Box p \vee \Box q \vee (p \wedge q) \quad \blacksquare$

Lemma B: $\Box R \wedge \neg \Box p \wedge \neg \Box q \Rightarrow \circ (\Box R \wedge \neg \Box p \wedge \neg \Box q)$

Proof:

$\quad \Box R \wedge \neg \Box p \wedge \neg \Box q$
$= \quad \langle (66) \text{ Expansion of } \Box \rangle$
$\quad R \wedge \circ \Box R \wedge \neg \Box p \wedge \neg \Box q$
$= \quad \langle (3.36) \text{ Symmetry of } \wedge, p \wedge q \equiv q \wedge p \rangle$
$\quad \circ \Box R \wedge R \wedge \neg \Box p \wedge \neg \Box q$
$= \quad \langle \text{Lemma A: } R \equiv \Box p \vee \Box q \vee (p \wedge q) \rangle$
$\quad \circ \Box R \wedge (\Box p \vee \Box q \vee (p \wedge q)) \wedge \neg \Box p \wedge \neg \Box q$
$= \quad \langle (3.44\text{a}) \text{ Absorption}, p \wedge (\neg p \vee q) \equiv p \wedge q, \text{twice} \rangle$
$\quad \circ \Box R \wedge (\neg \Box p \wedge \neg \Box q \wedge (p \wedge q))$
$= \quad \langle (66) \text{ Expansion of } \Box \text{ and } (3.47\text{a}) \text{ De Morgan}, \neg (p \wedge q) \equiv \neg p \vee \neg q, \text{twice} \rangle$
$\quad \circ \Box R \wedge (p \wedge q) \wedge (\neg p \vee \neg \circ \Box p) \wedge (\neg q \vee \neg \circ \Box q)$
$= \quad \langle (1) \text{ Self dual, twice} \rangle$
$\quad \circ \Box R \wedge (p \wedge q) \wedge (\neg p \vee \circ \neg \Box p) \wedge (\neg q \vee \circ \neg \Box q)$
$= \quad \langle (3.44\text{a}) \text{ Absorption}, p \wedge (\neg p \vee q) \equiv p \wedge q, \text{twice} \rangle$
$\quad \circ \Box R \wedge p \wedge \circ \neg \Box p \wedge q \wedge \circ \neg \Box q$
$\Rightarrow \quad \langle (3.76\text{b}) \text{ Strengthening}, p \wedge q \Rightarrow p \rangle$
$\quad \circ \Box R \wedge \circ \neg \Box p \wedge \circ \neg \Box q$
$= \quad \langle (5) \text{ Distributivity of } \circ \text{ over } \wedge \rangle$
$\quad \circ (\Box R \wedge \neg \Box p \wedge \neg \Box q) \quad \blacksquare$

Lemma C: $\Box R \wedge \neg \Box p \wedge \neg \Box q \Rightarrow \Box (\Box R \wedge \neg \Box p \wedge \neg \Box q)$

Proof: The proof is by (4.7.1) Truth implication.

    $true$
$=$ $\langle$Lemma B: $\Box R \wedge \neg \Box p \wedge \neg \Box q \Rightarrow \bigcirc (\Box R \wedge \neg \Box p \wedge \neg \Box q)\rangle$
    $\Box R \wedge \neg \Box p \wedge \neg \Box q \Rightarrow \bigcirc (\Box R \wedge \neg \Box p \wedge \neg \Box q)$
$=$ $\langle$(136) Metatheorem with the above theorem$\rangle$
    $\Box (\Box R \wedge \neg \Box p \wedge \neg \Box q \Rightarrow \bigcirc (\Box R \wedge \neg \Box p \wedge \neg \Box q))$
$\Rightarrow$ $\langle$(57) $\Box$ induction$\rangle$
    $\Box R \wedge \neg \Box p \wedge \neg \Box q \Rightarrow \Box (\Box R \wedge \neg \Box p \wedge \neg \Box q)$ ∎

Lemma D: $\Box (\Box R \wedge \neg \Box p \wedge \neg \Box q) \Rightarrow \Box p \wedge \Box q$

Proof: The proof is by (4.7.1) Truth implication.

    $true$
$=$ $\langle$(3.71) Reflexivity of $\Rightarrow$, $p \Rightarrow p\rangle$
    $\Box R \wedge \neg \Box p \wedge \neg \Box q \Rightarrow \Box R \wedge \neg \Box p \wedge \neg \Box q$
$\Rightarrow$ $\langle$(76) Strengthening of $\Box$ and (4.3) Monotonicity of $\wedge$
   and (3.82a) Transitivity$\rangle$
    $\Box R \wedge \neg \Box p \wedge \neg \Box q \Rightarrow R \wedge \neg \Box p \wedge \neg \Box q$
$=$ $\langle$Lemma A: $R \equiv \Box p \vee \Box q \vee (p \wedge q)\rangle$
    $\Box R \wedge \neg \Box p \wedge \neg \Box q \Rightarrow (\Box p \vee \Box q \vee (p \wedge q)) \wedge \neg \Box p \wedge \neg \Box q$
$=$ $\langle$(3.44a) Absorption, $p \wedge (\neg p \vee q) \equiv p \wedge q$, twice$\rangle$
    $\Box R \wedge \neg \Box p \wedge \neg \Box q \Rightarrow \neg \Box p \wedge \neg \Box q \wedge (p \wedge q)$
$\Rightarrow$ $\langle$(3.76b) Strengthening, $p \wedge q \Rightarrow p$ and (3.82a) Transitivity$\rangle$
    $\Box R \wedge \neg \Box p \wedge \neg \Box q \Rightarrow p \wedge q$
$=$ $\langle$(136) Metatheorem with the above theorem$\rangle$
    $\Box (\Box R \wedge \neg \Box p \wedge \neg \Box q \Rightarrow p \wedge q)$
$\Rightarrow$ $\langle$(120) Monotonicity of $\Box\rangle$
    $\Box (\Box R \wedge \neg \Box p \wedge \neg \Box q) \Rightarrow \Box (p \wedge q)$
$=$ $\langle$(99) Distributivity of $\Box$ over $\wedge\rangle$
    $\Box (\Box R \wedge \neg \Box p \wedge \neg \Box q) \Rightarrow \Box p \wedge \Box q$ ∎

Lemma E: $\Box R \wedge \neg \Box p \wedge \neg \Box q \Rightarrow \Box p \wedge \Box q$

Proof:

    $\Box R \wedge \neg \Box p \wedge \neg \Box q$
$\Rightarrow$ $\langle$Lemma C: $\Box R \wedge \neg \Box p \wedge \neg \Box q \Rightarrow \Box (\Box R \wedge \neg \Box p \wedge \neg \Box q)\rangle$
    $\Box (\Box R \wedge \neg \Box p \wedge \neg \Box q)$
$\Rightarrow$ $\langle$Lemma D: $\Box (\Box R \wedge \neg \Box p \wedge \neg \Box q) \Rightarrow \Box p \wedge \Box q\rangle$
    $\Box p \wedge \Box q$ ∎

Lemma F: $\Box p \vee \Box q \Rightarrow \bigcirc (\Box p \vee \Box q)$

Proof:

$\quad\quad \Box p \vee \Box q$
$= \quad \langle (66) \text{ Expansion of } \Box, \text{ twice} \rangle$
$\quad\quad (p \wedge \bigcirc \Box p) \vee (q \wedge \bigcirc \Box q)$
$\Rightarrow \quad \langle (3.76b) \text{ Strengthening}, p \wedge q \Rightarrow p \text{ and } (4.2) \text{ Monotonicity of } \vee, \text{ twice} \rangle$
$\quad\quad \bigcirc \Box p \vee \bigcirc \Box q$
$= \quad \langle (4) \text{ Distributivity of } \bigcirc \text{ over } \vee \rangle$
$\quad\quad \bigcirc (\Box p \vee \Box q) \quad \blacksquare$

Lemma G: $\Box (\Box p \vee \Box q) \Rightarrow \Box R$

Proof: The proof is by (4.7.1) Truth implication.

$\quad\quad true$
$= \quad \langle (3.76a) \text{ Weakening}, p \Rightarrow p \vee q \rangle$
$\quad\quad \Box p \vee \Box q \Rightarrow \Box p \vee \Box q \vee (p \wedge q)$
$= \quad \langle \text{Lemma A: } R \equiv \Box p \vee \Box q \vee (p \wedge q) \rangle$
$\quad\quad \Box p \vee \Box q \Rightarrow R$
$= \quad \langle (136) \text{ Metatheorem with the above theorem} \rangle$
$\quad\quad \Box (\Box p \vee \Box q \Rightarrow R)$
$\Rightarrow \quad \langle (120) \text{ Monotonicity of } \Box \rangle$
$\quad\quad \Box (\Box p \vee \Box q) \Rightarrow \Box R \quad \blacksquare$

The proof of (165) in the first direction is by (4.9) Proof by contradiction. To prove

$$\Box R \Rightarrow \Box p \vee \Box q$$

by contradiction, prove that

$$\neg (\Box R \Rightarrow \Box p \vee \Box q) \Rightarrow false$$

Proof:

$\quad\quad \neg (\Box R \Rightarrow \Box p \vee \Box q)$
$= \quad \langle (3.59) \text{ Implication}, p \Rightarrow q \equiv \neg p \vee q \rangle$
$\quad\quad \neg (\neg \Box R \vee \Box p \vee \Box q)$
$= \quad \langle (3.47b) \text{ De Morgan}, \neg (p \vee q) \equiv \neg p \wedge \neg q \rangle$
$\quad\quad \Box R \wedge \neg \Box p \wedge \neg \Box q$
$= \quad \langle (3.38) \text{ Idempotency of } \wedge, p \wedge p \equiv p \rangle$

$(\Box R \wedge \neg \Box p \wedge \neg \Box q) \wedge (\Box R \wedge \neg \Box p \wedge \neg \Box q)$
$\Rightarrow$ ⟨Lemma E: $\Box R \wedge \neg \Box p \wedge \neg \Box q \Rightarrow \Box p \wedge \Box q$
  and (4.3) Monotonicity of $\wedge$⟩
$\Box R \wedge \neg \Box p \wedge \neg \Box q \wedge \Box p \wedge \Box q$
$=$ ⟨(3.42) Contradiction, $p \wedge \neg p \equiv false$, twice⟩
$\Box R \wedge false \wedge false$
$=$ ⟨(3.40) Zero of $\wedge$, $p \wedge false \equiv false$⟩
$false$

The proof of (165) in the second direction

$$\Box p \vee \Box q \Rightarrow \Box R$$

is by (4.7.1) Truth implication.

Proof:

$true$
$=$ ⟨Lemma F: $\Box p \vee \Box q \Rightarrow \bigcirc (\Box p \vee \Box q)$⟩
$\Box p \vee \Box q \Rightarrow \bigcirc (\Box p \vee \Box q)$
$=$ ⟨(136) Metatheorem with the above theorem⟩
$\Box (\Box p \vee \Box q \Rightarrow \bigcirc (\Box p \vee \Box q))$
$\Rightarrow$ ⟨(57) $\Box$ induction⟩
$\Box p \vee \Box q \Rightarrow \Box (\Box p \vee \Box q)$
$\Rightarrow$ ⟨Lemma G: $\Box (\Box p \vee \Box q) \Rightarrow \Box R$ and (3.82a) Transitivity⟩
$\Box p \vee \Box q \Rightarrow \Box R$ ∎

The metatheorems and absorption laws imply the following intuitive theorem. If p will eventually be always true, and it is always the case that q will be eventually true, then it is always the case that $p \wedge q$ will eventually be true.

(166) $\Diamond \Box p \wedge \Box \Diamond q \Rightarrow \Box \Diamond (p \wedge q)$

Proof:

$true$
$=$ ⟨(88) Distributivity of $\Diamond$ over $\wedge$ and (139) Metatheorem $\Box$⟩
$\Box (\Box p \wedge \Diamond q) \Rightarrow \Box \Diamond (p \wedge q)$
$=$ ⟨(99) Distributivity of $\Box$ over $\wedge$⟩
$\Box \Box p \wedge \Box \Diamond q \Rightarrow \Box \Diamond (p \wedge q)$
$=$ ⟨(72) Absorption of $\Box$⟩

$$\square p \wedge \square \Diamond q \Rightarrow \square \Diamond (p \wedge q)$$
$= \quad \langle (3.65) \text{ Shunting}, p \wedge q \Rightarrow r \equiv p \Rightarrow (q \Rightarrow r) \rangle$
$$\square p \Rightarrow (\square \Diamond q \Rightarrow \square \Diamond (p \wedge q))$$
$= \quad \langle (138) \text{ Metatheorem } \Diamond \text{ with the above theorem} \rangle$
$$\Diamond \square p \Rightarrow \Diamond (\square \Diamond q \Rightarrow \square \Diamond (p \wedge q))$$
$= \quad \langle (104) \text{ Distributivity of } \Diamond \text{ over } \Rightarrow \rangle$
$$\Diamond \square p \Rightarrow (\Diamond \square \Diamond q \Rightarrow \Diamond \square \Diamond (p \wedge q))$$
$= \quad \langle (72) \text{ Absorption of } \square \text{ and } (151) \text{ Absorption of } \Diamond \text{ into } \square \rangle$
$$\Diamond \square p \Rightarrow (\square \Diamond q \Rightarrow \square \Diamond (p \wedge q))$$
$= \quad \langle (3.65) \text{ Shunting}, p \wedge q \Rightarrow r \equiv p \Rightarrow (q \Rightarrow r) \rangle$
$$\Diamond \square p \wedge \square \Diamond q \Rightarrow \square \Diamond (p \wedge q) \quad \blacksquare$$

(167) $\square ((\square p \Rightarrow \Diamond q) \wedge (q \Rightarrow \circ r)) \Rightarrow (\square p \Rightarrow \circ \square \Diamond r)$

Proof:

$$\square ((\square p \Rightarrow \Diamond q) \wedge (q \Rightarrow \circ r)) \Rightarrow (\square p \Rightarrow \circ \square \Diamond r)$$
$= \quad \langle (3.65) \text{ Shunting}, p \wedge q \Rightarrow r \equiv p \Rightarrow (q \Rightarrow r) \rangle$
$$\square ((\square p \Rightarrow \Diamond q) \wedge (q \Rightarrow \circ r)) \wedge \square p \Rightarrow \circ \square \Diamond r$$

And now,

$$\square ((\square p \Rightarrow \Diamond q) \wedge (q \Rightarrow \circ r)) \wedge \square p$$
$= \quad \langle (99) \text{ Distributivity of } \square \text{ over } \wedge \rangle$
$$\square (\square p \Rightarrow \Diamond q) \wedge \square (q \Rightarrow \circ r) \wedge \square p$$
$\Rightarrow \quad \langle (46) \text{ Weakening of } \Diamond \text{ with } p := \square p \text{ and } (4.3) \text{ Monotonicity of } \wedge \rangle$
$$\square (\square p \Rightarrow \Diamond q) \wedge \square (q \Rightarrow \circ r) \wedge \Diamond \square p$$
$\Rightarrow \quad \langle (157) \text{ Monotonicity of } \square \Diamond \text{ and } (4.3) \text{ Monotonicity of } \wedge \rangle$
$$(\square \Diamond \square p \Rightarrow \square \Diamond \Diamond q) \wedge \square (q \Rightarrow \circ r) \wedge \Diamond \square p$$
$= \quad \langle (152) \text{ Absorption of } \square \text{ into } \Diamond \square \text{ and } (50) \text{ Absorption of } \Diamond \rangle$
$$(\Diamond \square p \Rightarrow \square \Diamond q) \wedge \square (q \Rightarrow \circ r) \wedge \Diamond \square p$$
$\Rightarrow \quad \langle (3.77) \text{ Modus ponens}, p \wedge (p \Rightarrow q) \Rightarrow q \text{ and } (4.3) \text{ Monotonicity of } \wedge \rangle$
$$\square \Diamond q \wedge \square (q \Rightarrow \circ r)$$
$\Rightarrow \quad \langle (157) \text{ Monotonicity of } \square \Diamond \text{ and } (4.3) \text{ Monotonicity of } \wedge \rangle$
$$\square \Diamond q \wedge (\square \Diamond q \Rightarrow \square \Diamond \circ r)$$
$\Rightarrow \quad \langle (3.77) \text{ Modus ponens}, p \wedge (p \Rightarrow q) \Rightarrow q \rangle$
$$\square \Diamond \circ r$$
$= \quad \langle (51) \text{ Exchange of } \circ \text{ and } \Diamond \text{ and } (73) \text{ Exchange of } \circ \text{ and } \square \rangle$
$$\circ \square \Diamond r \quad \blacksquare$$

(168) **Progress proof rule:** $\Diamond\Box p \land \Box(\Box p \Rightarrow \Diamond q) \Rightarrow \Diamond q$

Proof:

$\quad \Diamond\Box p \land \Box(\Box p \Rightarrow \Diamond q)$
$\Rightarrow \quad \langle(119)\text{ Monotonicity of }\Diamond\text{ and }(4.3)\text{ Monotonicity of }\land\rangle$
$\quad \Diamond\Box p \land (\Diamond\Box p \Rightarrow \Diamond\Diamond q)$
$= \quad \langle(50)\text{ Absorption of }\Diamond\rangle$
$\quad \Diamond\Box p \land (\Diamond\Box p \Rightarrow \Diamond q)$
$\Rightarrow \quad \langle(3.77)\text{ Modus ponens, }p \land (p \Rightarrow q) \Rightarrow q\rangle$
$\quad \Diamond q \quad \blacksquare$

3.9 Wait

$p\,\mathcal{U}\,q$ requires p to be true until q, which is guaranteed to eventually be true. $p\,\mathcal{W}\,q$ has no such guarantee. That is, if q is eventually true then p is true until that time. But, if q is not eventually true then p must be true always. Accordingly, Kröger and Merz [20] refer to $\mathcal{W}$ as the *weak* version of $\mathcal{U}$. Equations (169) and (170) are the only defining axioms for the *wait* operator.

(169) **Definition of** $\mathcal{W}$: $\quad p\,\mathcal{W}\,q \equiv \Box p \lor p\,\mathcal{U}\,q$

(170) **Axiom, Distributivity of** $\neg$ **over** $\mathcal{W}$: $\quad \neg(p\,\mathcal{W}\,q) \equiv \neg q\,\mathcal{U}\,(\neg p \land \neg q)$

The defining equation gives $\mathcal{W}$ in terms of $\mathcal{U}$. The following theorem gives $\mathcal{U}$ in terms of $\mathcal{W}$.

(171) $\mathcal{U}$ **in terms of** $\mathcal{W}$: $\quad p\,\mathcal{U}\,q \equiv p\,\mathcal{W}\,q \land \Diamond q$

Proof:

$\quad p\,\mathcal{U}\,q \equiv p\,\mathcal{W}\,q \land \Diamond q$
$= \quad \langle(169)\text{ Definition of }\mathcal{W}\rangle$
$\quad p\,\mathcal{U}\,q \equiv (\Box p \lor p\,\mathcal{U}\,q) \land \Diamond q$
$= \quad \langle(3.46)\text{ Distributivity of }\land\text{ over }\lor, p \land (q \lor r) \equiv (p \land q) \lor (p \land r)\rangle$
$\quad p\,\mathcal{U}\,q \equiv (\Box p \land \Diamond q) \lor (p\,\mathcal{U}\,q \land \Diamond q)$
$= \quad \langle(39)\text{ Absorption of }\Diamond\text{ into }\mathcal{U}\rangle$
$\quad p\,\mathcal{U}\,q \equiv (\Box p \land \Diamond q) \lor p\,\mathcal{U}\,q$
$= \quad \langle(3.57)\text{ Definition of implication, }p \Rightarrow q \equiv p \lor q \equiv q\rangle$
$\quad \Box p \land \Diamond q \Rightarrow p\,\mathcal{U}\,q \quad -(84)\ \mathcal{U}\text{ implication} \quad \blacksquare$

(172) $p\,\mathcal{W}\,q \equiv \Box(p \land \neg q) \lor p\,\mathcal{U}\,q$

Proof:

$\square(p \wedge \neg q) \vee p\,\mathcal{U}\,q$
$= \quad \langle (99)\ \text{Distributivity of } \square \text{ over } \wedge \rangle$
$(\square p \wedge \square \neg q) \vee p\,\mathcal{U}\,q$
$= \quad \langle (3.45)\ \text{Distributivity of } \vee \text{ over } \wedge,\ p \vee (q \wedge r) \equiv (p \vee q) \wedge (p \vee r) \rangle$
$(\square p \vee p\,\mathcal{U}\,q) \wedge (\square \neg q \vee p\,\mathcal{U}\,q)$
$= \quad \langle (169)\ \text{Definition of } \mathcal{W} \rangle$
$p\,\mathcal{W}\,q \wedge (\square \neg q \vee p\,\mathcal{U}\,q)$
$= \quad \langle (171)\ \mathcal{U} \text{ in terms of } \mathcal{W} \rangle$
$p\,\mathcal{W}\,q \wedge (\square \neg q \vee (p\,\mathcal{W}\,q \wedge \Diamond q))$
$= \quad \langle (3.45)\ \text{Distributivity of } \vee \text{ over } \wedge,\ p \vee (q \wedge r) \equiv (p \vee q) \wedge (p \vee r) \rangle$
$p\,\mathcal{W}\,q \wedge ((\square \neg q \vee p\,\mathcal{W}\,q) \wedge (\square \neg q \vee \Diamond q))$
$= \quad \langle (89)\ \Diamond \text{ excluded middle and } (3.39)\ \text{Identity of } \wedge,\ p \wedge \textit{true} \equiv p \rangle$
$p\,\mathcal{W}\,q \wedge (\square \neg q \vee p\,\mathcal{W}\,q)$
$= \quad \langle (3.43a)\ \text{Absorption},\ p \wedge (p \vee q) \equiv p \rangle$
$p\,\mathcal{W}\,q \quad \blacksquare$

The proof of (173) Distributivity of $\neg$ over $\mathcal{U}$ uses (170) Axiom, Distributivity of $\neg$ over $\mathcal{W}$. Apparently, the reverse is not possible. That is, if (173) is taken as the axiom instead of (170), we were not able to prove (170) is a theorem.

(173) **Distributivity of $\neg$ over $\mathcal{U}$:** $\neg(p\,\mathcal{U}\,q) \equiv \neg q\,\mathcal{W}\,(\neg p \wedge \neg q)$

Proof: (Michael Ortiz)

$\neg q\,\mathcal{W}\,(\neg p \wedge \neg q)$
$= \quad \langle (169)\ \text{Definition of } \mathcal{W} \rangle$
$\square \neg q \vee \neg q\,\mathcal{U}\,(\neg p \wedge \neg q)$
$= \quad \langle (170)\ \text{Distributivity of } \neg \text{ over } \mathcal{W} \rangle$
$\square \neg q \vee \neg(p\,\mathcal{W}\,q)$
$= \quad \langle (61)\ \text{Dual of } \Diamond \rangle$
$\neg(p\,\mathcal{W}\,q) \vee \neg \Diamond q$
$= \quad \langle (3.47a)\ \text{De Morgan},\ \neg(p \wedge q) \equiv \neg p \vee \neg q \rangle$
$\neg(p\,\mathcal{W}\,q \wedge \Diamond q)$
$= \quad \langle (171)\ \mathcal{U} \text{ in terms of } \mathcal{W} \rangle$
$\neg(p\,\mathcal{U}\,q) \quad \blacksquare$

Theorem (174) $\mathcal{U}$ implication shows that $p\,\mathcal{U}\,q$ is stronger than $p\,\mathcal{W}\,q$. Theorem (175) Distributivity of $\wedge$ over $\mathcal{W}$ corresponds to, and is derived from, (83) Distributivity of $\wedge$ over $\mathcal{U}$. Theorem (176) $\mathcal{W}\Diamond$ equivalence comes from Manna and Pnueli [21] where it is used as the normal form for simple obligation formulas.

(174) **$\mathcal{U}$ implication:** $p\,\mathcal{U}\,q \Rightarrow p\,\mathcal{W}\,q$

Proof:

Pepperdine Papers on LTL

$$p\,\mathcal{U}\,q$$
= ⟨(171) $\mathcal{U}$ in terms of $\mathcal{W}$⟩
$$p\,\mathcal{W}\,q \wedge \Diamond q$$
⇒ ⟨(3.76b) Strengthening, $p \wedge q \Rightarrow p$⟩
$$p\,\mathcal{W}\,q \quad \blacksquare$$

(175) **Distributivity of $\wedge$ over $\mathcal{W}$:** $\Box p \wedge q\,\mathcal{W}\,r \Rightarrow (p \wedge q)\,\mathcal{W}\,(p \wedge r)$

Proof:

$$\Box p \wedge q\,\mathcal{W}\,r$$
= ⟨(169) Definition of $\mathcal{W}$⟩
$$\Box p \wedge (\Box q \vee q\,\mathcal{U}\,r)$$
= ⟨(3.46) Distributivity of $\wedge$ over $\vee$, $p \wedge (q \vee r) \equiv (p \wedge q) \vee (p \wedge r)$⟩
$$(\Box p \wedge \Box q) \vee (\Box p \wedge q\,\mathcal{U}\,r)$$
⇒ ⟨(83) Distributivity of $\wedge$ over $\mathcal{U}$ and (4.2) Monotonicity of $\vee$⟩
$$(\Box p \wedge \Box q) \vee (p \wedge q)\,\mathcal{U}\,(p \wedge r)$$
= ⟨(99) Distributivity of $\Box$ over $\wedge$⟩
$$\Box (p \wedge q) \vee (p \wedge q)\,\mathcal{U}\,(p \wedge r)$$
= ⟨(169) Definition of $\mathcal{W}$⟩
$$(p \wedge q)\,\mathcal{W}\,(p \wedge r) \quad \blacksquare$$

(176) **$\mathcal{W}\Diamond$ equivalence:** $p\,\mathcal{W}\,\Diamond q \equiv \Box p \vee \Diamond q$

Proof:

$$p\,\mathcal{W}\,\Diamond q$$
= ⟨(169) Definition of $\mathcal{W}$⟩
$$\Box p \vee p\,\mathcal{U}\,\Diamond q$$
= ⟨(41) Absorption of $\mathcal{U}$ into $\Diamond$⟩
$$\Box p \vee \Diamond q \quad \blacksquare$$

Theorem (177) $\mathcal{W}\Box$ implication corresponds to (140) $\mathcal{U}\Box$ implication. Theorem (178) Absorption of $\mathcal{W}$ into $\Box$ corresponds to (141) Absorption of $\mathcal{U}$ into $\Box$. Theorem (179) Perpetuity for the *wait* operator corresponds to (42) Eventuality for the *until* operator. The *always* operator, which is universal, is in the antecedent of the implication in Perpetuity, while the *eventually* operator, which is existential, is in the consequent of the implication in Eventuality.

(177) **$\mathcal{W}\Box$ implication:** $p\,\mathcal{W}\,\Box q \Rightarrow \Box(p\,\mathcal{W}\,q)$

Proof:

$\square (p \mathcal{W} q)$
= ⟨(169) Definition of $\mathcal{W}$⟩
$\square (\square p \vee p \mathcal{U} q)$
⇐ ⟨(100) Distributivity of $\square$ over $\vee$⟩
$\square \square p \vee \square (p \mathcal{U} q)$
= ⟨(72) Absorption of $\square$⟩
$\square p \vee \square (p \mathcal{U} q)$
⇐ ⟨(140) $\mathcal{U} \square$ implication and (4.2) Monotonicity of $\vee$⟩
$\square p \vee p \mathcal{U} \square q$
= ⟨(169) Definition of $\mathcal{W}$ with $q := \square q$⟩
$p \mathcal{W} \square q$ ∎

(178) **Absorption of** $\mathcal{W}$ **into** $\square$: $\quad p \mathcal{W} \square p \equiv \square p$

Proof:

$p \mathcal{W} \square p$
= ⟨(169) Definition of $\mathcal{W}$⟩
$\square p \vee p \mathcal{U} \square p$
= ⟨(141) Absorption of $\mathcal{U}$ into $\square$⟩
$\square p \vee \square p$
= ⟨(3.26) Idempotency of $\vee$, $p \vee p \equiv p$⟩
$\square p$ ∎

(179) **Perpetuity**: $\quad \square p \Rightarrow p \mathcal{W} q$

Proof:

$\square p$
⇒ ⟨(3.76a) Weakening, $p \Rightarrow p \vee q$⟩
$\square p \vee p \mathcal{U} q$
= ⟨(169) Definition of $\mathcal{W}$⟩
$p \mathcal{W} q$ ∎

(180) **Distributivity of** $\circ$ **over** $\mathcal{W}$: $\quad \circ (p \mathcal{W} q) \equiv \circ p \mathcal{W} \circ q$

Proof:

$\circ (p \mathcal{W} q)$
= ⟨(169) Definition of $\mathcal{W}$⟩
$\circ (\square p \vee p \mathcal{U} q)$
= ⟨(4) Distributivity of $\circ$ over $\vee$⟩

$$\bigcirc \Box p \vee \bigcirc (p \,\mathcal{U}\, q)$$
= ⟨(73) Exchange of $\bigcirc$ and $\Box$ and (9) Distributivity of $\bigcirc$ over $\mathcal{U}$⟩
$$\Box \bigcirc p \vee \bigcirc p \,\mathcal{U}\, \bigcirc q$$
= ⟨(169) Definition of $\mathcal{W}$ with $p, q := \bigcirc p, \bigcirc q$⟩
$$\bigcirc p \,\mathcal{W}\, \bigcirc q \quad \blacksquare$$

Expansion of the *wait* operator (181) corresponds to expansion of the *until* operator (10). Excluded middle for the *wait* operator (182) corresponds to excluded middle for the *until* operator (23). Left zero of the *wait* operator (183) corresponds to right zero of the *until* operator (11).

(181) **Expansion of** $\mathcal{W}$: $\quad p \,\mathcal{W}\, q \equiv q \vee (p \wedge \bigcirc(p \,\mathcal{W}\, q))$

Proof:

$$q \vee (p \wedge \bigcirc (p \,\mathcal{W}\, q))$$
= ⟨(169) Definition of $\mathcal{W}$⟩
$$q \vee (p \wedge \bigcirc (\Box p \vee p \,\mathcal{U}\, q))$$
= ⟨(4) Distributivity of $\bigcirc$ over $\vee$⟩
$$q \vee (p \wedge (\bigcirc \Box p \vee \bigcirc (p \,\mathcal{U}\, q)))$$
= ⟨(3.46) Distributivity of $\wedge$ over $\vee$, $p \wedge (q \vee r) \equiv (p \wedge q) \vee (p \wedge r)$⟩
$$q \vee (p \wedge \bigcirc \Box p) \vee (p \wedge \bigcirc (p \,\mathcal{U}\, q))$$
= ⟨(66) Expansion of $\Box$⟩
$$q \vee \Box p \vee (p \wedge \bigcirc (p \,\mathcal{U}\, q))$$
= ⟨(10) Expansion of $\mathcal{U}$⟩
$$\Box p \vee p \,\mathcal{U}\, q$$
= ⟨(169) Definition of $\mathcal{W}$⟩
$$p \,\mathcal{W}\, q \quad \blacksquare$$

(182) $\mathcal{W}$ **excluded middle:** $\quad p \,\mathcal{W}\, q \vee p \,\mathcal{W}\, \neg q$

Proof:

$$p \,\mathcal{W}\, q \vee p \,\mathcal{W}\, \neg q$$
= ⟨(181) Expansion, twice⟩
$$q \vee (p \wedge \bigcirc (p \,\mathcal{W}\, q)) \vee \neg q \vee (p \wedge \bigcirc (p \,\mathcal{W}\, \neg q))$$
= ⟨(3.28) Excluded middle, $p \vee \neg p$⟩
$$true \vee (p \wedge \bigcirc (p \,\mathcal{W}\, q)) \vee (p \wedge \bigcirc (p \,\mathcal{W}\, \neg q))$$
= ⟨(3.29) Zero of $\vee$, $p \vee true \equiv true$⟩
$$true \quad \blacksquare$$

(183) **Left zero of** $\mathcal{W}$: $\quad true \, \mathcal{W} \, q \equiv true$

Proof:

$\quad true \, \mathcal{W} \, q$
$= \quad \langle(169) \text{ Definition of } \mathcal{W}\rangle$
$\quad \Box \, true \lor true \, \mathcal{U} \, q$
$= \quad \langle(64) \text{ Truth of } \Box\rangle$
$\quad true \lor true \, \mathcal{U} \, q$
$= \quad \langle(3.29) \text{ Zero of } \lor, p \lor true \equiv true\rangle$
$\quad true \quad \blacksquare$

The next four distributive theorems for the *wait* operator correspond to, and are proved from, the distributive axioms for the *until* operator (12), (13), (14), and (15). The proof of (187) Right distributivity of $\mathcal{W}$ over $\land$ is an example of one benefit of $\mathcal{E}$, with its emphasis on equality, over $\mathcal{H}$, with its emphasis on implication and modus ponens. Before we formulated the calculational 8-step proof of (187), we had an earlier proof based on mutual implication that required 20 steps.

(184) **Left distributivity of** $\mathcal{W}$ **over** $\lor$: $\quad p \, \mathcal{W} \, (q \lor r) \equiv p \, \mathcal{W} \, q \lor p \, \mathcal{W} \, r$

Proof:

$\quad p \, \mathcal{W} \, q \lor p \, \mathcal{W} \, r$
$= \quad \langle(169) \text{ Definition of } \mathcal{W}, \text{twice}\rangle$
$\quad \Box p \lor p \, \mathcal{U} \, q \lor \Box p \lor p \, \mathcal{U} \, r$
$= \quad \langle(3.26) \text{ Idempotency of } \lor, p \lor p \equiv p\rangle$
$\quad \Box p \lor p \, \mathcal{U} \, q \lor p \, \mathcal{U} \, r$
$= \quad \langle(12) \text{ Left Distributivity of } \mathcal{U} \text{ over } \lor\rangle$
$\quad \Box p \lor p \, \mathcal{U} \, (q \lor r)$
$= \quad \langle(169) \text{ Definition of } \mathcal{W}\rangle$
$\quad p \, \mathcal{W} \, (q \lor r) \quad \blacksquare$

(185) **Right distributivity of** $\mathcal{W}$ **over** $\lor$: $\quad p \, \mathcal{W} \, r \lor q \, \mathcal{W} \, r \Rightarrow (p \lor q) \, \mathcal{W} \, r$

Proof:

$\quad p \, \mathcal{W} \, r \lor q \, \mathcal{W} \, r$
$= \quad \langle(169) \text{ Definition of } \mathcal{W}\rangle$
$\quad \Box p \lor \Box q \lor p \, \mathcal{U} \, r \lor q \, \mathcal{U} \, r$
$\Rightarrow \quad \langle(100) \text{ Distributivity of } \Box \text{ over } \lor \text{ and } (4.2) \text{ Monotonicity of } \lor,$
$\quad \quad (p \Rightarrow q) \Rightarrow (p \lor r \Rightarrow q \lor r)\rangle$

$\quad\quad \Box(p \vee q) \vee p\,\mathcal{U}\,r \vee q\,\mathcal{U}\,r$
$\Rightarrow \quad \langle\text{(13) Right distributivity of } \mathcal{U} \text{ over } \vee \text{ and (4.2) Monotonicity of } \vee\rangle$
$\quad\quad \Box(p \vee q) \vee (p \vee q)\,\mathcal{U}\,r$
$= \quad \langle\text{(169) Definition of } \mathcal{W}\rangle$
$\quad\quad (p \vee q)\,\mathcal{W}\,r \quad \blacksquare$

(186) **Left distributivity of** $\mathcal{W}$ **over** $\wedge$: $\quad p\,\mathcal{W}\,(q \wedge r) \Rightarrow p\,\mathcal{W}\,q \wedge p\,\mathcal{W}\,r$

Proof:

$\quad\quad p\,\mathcal{W}\,(q \wedge r)$
$= \quad \langle\text{(169) Definition of } \mathcal{W}\rangle$
$\quad\quad \Box p \vee p\,\mathcal{U}\,(q \wedge r)$
$\Rightarrow \quad \langle\text{(14) Left Distributivity of } \mathcal{U} \text{ over } \wedge \text{ and (4.2) Monotonicity of } \vee\rangle$
$\quad\quad \Box p \vee (p\,\mathcal{U}\,q \wedge p\,\mathcal{U}\,r)$
$= \quad \langle\text{(3.45) Distributivity of } \vee \text{ over } \wedge, p \vee (q \wedge r) \equiv (p \vee q) \wedge (p \vee r)\rangle$
$\quad\quad (\Box p \vee p\,\mathcal{U}\,q) \wedge (\Box p \vee p\,\mathcal{U}\,r)$
$= \quad \langle\text{(169) Definition of } \mathcal{W}, \text{twice}\rangle$
$\quad\quad p\,\mathcal{W}\,q \wedge p\,\mathcal{W}\,r \quad \blacksquare$

(187) **Right distributivity of** $\mathcal{W}$ **over** $\wedge$: $\quad (p \wedge q)\,\mathcal{W}\,r \equiv p\,\mathcal{W}\,r \wedge q\,\mathcal{W}\,r$

Proof:

$\quad\quad p\,\mathcal{W}\,r \wedge q\,\mathcal{W}\,r$
$= \quad \langle\text{(3.12) Double negation, } \neg\neg p \equiv p\rangle$
$\quad\quad \neg\neg(p\,\mathcal{W}\,r \wedge q\,\mathcal{W}\,r)$
$= \quad \langle\text{(3.47a) De Morgan, } \neg(p \wedge q) \equiv \neg p \vee \neg q\rangle$
$\quad\quad \neg(\neg(p\,\mathcal{W}\,r) \vee \neg(q\,\mathcal{W}\,r))$
$= \quad \langle\text{(170) Distributivity of } \neg \text{ over } \mathcal{W}, \text{twice}\rangle$
$\quad\quad \neg(\neg r\,\mathcal{U}\,(\neg p \wedge \neg r) \vee \neg r\,\mathcal{U}\,(\neg q \wedge \neg r))$
$= \quad \langle\text{(12) Left Distributivity of } \mathcal{U} \text{ over } \vee\rangle$
$\quad\quad \neg(\neg r\,\mathcal{U}\,((\neg p \wedge \neg r) \vee (\neg q \wedge \neg r)))$
$= \quad \langle\text{(3.46) Distributivity of } \wedge \text{ over } \vee, p \wedge (q \vee r) \equiv (p \wedge q) \vee (p \wedge r)\rangle$
$\quad\quad \neg(\neg r\,\mathcal{U}\,(\neg r \wedge (\neg p \vee \neg q)))$
$= \quad \langle\text{(3.47a) De Morgan, } \neg(p \wedge q) \equiv \neg p \vee \neg q\rangle$
$\quad\quad \neg(\neg r\,\mathcal{U}\,(\neg r \wedge \neg(p \wedge q)))$
$= \quad \langle\text{(170) Distributivity of } \neg \text{ over } \mathcal{W} \text{ with } p, q := p \wedge q, r\rangle$
$\quad\quad \neg\neg((p \wedge q)\,\mathcal{W}\,r)$
$= \quad \langle\text{(3.12) Double negation, } \neg\neg p \equiv p\rangle$
$\quad\quad (p \wedge q)\,\mathcal{W}\,r \quad \blacksquare$

Theorem (188) for $\mathcal{W}$ is identical to (19) for $\mathcal{U}$. Both $\mathcal{W}$ and $\mathcal{U}$ obey the disjunction and conjunction rules—(189), (190), (195), and (196)—which give rise, in turn, to expanded distributivity theorems of $\neg$ over $\mathcal{W}$ and $\mathcal{U}$, (197) to (202). The conjunction and disjunction rules do not appear in other LTL systems.

(188) **Right distributivity of $\mathcal{W}$ over $\Rightarrow$:** $(p \Rightarrow q) \mathcal{W} r \Rightarrow (p \mathcal{W} r \Rightarrow q \mathcal{W} r)$

Proof:

$\ (p \Rightarrow q) \mathcal{W} r \Rightarrow (p \mathcal{W} r \Rightarrow q \mathcal{W} r)$
$= \quad \langle (3.65)\ \text{Shunting},\ p \wedge q \Rightarrow r \equiv p \Rightarrow (q \Rightarrow r) \rangle$
$\ (p \Rightarrow q) \mathcal{W} r \wedge p \mathcal{W} r \Rightarrow q \mathcal{W} r$

And now,

$\ (p \Rightarrow q) \mathcal{W} r \wedge p \mathcal{W} r$
$= \quad \langle (187)\ \text{Right distributivity of } \mathcal{W} \text{ over } \wedge \rangle$
$\ (p \wedge (p \Rightarrow q)) \mathcal{W} r$
$= \quad \langle (3.66)\ p \wedge (p \Rightarrow q) \equiv p \wedge q \rangle$
$\ (p \wedge q) \mathcal{W} r$
$= \quad \langle (187)\ \text{Right distributivity of } \mathcal{W} \text{ over } \wedge \rangle$
$\ p \mathcal{W} r \wedge q \mathcal{W} r$
$\Rightarrow \quad \langle (3.76\text{b})\ \text{Strengthening},\ p \wedge q \Rightarrow p \rangle$
$\ q \mathcal{W} r\ \blacksquare$

(189) **Disjunction rule of $\mathcal{W}$:** $p \mathcal{W} q \equiv (p \vee q) \mathcal{W} q$

Proof:

$\ p \mathcal{W} q$
$= \quad \langle (3.12)\ \text{Double negation},\ \neg\neg p \equiv p \rangle$
$\ \neg\neg(p \mathcal{W} q)$
$= \quad \langle (170)\ \text{Distributivity of } \neg \text{ over } \mathcal{W} \rangle$
$\ \neg(\neg q\ \mathcal{U}\ (\neg p \wedge \neg q))$
$= \quad \langle (173)\ \text{Distributivity of } \neg \text{ over } \mathcal{U} \rangle$
$\ \neg(\neg p \wedge \neg q) \mathcal{W} (q \wedge \neg(\neg p \wedge \neg q))$
$= \quad \langle (3.47\text{a})\ \text{De Morgan},\ \neg(p \wedge q) \equiv \neg p \vee \neg q,\ \text{twice} \rangle$
$\ (p \vee q) \mathcal{W} (q \wedge (p \vee q))$
$= \quad \langle (3.43\text{a})\ \text{Absorption},\ p \wedge (p \vee q) \equiv p \rangle$
$\ (p \vee q) \mathcal{W} q\ \blacksquare$

(190) **Disjunction rule of** $\mathcal{U}$: $\quad p\,\mathcal{U}\,q \equiv (p \vee q)\,\mathcal{U}\,q$

Proof:

$$p\,\mathcal{U}\,q$$
$= \quad \langle(3.12)\text{ Double negation}, \neg\neg p \equiv p\rangle$
$$\neg\neg(p\,\mathcal{U}\,q)$$
$= \quad \langle(173)\text{ Distributivity of } \neg \text{ over } \mathcal{U}\rangle$
$$\neg(\neg q\,\mathcal{W}\,(\neg p \wedge \neg q))$$
$= \quad \langle(170)\text{ Distributivity of } \neg \text{ over } \mathcal{W}\rangle$
$$\neg(\neg p \wedge \neg q)\,\mathcal{U}\,(q \wedge \neg(\neg p \wedge \neg q))$$
$= \quad \langle(3.47\text{a})\text{ De Morgan}, \neg(p \wedge q) \equiv \neg p \vee \neg q, \text{twice}\rangle$
$$(p \vee q)\,\mathcal{U}\,(q \wedge (p \vee q))$$
$= \quad \langle(3.43\text{a})\text{ Absorption}, p \wedge (p \vee q) \equiv p\rangle$
$$(p \vee q)\,\mathcal{U}\,q \quad \blacksquare$$

(191) **Rule of** $\mathcal{W}$: $\quad \neg q\,\mathcal{W}\,q$

Proof:

$$\neg q\,\mathcal{W}\,q$$
$= \quad \langle(189)\text{ Disjunction rule of } \mathcal{W}\rangle$
$$(\neg q \vee q)\,\mathcal{W}\,q$$
$= \quad \langle(3.28)\text{ Excluded middle}, p \vee \neg p\rangle$
$$true\,\mathcal{W}\,q$$
$= \quad \langle(183)\text{ Left zero of } \mathcal{W}\rangle$
$$true \quad \blacksquare$$

(192) **Rule of** $\mathcal{U}$: $\quad \neg q\,\mathcal{U}\,q \equiv \Diamond q$

Proof:

$$\neg q\,\mathcal{U}\,q$$
$= \quad \langle(190)\text{ Disjunction rule of } \mathcal{U}\rangle$
$$(\neg q \vee q)\,\mathcal{U}\,q$$
$= \quad \langle(3.28)\text{ Excluded middle}, p \vee \neg p\rangle$
$$true\,\mathcal{U}\,q$$
$= \quad \langle(38)\text{ Definition of } \Diamond\rangle$
$$\Diamond q \quad \blacksquare$$

(193) $(p \Rightarrow q)\,\mathcal{W}\,p$

Proof: The proof is by (4.7.1) Truth implication.

$$(p \Rightarrow q) \, \mathcal{W} \, p$$
= ⟨(3.59) Implication, $p \Rightarrow q \equiv \neg p \vee q$⟩
$$(\neg p \vee q) \, \mathcal{W} \, p$$
⇐ ⟨(185) Right distributivity of $\mathcal{W}$ over $\vee$⟩
$$\neg p \, \mathcal{W} \, p \vee q \, \mathcal{W} \, p$$
= ⟨(191) Rule of $\mathcal{W}$⟩
$$true \vee q \, \mathcal{W} \, p$$
= ⟨(3.29) Zero of $\vee$, $p \vee true \equiv true$⟩
$$true \quad \blacksquare$$

(194) $\Diamond p \Rightarrow (p \Rightarrow q) \, \mathcal{U} \, p$

Proof:

$$(p \Rightarrow q) \, \mathcal{U} \, p$$
= ⟨(3.59) Implication, $p \Rightarrow q \equiv \neg p \vee q$⟩
$$(\neg p \vee q) \, \mathcal{U} \, p$$
⇐ ⟨(13) Right distributivity of $\mathcal{U}$ over $\vee$⟩
$$\neg p \, \mathcal{U} \, p \vee q \, \mathcal{U} \, p$$
⇐ ⟨(3.76a) Weakening, $p \Rightarrow p \vee q$⟩
$$\neg p \, \mathcal{U} \, p$$
= ⟨(192) Rule of $\mathcal{U}$⟩
$$\Diamond p \quad \blacksquare$$

(195) **Conjunction rule of** $\mathcal{W}$: $p \, \mathcal{W} \, q \equiv (p \wedge \neg q) \, \mathcal{W} \, q$

Proof:

$$(p \wedge \neg q) \, \mathcal{W} \, q$$
= ⟨(187) Right distributivity of $\mathcal{W}$ over $\wedge$⟩
$$p \, \mathcal{W} \, q \wedge \neg q \, \mathcal{W} \, q$$
= ⟨(191) Rule of $\mathcal{W}$⟩
$$p \, \mathcal{W} \, q \wedge true$$
= ⟨(3.39) Identity of $\wedge$, $p \wedge true \equiv p$⟩
$$p \, \mathcal{W} \, q \quad \blacksquare$$

(196) **Conjunction rule of** $\mathcal{U}$: $p \, \mathcal{U} \, q \equiv (p \wedge \neg q) \, \mathcal{U} \, q$

Proof:

$(p \wedge \neg q) \, \mathcal{U} \, q$
$= \quad \langle (171) \; \mathcal{U} \text{ in terms of } \mathcal{W} \rangle$
$(p \wedge \neg q) \, \mathcal{W} \, q \wedge \Diamond q$
$= \quad \langle (195) \text{ Conjunction rule of } \mathcal{W} \rangle$
$p \, \mathcal{W} \, q \wedge \Diamond q$
$= \quad \langle (171) \; \mathcal{U} \text{ in terms of } \mathcal{W} \rangle$
$p \, \mathcal{U} \, q$ ∎

(197) **Distributivity of $\neg$ over $\mathcal{W}$:** $\quad \neg(p \, \mathcal{W} \, q) \equiv (p \wedge \neg q) \, \mathcal{U} \, (\neg p \wedge \neg q)$

Proof:

$\neg(p \, \mathcal{W} \, q)$
$= \quad \langle (170) \text{ Distributivity of } \neg \text{ over } \mathcal{W} \rangle$
$\neg q \, \mathcal{U} \, (\neg p \wedge \neg q)$
$= \quad \langle (196) \text{ Conjunction rule of } \mathcal{U} \text{ with } p, q := \neg q, \neg p \wedge \neg q \rangle$
$(\neg q \wedge \neg(\neg p \wedge \neg q)) \, \mathcal{U} \, (\neg p \wedge \neg q)$
$= \quad \langle (3.47\text{a}) \text{ De Morgan}, \neg(p \wedge q) \equiv \neg p \vee \neg q \rangle$
$(\neg q \wedge (p \vee q)) \, \mathcal{U} \, (\neg p \wedge \neg q)$
$= \quad \langle (3.44\text{a}) \text{ Absorption}, p \wedge (\neg p \vee q) \equiv p \wedge q \text{ with } p, q := \neg q, p \rangle$
$(p \wedge \neg q) \, \mathcal{U} \, (\neg p \wedge \neg q)$ ∎

(198) **Distributivity of $\neg$ over $\mathcal{U}$:** $\quad \neg(p \, \mathcal{U} \, q) \equiv (p \wedge \neg q) \, \mathcal{W} \, (\neg p \wedge \neg q)$

Proof:

$\neg(p \, \mathcal{U} \, q)$
$= \quad \langle (173) \text{ Distributivity of } \neg \text{ over } \mathcal{U} \rangle$
$\neg q \, \mathcal{W} \, (\neg p \wedge \neg q)$
$= \quad \langle (195) \text{ Conjunction rule of } \mathcal{W} \text{ with } p, q := \neg q, \neg p \wedge \neg q \rangle$
$(\neg q \wedge \neg(\neg p \wedge \neg q)) \, \mathcal{W} \, (\neg p \wedge \neg q)$
$= \quad \langle (3.47\text{a}) \text{ De Morgan}, \neg(p \wedge q) \equiv \neg p \vee \neg q \rangle$
$(\neg q \wedge (p \vee q)) \, \mathcal{W} \, (\neg p \wedge \neg q)$
$= \quad \langle (3.44\text{a}) \text{ Absorption}, p \wedge (\neg p \vee q) \equiv p \wedge q \text{ with } p, q := \neg q, p \rangle$
$(p \wedge \neg q) \, \mathcal{W} \, (\neg p \wedge \neg q)$ ∎

(199) **Dual of $\mathcal{U}$:** $\quad \neg(\neg p \, \mathcal{U} \, \neg q) \equiv q \, \mathcal{W} \, (p \wedge q)$

Proof:

$\neg(\neg p \, \mathcal{U} \, \neg q)$
$= \quad \langle (173) \text{ Distributivity of } \neg \text{ over } \mathcal{U} \text{ with } p, q := \neg p, \neg q \rangle$
$q \, \mathcal{W} \, (p \wedge q)$ ∎

(200) **Dual of** $\mathcal{U}$: $\neg(\neg p \, \mathcal{U} \, \neg q) \equiv (\neg p \wedge q) \, \mathcal{W} \, (p \wedge q)$

Proof:

$\quad \neg(\neg p \, \mathcal{U} \, \neg q)$
$= \quad \langle (198)$ Distributivity of $\neg$ over $\mathcal{U}$ with $p, q := \neg p, \neg q \rangle$
$\quad (\neg p \wedge q) \, \mathcal{W} \, (p \wedge q)$ ∎

(201) **Dual of** $\mathcal{W}$: $\neg(\neg p \, \mathcal{W} \, \neg q) \equiv q \, \mathcal{U} \, (p \wedge q)$

Proof:

$\quad \neg(\neg p \, \mathcal{W} \, \neg q)$
$= \quad \langle (170)$ Distributivity of $\neg$ over $\mathcal{W}$ with $p, q := \neg p, \neg q \rangle$
$\quad q \, \mathcal{U} \, (p \wedge q)$ ∎

(202) **Dual of** $\mathcal{W}$: $\neg(\neg p \, \mathcal{W} \, \neg q) \equiv (\neg p \wedge q) \, \mathcal{U} \, (p \wedge q)$

Proof:

$\quad \neg(\neg p \, \mathcal{W} \, \neg q)$
$= \quad \langle (197)$ Distributivity of $\neg$ over $\mathcal{W}$ with $p, q := \neg p, \neg q \rangle$
$\quad (\neg p \wedge q) \, \mathcal{U} \, (p \wedge q)$ ∎

Theorem (203) shows that the *wait* operator, like the *until* operator, is idempotent. Theorem (204) shows that *true* is the right zero of *wait*, as it is for *until*. Theorem (205) shows that *false* is the left identity of *wait*, as it is for *until*. Theorem (206) for the *wait* operator corresponds to theorem (28) for the *until* operator and shows that $p \, \mathcal{W} \, q$ is stronger than $p \vee q$. Theorem (207) shows that $\Box (p \vee q)$ is stronger than $p \, \mathcal{W} \, q$. Theorem (209), insertion for the *wait* operator, corresponds to (29), insertion for the *until* operator.

(203) **Idempotency of** $\mathcal{W}$: $p \, \mathcal{W} \, p \equiv p$

Proof:

$\quad p \, \mathcal{W} \, p$
$= \quad \langle (169)$ Definition of $\mathcal{W} \rangle$
$\quad \Box p \vee p \, \mathcal{U} \, p$
$= \quad \langle (22)$ Idempotency of $\mathcal{U} \rangle$
$\quad \Box p \vee p$
$= \quad \langle (69)$ Absorption of $\Box$ into $\vee \rangle$
$\quad p$ ∎

(204) **Right zero of** $\mathcal{W}$: $p \,\mathcal{W}\, true \equiv true$

Proof:

$\quad p \,\mathcal{W}\, true$
$= \quad \langle (169) \text{ Definition of } \mathcal{W} \rangle$
$\quad \Box p \vee p \,\mathcal{U}\, true$
$= \quad \langle (20) \text{ Right zero of } \mathcal{U} \rangle$
$\quad \Box p \vee true$
$= \quad \langle (3.29) \text{ Zero of } \vee, p \vee true \equiv true \rangle$
$\quad true \quad\blacksquare$

(205) **Left identity of** $\mathcal{W}$: $false \,\mathcal{W}\, q \equiv q$

Proof:

$\quad false \,\mathcal{W}\, q$
$= \quad \langle (169) \text{ Definition of } \mathcal{W} \rangle$
$\quad \Box false \vee false \,\mathcal{U}\, q$
$= \quad \langle (21) \text{ Left identity of } \mathcal{U} \rangle$
$\quad \Box false \vee q$
$= \quad \langle (65) \text{ Falsehood of } \Box \rangle$
$\quad false \vee q$
$= \quad \langle (3.30) \text{ Identity of } \vee, p \vee false \equiv p \rangle$
$\quad q \quad\blacksquare$

(206) $p \,\mathcal{W}\, q \Rightarrow p \vee q$

Proof:

$\quad p \,\mathcal{W}\, q$
$= \quad \langle (181) \text{ Expansion of } \mathcal{W} \rangle$
$\quad q \vee (p \wedge \bigcirc (p \,\mathcal{W}\, q))$
$\Rightarrow \quad \langle (3.76\text{d}) \, p \vee (q \wedge r) \Rightarrow p \vee q \rangle$
$\quad p \vee q \quad\blacksquare$

(207) $\Box (p \vee q) \Rightarrow p \,\mathcal{W}\, q$

Proof:

$\quad \Box (p \vee q)$
$\Rightarrow \quad \langle (179) \text{ Perpetuity} \rangle$
$\quad (p \vee q) \,\mathcal{W}\, q$
$= \quad \langle (189) \text{ Disjunction rule of } \mathcal{W} \rangle$
$\quad p \,\mathcal{W}\, q \quad\blacksquare$

(208) $\Box(\neg q \Rightarrow p) \Rightarrow p \mathcal{W} q$

Proof:

$\quad\quad \Box(\neg q \Rightarrow p)$
$= \quad \langle(3.59) \text{ Implication}, p \Rightarrow q \equiv \neg p \vee q \text{ and } (3.24) \text{ Symmetry of } \vee, p \vee q \equiv q \vee p\rangle$
$\quad\quad \Box(p \vee q)$
$\Rightarrow \quad \langle(207) \Box(p \vee q) \Rightarrow p \mathcal{W} q\rangle$
$\quad\quad p \mathcal{W} q \quad \blacksquare$

(209) $\mathcal{W}$ **insertion**: $\quad q \Rightarrow p \mathcal{W} q$

Proof:

$\quad\quad p \mathcal{W} q$
$= \quad \langle(181) \text{ Expansion of } \mathcal{W}\rangle$
$\quad\quad q \vee (p \wedge \circ(p \mathcal{W} q))$
$\Leftarrow \quad \langle(3.76a) \text{ Weakening}, p \Rightarrow p \vee q\rangle$
$\quad\quad q \quad \blacksquare$

The next three theorems (210), (211), and (212) complete the set of frame laws which included the theorems from (106) to (117) for binary propositional operators, and the $\mathcal{U}$ frame laws from theorems (148), (149), and (150).

(210) $\mathcal{W}$ **frame law of** $\circ$: $\quad \Box p \Rightarrow (\circ q \Rightarrow \circ(p \mathcal{W} q))$

Proof:

$\quad\quad \Box p \Rightarrow (\circ q \Rightarrow \circ(p \mathcal{W} q))$
$= \quad \langle(2) \text{ Distributivity of } \circ \text{ over } \Rightarrow\rangle$
$\quad\quad \Box p \Rightarrow \circ(q \Rightarrow p \mathcal{W} q)$
$= \quad \langle(209) \mathcal{W} \text{ insertion}\rangle$
$\quad\quad \Box p \Rightarrow \circ \textit{true}$
$= \quad \langle(7) \text{ Truth of } \circ\rangle$
$\quad\quad \Box p \Rightarrow \textit{true} \quad -(3.72) \text{ Right zero of } \Rightarrow, p \Rightarrow \textit{true} \equiv \textit{true} \quad \blacksquare$

(211) $\mathcal{W}$ **frame law of** $\Diamond$: $\quad \Box p \Rightarrow (\Diamond q \Rightarrow \Diamond(p \mathcal{W} q))$

Proof:

$\quad\quad \Box p \Rightarrow (\Diamond q \Rightarrow \Diamond(p \mathcal{W} q))$
$= \quad \langle(3.65) \text{ Shunting}, p \wedge q \Rightarrow r \equiv p \Rightarrow (q \Rightarrow r)\rangle$
$\quad\quad \Box p \wedge \Diamond q \Rightarrow \Diamond(p \mathcal{W} q)$

And now,

$\quad \square p \wedge \diamond q$
$\Rightarrow \quad \langle (84)\ \mathcal{U}\ \text{implication} \rangle$
$\quad p\ \mathcal{U}\ q$
$\Rightarrow \quad \langle (174)\ p\ \mathcal{U}\ q \Rightarrow p\ \mathcal{W}\ q \rangle$
$\quad p\ \mathcal{W}\ q$
$\Rightarrow \quad \langle (46)\ \text{Weakening of}\ \diamond,\ p \Rightarrow \diamond p\ \text{with}\ p := p\ \mathcal{W}\ q \rangle$
$\quad \diamond(p\ \mathcal{W}\ q) \quad \blacksquare$

(212) $\mathcal{W}$ **frame law of** $\square$: $\quad \square p \Rightarrow (\square q \Rightarrow \square(p\ \mathcal{W}\ q))$

Proof:

$\quad \square p \Rightarrow (\square q \Rightarrow \square(p\ \mathcal{W}\ q))$
$= \quad \langle (3.65)\ \text{Shunting},\ p \wedge q \Rightarrow r \equiv p \Rightarrow (q \Rightarrow r) \rangle$
$\quad \square p \wedge \square q \Rightarrow \square(p\ \mathcal{W}\ q))$

And now,

$\quad \square p \wedge \square q$
$\Rightarrow \quad \langle (3.76b)\ \text{Strengthening},\ p \wedge q \Rightarrow p\ \text{with}\ p, q := \square(p\ \mathcal{W}\ q), \square p \rangle$
$\quad \square q$
$\Rightarrow \quad \langle (209)\ \mathcal{W}\ \text{Insertion with}\ q := \square q \rangle$
$\quad p\ \mathcal{W}\ \square q$
$\Rightarrow \quad \langle (177)\ \mathcal{W}\ \square\ \text{implication} \rangle$
$\quad \square(p\ \mathcal{W}\ q) \quad \blacksquare$

Here are four induction theorems for $\mathcal{W}$, the last two of which are unique to this system.

(213) $\mathcal{W}$ **induction:** $\quad \square(p \Rightarrow (\circ p \wedge q) \vee r) \Rightarrow (p \Rightarrow q\ \mathcal{W}\ r)$

Proof:

$\quad \square(p \Rightarrow (\circ p \wedge q) \vee r) \Rightarrow (p \Rightarrow q\ \mathcal{W}\ r)$
$= \quad \langle (169)\ \text{Definition of}\ \mathcal{W} \rangle$
$\quad \square(p \Rightarrow (\circ p \wedge q) \vee r) \Rightarrow (p \Rightarrow \square q \vee q\ \mathcal{U}\ r) \quad -(55)\ \mathcal{U}\ \text{induction} \quad \blacksquare$

(214) $\mathcal{W}$ **induction:** $\quad \square(p \Rightarrow \circ(p \vee q)) \Rightarrow (p \Rightarrow p\ \mathcal{W}\ q)$

Proof:

$$\square(p \Rightarrow \circ (p \vee q))$$
$\Rightarrow \quad \langle (56) \; \mathcal{U} \text{ induction} \rangle$
$$p \Rightarrow \square p \vee p \, \mathcal{U} \, q$$
$= \quad \langle (169) \text{ Definition of } \mathcal{W} \rangle$
$$p \Rightarrow p \, \mathcal{W} \, q \quad \blacksquare$$

(215) $\mathcal{W}$ **induction:** $\quad \square(p \Rightarrow \circ p) \Rightarrow (p \Rightarrow p \, \mathcal{W} \, q)$

Proof:

$$\square(p \Rightarrow \circ p)$$
$\Rightarrow \quad \langle (57) \; \square \text{ induction} \rangle$
$$p \Rightarrow \square p$$
$\Rightarrow \quad \langle (179) \text{ Perpetuity and (3.82a) Transitivity} \rangle$
$$p \Rightarrow p \, \mathcal{W} \, q \quad \blacksquare$$

(216) $\mathcal{W}$ **induction:** $\quad \square(p \Rightarrow q \wedge \circ p) \Rightarrow (p \Rightarrow p \, \mathcal{W} \, q)$

Proof:

$$\square(p \Rightarrow q \wedge \circ p)$$
$= \quad \langle (3.63.1) \text{ Distributivity of } \Rightarrow \text{ over } \wedge, \; p \Rightarrow q \wedge r \equiv (p \Rightarrow q) \wedge (p \Rightarrow r) \rangle$
$$\square((p \Rightarrow q) \wedge (p \Rightarrow \circ p))$$
$= \quad \langle (99) \text{ Distributivity of } \square \text{ over } \wedge \rangle$
$$\square(p \Rightarrow q) \wedge \square(p \Rightarrow \circ p)$$
$\Rightarrow \quad \langle (3.76b) \text{ Strengthening}, \; p \wedge q \Rightarrow p \rangle$
$$\square(p \Rightarrow \circ p)$$
$\Rightarrow \quad \langle (57) \; \square \text{ induction} \rangle$
$$p \Rightarrow \square p$$
$\Rightarrow \quad \langle (179) \text{ Perpetuity and (3.82a) Transitivity} \rangle$
$$p \Rightarrow p \, \mathcal{W} \, q \quad \blacksquare$$

The next five absorption theorems correspond to the five absorption theorems for *until*, (31), (32), (33), (35), and (34) respectively. None appear in other LTL systems.

(217) **Absorption:** $\quad p \vee p \, \mathcal{W} \, q \equiv p \vee q$

Proof:

$$p \vee p \, \mathcal{W} \, q$$
$= \quad \langle (169) \text{ Definition of } \mathcal{W} \rangle$
$$p \vee p \, \mathcal{U} \, q \vee \square p$$
$= \quad \langle (69) \text{ Absorption of } \vee \text{ into } \square \rangle$
$$p \vee p \, \mathcal{U} \, q$$
$= \quad \langle (31) \text{ Absorption} \rangle$
$$p \vee q \quad \blacksquare$$

(218) **Absorption:** $p\,\mathcal{W}\,q \vee q \equiv p\,\mathcal{W}\,q$

Proof: (Ravi Mohan)

$\quad p\,\mathcal{W}\,q \vee q \equiv p\,\mathcal{W}\,q$
$= \quad \langle (3.57) \text{ Definition of implication}, p \Rightarrow q \equiv p \vee q \equiv q \rangle$
$\quad q \Rightarrow p\,\mathcal{W}\,q \quad -(209)\ \mathcal{W}\ \text{Insertion} \quad \blacksquare$

(219) **Absorption:** $p\,\mathcal{W}\,q \wedge q \equiv q$

Proof: (Ravi Mohan)

$\quad p\,\mathcal{W}\,q \wedge q \equiv q$
$= \quad \langle (3.60) \text{ Implication}, p \Rightarrow q \equiv p \wedge q \equiv p \rangle$
$\quad q \Rightarrow p\,\mathcal{W}\,q \quad -(209)\ \mathcal{W}\ \text{Insertion} \quad \blacksquare$

(220) **Absorption:** $p\,\mathcal{W}\,q \wedge (p \vee q) \equiv p\,\mathcal{W}\,q$

Proof: (Ravi Mohan)

$\quad p\,\mathcal{W}\,q \wedge (p \vee q) \equiv p\,\mathcal{W}\,q$
$= \quad \langle (3.60) \text{ Implication}, p \Rightarrow q \equiv p \wedge q \equiv p \rangle$
$\quad p\,\mathcal{W}\,q \Rightarrow p \vee q \quad -(206) \quad \blacksquare$

(221) **Absorption:** $p\,\mathcal{W}\,q \vee (p \wedge q) \equiv p\,\mathcal{W}\,q$

Proof:

$\quad p\,\mathcal{W}\,q \vee (p \wedge q)$
$= \quad \langle (181) \text{ Expansion of } \mathcal{W} \rangle$
$\quad q \vee (p \wedge \circ(p\,\mathcal{W}\,q)) \vee (p \wedge q)$
$= \quad \langle (3.43b) \text{ Absorption}, p \vee (p \wedge q) \equiv p \text{ with } p,q := q,p \rangle$
$\quad q \vee (p \wedge \circ(p\,\mathcal{W}\,q))$
$= \quad \langle (181) \text{ Expansion of } \mathcal{W} \rangle$
$\quad p\,\mathcal{W}\,q \quad \blacksquare$

The left and right absorption theorems for $\mathcal{W}$, (222) and (223), correspond to, and are proved from, the left and right absorption theorems for $\mathcal{U}$, (36) and (37). Theorem (224) corresponds to the definition of the $\Diamond$ operator (38). Figure 4 summarizes the eight possibilities of *true* and *false* on the left hand side and right hand side of $\mathcal{U}$ and $\mathcal{W}$.

(11) Right zero of $\mathcal{U}$: $p\,\mathcal{U}\,\mathit{false} \equiv \mathit{false}$	(183) Left zero of $\mathcal{W}$: $\mathit{true}\,\mathcal{W}\,q \equiv \mathit{true}$
(20) Right zero of $\mathcal{U}$: $p\,\mathcal{U}\,\mathit{true} \equiv \mathit{true}$	(204) Right zero of $\mathcal{W}$: $p\,\mathcal{W}\,\mathit{true} \equiv \mathit{true}$
(21) Left identity of $\mathcal{U}$: $\mathit{false}\,\mathcal{U}\,q \equiv q$	(205) Left identity of $\mathcal{W}$: $\mathit{false}\,\mathcal{W}\,q \equiv q$
(38) Definition of $\Diamond$: $\Diamond q \equiv \mathit{true}\,\mathcal{U}\,q$	(224) $\Box$ to $\mathcal{W}$ law: $\Box p \equiv p\,\mathcal{W}\,\mathit{false}$

Figure 4: The eight possibilities of *true* and *false* on the left hand side and right hand side of $\mathcal{U}$ and $\mathcal{W}$.

(222) **Left absorption of** $\mathcal{W}$: $p\,\mathcal{W}\,(p\,\mathcal{W}\,q) \equiv p\,\mathcal{W}\,q$

Proof:

$\quad p\,\mathcal{W}\,(p\,\mathcal{W}\,q)$
$= \quad \langle(169)\text{ Definition of }\mathcal{W}\text{, twice}\rangle$
$\quad p\,\mathcal{U}\,(p\,\mathcal{U}\,q \vee \Box p) \vee \Box p$
$= \quad \langle(12)\text{ Left distributivity of }\mathcal{U}\text{ over }\vee\rangle$
$\quad p\,\mathcal{U}\,(p\,\mathcal{U}\,q) \vee p\,\mathcal{U}\,\Box p \vee \Box p$
$= \quad \langle(141)\text{ Absorption of }\mathcal{U}\text{ into }\Box\text{ and }(3.26)\text{ Idempotency of }\vee, p \vee p \equiv p\rangle$
$\quad p\,\mathcal{U}\,(p\,\mathcal{U}\,q) \vee \Box p$
$= \quad \langle(36)\text{ Left absorption of }\mathcal{U}\rangle$
$\quad p\,\mathcal{U}\,q \vee \Box p$
$= \quad \langle(169)\text{ Definition of }\mathcal{W}\rangle$
$\quad p\,\mathcal{W}\,q \quad \blacksquare$

(223) **Right absorption of** $\mathcal{W}$: $(p\,\mathcal{W}\,q)\,\mathcal{W}\,q \equiv p\,\mathcal{W}\,q$

Proof: The proof is by (4.7) Mutual implication.
The proof in the first direction follows.

$\quad (p\,\mathcal{W}\,q)\,\mathcal{W}\,q$
$= \quad \langle(169)\text{ Definition of }\mathcal{W}\rangle$
$\quad (p\,\mathcal{U}\,q \vee \Box p)\,\mathcal{W}\,q$
$\Leftarrow \quad \langle(185)\text{ Right distributivity of }\mathcal{W}\text{ over }\vee\rangle$
$\quad (p\,\mathcal{U}\,q)\,\mathcal{W}\,q \vee \Box p\,\mathcal{W}\,q$
$\Leftarrow \quad \langle(174)\,\mathcal{U}\text{ implication and }(4.2)\text{ Monotonicity of }\vee\rangle$
$\quad (p\,\mathcal{U}\,q)\,\mathcal{U}\,q \vee \Box p\,\mathcal{W}\,q$
$= \quad \langle(37)\text{ Right absorption of }\mathcal{U}\rangle$
$\quad p\,\mathcal{U}\,q \vee \Box p\,\mathcal{W}\,q$

Pepperdine Papers on LTL 109

$\quad\Leftarrow\quad \langle(179)\text{ Perpetuity and }(4.2)\text{ Monotonicity of }\vee\rangle$
$\quad\quad p\,\mathcal{U}\,q \vee \square\square p$
$\quad=\quad \langle(72)\text{ Absorption of }\square\rangle$
$\quad\quad p\,\mathcal{U}\,q \vee \square p$
$\quad=\quad \langle(169)\text{ Definition of }\mathcal{W}\rangle$
$\quad\quad p\,\mathcal{W}\,q$

The proof in the second direction follows.

$\quad (p\,\mathcal{W}\,q)\,\mathcal{W}\,q$
$\Rightarrow\quad \langle(206)\ p\,\mathcal{W}\,q \Rightarrow p \vee q\rangle$
$\quad\quad p\,\mathcal{W}\,q \vee q$
$\quad=\quad \langle(218)\text{ Absorption}\rangle$
$\quad\quad p\,\mathcal{W}\,q$ ∎

(224) $\square$ **to** $\mathcal{W}$ **law:** $\quad \square p \equiv p\,\mathcal{W}\,\mathit{false}$

Proof:

$\quad p\,\mathcal{W}\,\mathit{false}$
$=\quad \langle(169)\text{ Definition of }\mathcal{W}\rangle$
$\quad \square p \vee p\,\mathcal{U}\,\mathit{false}$
$=\quad \langle(11)\text{ Right zero of }\mathcal{U}\rangle$
$\quad \square p \vee \mathit{false}$
$=\quad \langle(3.30)\text{ Identity of }\vee, p\vee\mathit{false}\equiv p\rangle$
$\quad \square p$ ∎

(225) $\diamond$ **to** $\mathcal{W}$ **law:** $\quad \diamond p \equiv \neg(\neg p\,\mathcal{W}\,\mathit{false})$

Proof:

$\quad \neg(\neg p\,\mathcal{W}\,\mathit{false})$
$=\quad \langle(224)\ \square\text{ to }\mathcal{W}\text{ law with }p := \neg p\rangle$
$\quad \neg\square\neg p$
$=\quad \langle(59)\ \diamond p \equiv \neg\square\neg p\rangle$
$\quad \diamond p$ ∎

Theorem (226) for $\mathcal{W}$ corresponds to (84) $\mathcal{U}$ implication. It is also known as an entailment law for the *wait* operator. [24] The following four absorption theorems combine an *until* operation with a *wait* operation. Theorem (231) corresponds to (141) Absorption of $\mathcal{U}$ into $\square$. Theorem (232) corresponds to (40) Absorption of $\mathcal{U}$ into $\diamond$. Theorem (233) corresponds to (39) Absorption of $\diamond$ into $\mathcal{U}$.

(226) $\mathcal{W}$ **implication:** $\quad p\,\mathcal{W}\,q \Rightarrow \square p \vee \diamond q$

Proof:

$p \, \mathcal{W} \, q$
$= \quad \langle(169) \text{ Definition of } \mathcal{W}\rangle$
$\Box p \vee p \, \mathcal{U} \, q$
$\Rightarrow \quad \langle(42) \text{ Eventuality and } (4.2) \text{ Monotonicity of } \vee\rangle$
$\Box p \vee \Diamond q \quad \blacksquare$

(227) **Absorption:** $p \, \mathcal{W} \, (p \, \mathcal{U} \, q) \equiv p \, \mathcal{W} \, q$

Proof:

$p \, \mathcal{W} \, (p \, \mathcal{U} \, q)$
$= \quad \langle(169) \text{ Definition of } \mathcal{W}\rangle$
$\Box p \vee p \, \mathcal{U} \, (p \, \mathcal{U} \, q)$
$= \quad \langle(36) \text{ Left absorption of } \mathcal{U}\rangle$
$\Box p \vee p \, \mathcal{U} \, q$
$= \quad \langle(169) \text{ Definition of } \mathcal{W}\rangle$
$p \, \mathcal{W} \, q \quad \blacksquare$

(228) **Absorption:** $(p \, \mathcal{U} \, q) \, \mathcal{W} \, q \equiv p \, \mathcal{U} \, q$

Proof:

$(p \, \mathcal{U} \, q) \, \mathcal{W} \, q$
$= \quad \langle(169) \text{ Definition of } \mathcal{W}\rangle$
$\Box (p \, \mathcal{U} \, q) \vee (p \, \mathcal{U} \, q) \, \mathcal{U} \, q$
$= \quad \langle(37) \text{ Right absorption of } \mathcal{U}\rangle$
$\Box (p \, \mathcal{U} \, q) \vee p \, \mathcal{U} \, q$
$= \quad \langle(69) \text{ Absorption of } \Box \text{ into } \vee \text{ with } p := p \, \mathcal{U} \, q\rangle$
$p \, \mathcal{U} \, q \quad \blacksquare$

(229) **Absorption:** $p \, \mathcal{U} \, (p \, \mathcal{W} \, q) \equiv p \, \mathcal{W} \, q$

Proof:

$p \, \mathcal{U} \, (p \, \mathcal{W} \, q)$
$= \quad \langle(169) \text{ Definition of } \mathcal{W}\rangle$
$p \, \mathcal{U} \, (\Box p \vee p \, \mathcal{U} \, q)$
$= \quad \langle(12) \text{ Left Distributivity of } \mathcal{U} \text{ over } \vee\rangle$
$p \, \mathcal{U} \, \Box p \vee p \, \mathcal{U} \, (p \, \mathcal{U} \, q)$
$= \quad \langle(36) \text{ Left absorption of } \mathcal{U}\rangle$
$p \, \mathcal{U} \, \Box p \vee p \, \mathcal{U} \, q$
$= \quad \langle(141) \text{ Absorption}\rangle$
$\Box p \vee p \, \mathcal{U} \, q$
$= \quad \langle(169) \text{ Definition of } \mathcal{W}\rangle$
$p \, \mathcal{W} \, q \quad \blacksquare$

(230) **Absorption:** $(p \, \mathcal{W} \, q) \, \mathcal{U} \, q \equiv p \, \mathcal{U} \, q$

Proof:

$\quad (p \, \mathcal{W} \, q) \, \mathcal{U} \, q$
$= \quad \langle (171) \; \mathcal{U} \text{ in terms of } \mathcal{W} \rangle$
$\quad (p \, \mathcal{W} \, q) \, \mathcal{W} \, q \wedge \Diamond q$
$= \quad \langle (223) \text{ Right absorption of } \mathcal{W} \rangle$
$\quad p \, \mathcal{W} \, q \wedge \Diamond q$
$= \quad \langle (171) \; \mathcal{U} \text{ in terms of } \mathcal{W} \rangle$
$\quad p \, \mathcal{U} \, q \quad \blacksquare$

(231) **Absorption of $\mathcal{W}$ into $\Diamond$:** $\Diamond q \, \mathcal{W} \, q \equiv \Diamond q$

Proof:

$\quad \Diamond q \, \mathcal{W} \, q$
$= \quad \langle (38) \text{ Definition of } \Diamond \rangle$
$\quad (\text{true} \, \mathcal{U} \, q) \, \mathcal{W} \, q$
$= \quad \langle (228) \text{ Absorption} \rangle$
$\quad \text{true} \, \mathcal{U} \, q$
$= \quad \langle (38) \text{ Definition of } \Diamond \rangle$
$\quad \Diamond q \quad \blacksquare$

(232) **Absorption of $\mathcal{W}$ into $\Box$:** $\Box p \wedge p \, \mathcal{W} \, q \equiv \Box p$

Proof:

$\quad \Box p \wedge p \, \mathcal{W} \, q$
$= \quad \langle (169) \text{ Definition of } \mathcal{W} \rangle$
$\quad \Box p \wedge (\Box p \vee p \, \mathcal{U} \, q)$
$= \quad \langle (3.43a) \text{ Absorption}, p \wedge (p \vee q) \equiv p \rangle$
$\quad \Box p \quad \blacksquare$

(233) **Absorption of $\Box$ into $\mathcal{W}$:** $\Box p \vee p \, \mathcal{W} \, q \equiv p \, \mathcal{W} \, q$

Proof:

$\quad \Box p \vee p \, \mathcal{W} \, q$
$= \quad \langle (169) \text{ Definition of } \mathcal{W} \rangle$
$\quad \Box p \vee \Box p \vee p \, \mathcal{U} \, q$
$= \quad \langle (3.26) \text{ Idempotency of } \vee, p \vee p \equiv p \rangle$
$\quad \Box p \vee p \, \mathcal{U} \, q$
$= \quad \langle (169) \text{ Definition of } \mathcal{W} \rangle$
$\quad p \, \mathcal{W} \, q \quad \blacksquare$

The next two pair of theorems correspond. The *wait* versions are in the temporal logic literature. The *until* versions are unique to this system. Theorem (235) has a near symmetry with theorem (238). The monotonicity theorems and the strengthening and catenation rules are common. The implication rules (243) to (246) are unique to this system.

(234) $p \mathcal{W} q \wedge \Box \neg q \Rightarrow \Box p$

Proof:

$\quad p \mathcal{W} q \wedge \Box \neg q$
$\Rightarrow \quad \langle(226) \; \mathcal{W} \text{ implication and (4.3) Monotonicity of } \wedge\rangle$
$\quad (\Box p \vee \Diamond q) \wedge \Box \neg q$
$= \quad \langle(61) \text{ Dual of } \Diamond \rangle$
$\quad (\Box p \vee \Diamond q) \wedge \neg \Diamond q$
$= \quad \langle(3.44\text{a}) \text{ Absorption}, p \wedge (\neg p \vee q) \equiv p \wedge q\rangle$
$\quad \Box p \wedge \neg \Diamond q$
$\Rightarrow \quad \langle(3.76\text{b}) \text{ Strengthening}, p \wedge q \Rightarrow p\rangle$
$\quad \Box p \quad \blacksquare$

(235) $\Box p \Rightarrow p \mathcal{U} q \vee \Box \neg q$

Proof:

$\quad p \mathcal{U} q \vee \Box \neg q$
$\Leftarrow \quad \langle(84) \; \mathcal{U} \text{ implication and (4.2) Monotonicity of } \vee\rangle$
$\quad (\Box p \wedge \Diamond q) \vee \Box \neg q$
$= \quad \langle(61) \text{ Dual of } \Diamond \rangle$
$\quad (\Box p \wedge \Diamond q) \vee \neg \Diamond q$
$= \quad \langle(3.44\text{b}) \text{ Absorption}, p \vee (\neg p \wedge q) \equiv p \vee q\rangle$
$\quad \Box p \vee \neg \Diamond q$
$\Leftarrow \quad \langle(3.76\text{a}) \text{ Weakening}, p \Rightarrow p \vee q\rangle$
$\quad \Box p \quad \blacksquare$

(236) $\neg \Box p \wedge p \mathcal{W} q \Rightarrow \Diamond q$

Proof:

$\quad \neg \Box p \wedge p \mathcal{W} q \Rightarrow \Diamond q$
$= \quad \langle(3.65) \text{ Shunting}, p \wedge q \Rightarrow r \equiv p \Rightarrow (q \Rightarrow r)\rangle$
$\quad p \mathcal{W} q \Rightarrow (\neg \Box p \Rightarrow \Diamond q)$
$= \quad \langle(3.59) \text{ Implication}, p \Rightarrow q \equiv \neg p \vee q\rangle$
$\quad p \mathcal{W} q \Rightarrow \Box p \vee \Diamond q \quad -(226) \; \mathcal{W} \text{ implication} \quad \blacksquare$

(237) $\Diamond q \Rightarrow \neg \Box p \vee p \, \mathcal{U} \, q$

Proof:

$\quad \Diamond q \Rightarrow \neg \Box p \vee p \, \mathcal{U} \, q$
$= \quad \langle (3.59) \text{ Implication}, p \Rightarrow q \equiv \neg p \vee q \rangle$
$\quad \Diamond q \Rightarrow (\Box p \Rightarrow p \, \mathcal{U} \, q)$
$= \quad \langle (3.65) \text{ Shunting}, p \wedge q \Rightarrow r \equiv p \Rightarrow (q \Rightarrow r) \rangle$
$\quad \Diamond q \wedge \Box p \Rightarrow p \, \mathcal{U} \, q \quad -(84) \, \mathcal{U} \text{ implication} \quad \blacksquare$

(238) $\Box \neg p \wedge p \, \mathcal{U} \, q \Rightarrow q$

Proof: (John Wiegley)

$\quad \Box \neg p \wedge p \, \mathcal{U} \, q$
$\Rightarrow \quad \langle (76) \text{ Strengthening of } \Box \text{ and } (28) \, p \, \mathcal{U} \, q \Rightarrow p \vee q$
$\quad\quad \text{ and } (4.3) \text{ Monotonicity of } \wedge, \text{ twice} \rangle$
$\quad \neg p \wedge (p \vee q)$
$= \quad \langle (3.44a) \text{ Absorption}, p \wedge (\neg p \vee q) \equiv p \wedge q \rangle$
$\quad \neg p \wedge q$
$\Rightarrow \quad \langle (3.76b) \text{ Strengthening}, p \wedge q \Rightarrow p \rangle$
$\quad q \quad \blacksquare$

(239) **Left monotonicity of** $\mathcal{W}$: $\Box(p \Rightarrow q) \Rightarrow (p \, \mathcal{W} \, r \Rightarrow q \, \mathcal{W} \, r)$

Proof:

$\quad \Box(p \Rightarrow q) \Rightarrow (p \, \mathcal{W} \, r \Rightarrow q \, \mathcal{W} \, r)$
$= \quad \langle (3.65) \text{ Shunting}, p \wedge q \Rightarrow r \equiv p \Rightarrow (q \Rightarrow r) \rangle$
$\quad \Box(p \Rightarrow q) \wedge p \, \mathcal{W} \, r \Rightarrow q \, \mathcal{W} \, r$

And now,

$\quad \Box(p \Rightarrow q) \wedge p \, \mathcal{W} \, r$
$\Rightarrow \quad \langle (175) \text{ Distributivity of } \wedge \text{ over } \mathcal{W} \rangle$
$\quad ((p \Rightarrow q) \wedge p) \, \mathcal{W} \, ((p \Rightarrow q) \wedge r)$
$= \quad \langle (3.66) \, p \wedge (p \Rightarrow q) \equiv p \wedge q \rangle$
$\quad ((p \wedge q) \, \mathcal{W} \, ((p \Rightarrow q) \wedge r)$
$\Rightarrow \quad \langle (186) \text{ Left Distributivity of } \mathcal{W} \text{ over } \wedge \rangle$
$\quad ((p \wedge q) \, \mathcal{W} \, (p \Rightarrow q) \wedge (p \wedge q) \, \mathcal{W} \, r$
$= \quad \langle (187) \text{ Right Distributivity of } \mathcal{W} \text{ over } \wedge \rangle$
$\quad ((p \wedge q) \, \mathcal{W} \, (p \Rightarrow q) \wedge p \, \mathcal{W} \, r \wedge q \, \mathcal{W} \, r$
$\Rightarrow \quad \langle (3.76b) \text{ Strengthening}, p \wedge q \Rightarrow p \rangle$
$\quad q \, \mathcal{W} \, r \quad \blacksquare$

(240) **Right monotonicity of** $\mathcal{W}$: $\Box(p \Rightarrow q) \Rightarrow (r\,\mathcal{W}\,p \Rightarrow r\,\mathcal{W}\,q)$

Proof:

$\quad r\,\mathcal{W}\,p \Rightarrow r\,\mathcal{W}\,q$
$=\quad \langle(169)\text{ Definition of }\mathcal{W}\text{, twice}\rangle$
$\quad \Box r \vee r\,\mathcal{U}\,p \Rightarrow \Box r \vee r\,\mathcal{U}\,q$
$\Leftarrow\quad \langle(4.2)\text{ Monotonicity of }\vee, (p \Rightarrow q) \Rightarrow (p \vee r \Rightarrow q \vee r)\rangle$
$\quad r\,\mathcal{U}\,p \Rightarrow r\,\mathcal{U}\,q$
$\Leftarrow\quad \langle(85)\text{ Right monotonicity of }\mathcal{U}\rangle$
$\quad \Box(p \Rightarrow q)$ ∎

(241) $\mathcal{W}$ **strengthening rule:** $\Box((p \Rightarrow r) \wedge (q \Rightarrow s)) \Rightarrow (p\,\mathcal{W}\,q \Rightarrow r\,\mathcal{W}\,s)$

Proof:

$\quad \Box((p \Rightarrow r) \wedge (q \Rightarrow s)) \Rightarrow (p\,\mathcal{W}\,q \Rightarrow r\,\mathcal{W}\,s)$
$=\quad \langle(3.65)\text{ Shunting}, p \wedge q \Rightarrow r \equiv p \Rightarrow (q \Rightarrow r)\rangle$
$\quad \Box((p \Rightarrow r) \wedge (q \Rightarrow s)) \wedge p\,\mathcal{W}\,q \Rightarrow r\,\mathcal{W}\,s$

And now,

$\quad \Box((p \Rightarrow r) \wedge (q \Rightarrow s)) \wedge p\,\mathcal{W}\,q$
$=\quad \langle(99)\text{ Distributivity of }\Box\text{ over }\wedge\rangle$
$\quad \Box(p \Rightarrow r) \wedge \Box(q \Rightarrow s) \wedge p\,\mathcal{W}\,q$
$\Rightarrow\quad \langle(240)\text{ Right monotonicity of }\mathcal{W}\text{ and }(4.3)\text{ Monotonicity of }\wedge\rangle$
$\quad \Box(p \Rightarrow r) \wedge (p\,\mathcal{W}\,q \Rightarrow p\,\mathcal{W}\,s) \wedge p\,\mathcal{W}\,q$
$\Rightarrow\quad \langle(3.77)\text{ Modus ponens}, p \wedge (p \Rightarrow q) \Rightarrow q\text{ and }(4.3)\text{ Monotonicity of }\wedge\rangle$
$\quad \Box(p \Rightarrow r) \wedge p\,\mathcal{W}\,s$
$\Rightarrow\quad \langle(239)\text{ Left monotonicity of }\mathcal{W}\text{ and }(4.3)\text{ Monotonicity of }\wedge\rangle$
$\quad (p\,\mathcal{W}\,s \Rightarrow r\,\mathcal{W}\,s) \wedge p\,\mathcal{W}\,s$
$\Rightarrow\quad \langle(3.77)\text{ Modus ponens}, p \wedge (p \Rightarrow q) \Rightarrow q\rangle$
$\quad r\,\mathcal{W}\,s$ ∎

(242) $\mathcal{W}$ **catenation rule:** $\Box((p \Rightarrow q\,\mathcal{W}\,r) \wedge (r \Rightarrow q\,\mathcal{W}\,s)) \Rightarrow (p \Rightarrow q\,\mathcal{W}\,s)$

Proof:

$\quad \Box((p \Rightarrow q\,\mathcal{W}\,r) \wedge (r \Rightarrow q\,\mathcal{W}\,s)) \Rightarrow (p \Rightarrow q\,\mathcal{W}\,s)$
$=\quad \langle(3.65)\text{ Shunting}, p \wedge q \Rightarrow r \equiv p \Rightarrow (q \Rightarrow r)\rangle$
$\quad \Box((p \Rightarrow q\,\mathcal{W}\,r) \wedge (r \Rightarrow q\,\mathcal{W}\,s)) \wedge p \Rightarrow q\,\mathcal{W}\,s$

And now,

$$\Box((p \Rightarrow q \mathcal{W} r) \land (r \Rightarrow q \mathcal{W} s)) \land p$$
$= \quad \langle(99) \text{ Distributivity of } \Box \text{ over } \land\rangle$
$$\Box(p \Rightarrow q \mathcal{W} r) \land \Box(r \Rightarrow q \mathcal{W} s) \land p$$
$\Rightarrow \quad \langle(76) \text{ Strengthening of } \Box \text{ and } (4.3) \text{ Monotonicity of } \land\rangle$
$$(p \Rightarrow q \mathcal{W} r) \land \Box(r \Rightarrow q \mathcal{W} s) \land p$$
$\Rightarrow \quad \langle(3.77) \text{ Modus ponens, } p \land (p \Rightarrow q) \Rightarrow q \text{ and } (4.3) \text{ Monotonicity of } \land\rangle$
$$q \mathcal{W} r \land \Box(r \Rightarrow q \mathcal{W} s)$$
$\Rightarrow \quad \langle(240) \text{ Right monotonicity of } \mathcal{W} \text{ with } p,q,r := r, q \mathcal{W} s, q\rangle$
$$q \mathcal{W} r \land (q \mathcal{W} r \Rightarrow q \mathcal{W} (q \mathcal{W} s))$$
$\Rightarrow \quad \langle(3.77) \text{ Modus ponens, } p \land (p \Rightarrow q) \Rightarrow q\rangle$
$$q \mathcal{W} (q \mathcal{W} s)$$
$= \quad \langle(222) \text{ Left absorption of } \mathcal{W}\rangle$
$$q \mathcal{W} s \quad \blacksquare$$

(243) Left $\mathcal{U}$ $\mathcal{W}$ implication: $(p \mathcal{U} q) \mathcal{W} r \Rightarrow (p \mathcal{W} q) \mathcal{W} r$

Proof: The proof is by (4.7.1) Truth implication.

$$true$$
$= \quad \langle(136) \text{ Metatheorem and } (174) \ \mathcal{U} \text{ implication}\rangle$
$$\Box(p \mathcal{U} q \Rightarrow p \mathcal{W} q)$$
$\Rightarrow \quad \langle(239) \text{ Left monotonicity of } \mathcal{W} \text{ with } p,q := p \mathcal{U} q, p \mathcal{W} q\rangle$
$$(p \mathcal{U} q) \mathcal{W} r \Rightarrow (p \mathcal{W} q) \mathcal{W} r \quad \blacksquare$$

(244) Right $\mathcal{W}$ $\mathcal{U}$ implication: $p \mathcal{W} (q \mathcal{U} r) \Rightarrow p \mathcal{W} (q \mathcal{W} r)$

Proof: The proof is by (4.7.1) Truth implication.

$$true$$
$= \quad \langle(136) \text{ Metatheorem and } (174) \ \mathcal{U} \text{ implication}\rangle$
$$\Box(q \mathcal{U} r \Rightarrow q \mathcal{W} r)$$
$\Rightarrow \quad \langle(240) \text{ Right monotonicity of } \mathcal{W} \text{ with } p,q,r := q \mathcal{U} r, q \mathcal{W} r, p\rangle$
$$p \mathcal{W} (q \mathcal{U} r) \Rightarrow p \mathcal{W} (q \mathcal{W} r) \quad \blacksquare$$

(245) Right $\mathcal{U}$ $\mathcal{U}$ implication: $p \mathcal{U} (q \mathcal{U} r) \Rightarrow p \mathcal{U} (q \mathcal{W} r)$

Proof: The proof is by (4.7.1) Truth implication.

$$true$$
$= \quad \langle(136) \text{ Metatheorem and } (174) \ \mathcal{U} \text{ implication}\rangle$
$$\Box(q \mathcal{U} r \Rightarrow q \mathcal{W} r)$$
$\Rightarrow \quad \langle(85) \text{ Right monotonicity of } \mathcal{U} \text{ with } p,q,r := q \mathcal{U} r, q \mathcal{W} r, p\rangle$
$$p \mathcal{U} (q \mathcal{U} r) \Rightarrow p \mathcal{U} (q \mathcal{W} r) \quad \blacksquare$$

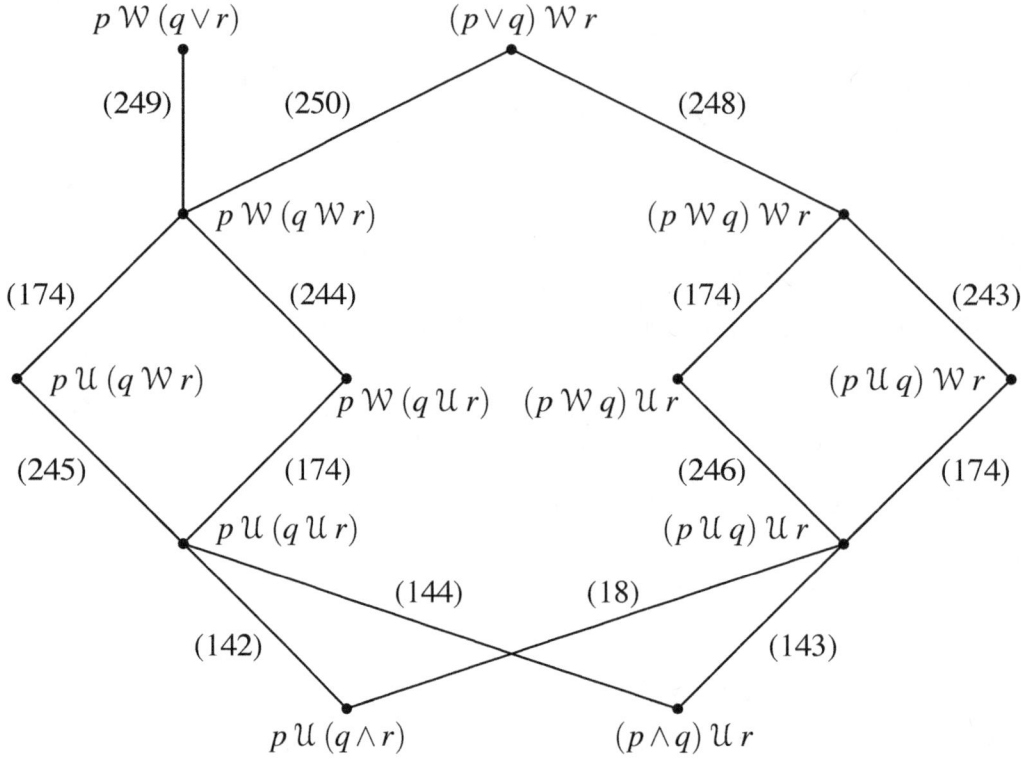

Figure 5: A Hasse diagram showing some implication relations of linear temporal logic.

(246) **Left $\mathcal{U}\,\mathcal{U}$ implication:** $(p\,\mathcal{U}\,q)\,\mathcal{U}\,r \Rightarrow (p\,\mathcal{W}\,q)\,\mathcal{U}\,r$

Proof: The proof is by (4.7.1) Truth implication.

$\quad true$
$= \quad \langle(136)\text{ Metatheorem and }(174)\ \mathcal{U}\text{ implication}\rangle$
$\quad \Box(p\,\mathcal{U}\,q \Rightarrow p\,\mathcal{W}\,q)$
$\Rightarrow \quad \langle(86)\text{ Left monotonicity of }\mathcal{U}\text{ with }p,q := p\,\mathcal{U}\,q, p\,\mathcal{W}\,q\rangle$
$\quad (p\,\mathcal{U}\,q)\,\mathcal{U}\,r \Rightarrow (p\,\mathcal{W}\,q)\,\mathcal{U}\,r \quad \blacksquare$

Of the three strengthening rules, only (248) appears in the LTL literature. All of the following ordering rules are in the literature except for (252) $\mathcal{U}$ ordering. Figure 5 is a Hasse diagram showing some implication relations of these theorems.

(247) **Left $\mathcal{U} \vee$ strengthening:** $(p\,\mathcal{U}\,q)\,\mathcal{U}\,r \Rightarrow (p \vee q)\,\mathcal{U}\,r$

Proof: The proof is by (4.7.1) Truth implication.

$\quad true$
$= \quad \langle(136)\text{ Metatheorem and }(28)\rangle$
$\quad \Box(p\,\mathcal{U}\,q \Rightarrow p \vee q)$
$\Rightarrow \quad \langle(86)\text{ Left monotonicity of }\mathcal{U}\text{ with }p,q := p\,\mathcal{U}\,q, p \vee q\rangle$
$\quad (p\,\mathcal{U}\,q)\,\mathcal{U}\,r \Rightarrow (p \vee q)\,\mathcal{U}\,r \quad \blacksquare$

(248) **Left $\mathcal{W} \vee$ strengthening:** $(p \mathcal{W} q) \mathcal{W} r \Rightarrow (p \vee q) \mathcal{W} r$

Proof: The proof is by (4.7.1) Truth implication.

$\quad$ *true*
$=\quad \langle(136)\text{ Metatheorem and }(206)\rangle$
$\quad \square (p \mathcal{W} q \Rightarrow p \vee q)$
$\Rightarrow \quad \langle(239)\text{ Left monotonicity of }\mathcal{W}\text{ with }p,q := p \mathcal{W} q, p \vee q\rangle$
$\quad (p \mathcal{W} q) \mathcal{W} r \Rightarrow (p \vee q) \mathcal{W} r \quad \blacksquare$

(249) **Right $\mathcal{W} \vee$ strengthening:** $p \mathcal{W} (q \mathcal{W} r) \Rightarrow p \mathcal{W} (q \vee r)$

Proof: The proof is by (4.7.1) Truth implication.

$\quad$ *true*
$=\quad \langle(136)\text{ Metatheorem and }(206)\rangle$
$\quad \square (q \mathcal{W} r \Rightarrow q \vee r)$
$\Rightarrow \quad \langle(240)\text{ Right monotonicity of }\mathcal{W}\text{ with }p,q,r := q \mathcal{W} r, q \vee r, p\rangle$
$\quad p \mathcal{W} (q \mathcal{W} r) \Rightarrow p \mathcal{W} (q \vee r) \quad \blacksquare$

(250) **Right $\mathcal{W} \vee$ ordering:** $p \mathcal{W} (q \mathcal{W} r) \Rightarrow (p \vee q) \mathcal{W} r$

Proof:

$\quad p \mathcal{W} (q \mathcal{W} r)$
$=\quad \langle(169)\text{ Definition of }\mathcal{W}\text{, twice}\rangle$
$\quad \square p \vee p \,\mathcal{U}\, (\square q \vee q \,\mathcal{U}\, r)$
$=\quad \langle(12)\text{ Left distributivity of }\mathcal{U}\text{ over }\vee\rangle$
$\quad \square p \vee p \,\mathcal{U}\, \square q \vee p \,\mathcal{U}\, (q \,\mathcal{U}\, r)$
$\Rightarrow \quad \langle(140)\ \mathcal{U}\square\text{ implication and }(4.2)\text{ Monotonicity of }\vee\rangle$
$\quad \square p \vee \square (p \,\mathcal{U}\, q) \vee p \,\mathcal{U}\, (q \,\mathcal{U}\, r)$
$\Rightarrow \quad \langle(17)\text{ Right }\mathcal{U} \vee\text{ ordering and }(4.2)\text{ Monotonicity of }\vee\rangle$
$\quad \square p \vee \square (p \,\mathcal{U}\, q) \vee (p \vee q) \,\mathcal{U}\, r$
$\Rightarrow \quad \langle(100)\text{ Distributivity of }\square\text{ over }\vee\text{ and }(4.2)\text{ Monotonicity of }\vee\rangle$
$\quad \square (p \vee p \,\mathcal{U}\, q) \vee (p \vee q) \,\mathcal{U}\, r$
$=\quad \langle(31)\text{ Absorption}\rangle$
$\quad \square (p \vee q) \vee (p \vee q) \,\mathcal{U}\, r$
$=\quad \langle(169)\text{ Definition of }\mathcal{W}\rangle$
$\quad (p \vee q) \mathcal{W} r \quad \blacksquare$

(251) **Right $\wedge \mathcal{W}$ ordering:** $p \mathcal{W} (q \wedge r) \Rightarrow (p \mathcal{W} q) \mathcal{W} r$

Proof:

$(p \,\mathcal{W}\, q) \,\mathcal{W}\, r$
$=\quad \langle(169)\text{ Definition of }\mathcal{W}\rangle$
$(\Box p \vee p \,\mathcal{U}\, q) \,\mathcal{W}\, r$
$\Leftarrow\quad \langle(185)\text{ Right distributivity of }\mathcal{W}\text{ over }\vee\rangle$
$(\Box p \,\mathcal{W}\, r) \vee (p \,\mathcal{U}\, q) \,\mathcal{W}\, r$
$=\quad \langle(169)\text{ Definition of }\mathcal{W}\text{, twice}\rangle$
$\Box\Box p \vee \Box p \,\mathcal{U}\, r \vee \Box(p \,\mathcal{U}\, q) \vee (p \,\mathcal{U}\, q) \,\mathcal{U}\, r$
$=\quad \langle(72)\text{ Absorption of }\Box\rangle$
$\Box p \vee \Box p \,\mathcal{U}\, r \vee \Box(p \,\mathcal{U}\, q) \vee (p \,\mathcal{U}\, q) \,\mathcal{U}\, r$
$\Leftarrow\quad \langle(3.76\text{a})\text{ Weakening, } p \Rightarrow p \vee q\rangle$
$\Box p \vee (p \,\mathcal{U}\, q) \,\mathcal{U}\, r$
$\Leftarrow\quad \langle(18)\text{ Right }\wedge\, \mathcal{U}\text{ ordering and }(4.2)\text{ Monotonicity of }\vee\rangle$
$\Box p \vee p \,\mathcal{U}\, (q \wedge r)$
$=\quad \langle(169)\text{ Definition of }\mathcal{W}\rangle$
$p \,\mathcal{W}\, (q \wedge r)\quad \blacksquare$

(252) **$\mathcal{U}$ ordering:** $\quad \neg p \,\mathcal{U}\, q \vee \neg q \,\mathcal{U}\, p \equiv \Diamond(p \vee q)$

Proof:
$\neg p \,\mathcal{U}\, q \vee \neg q \,\mathcal{U}\, p$
$=\quad \langle(196)\text{ Conjunction rule of }\mathcal{U}\text{, twice}\rangle$
$(\neg p \wedge \neg q) \,\mathcal{U}\, q \vee (\neg q \wedge \neg p) \,\mathcal{U}\, p$
$=\quad \langle(12)\text{ Left distributivity of }\mathcal{U}\text{ over }\vee\rangle$
$(\neg p \wedge \neg q) \,\mathcal{U}\, (p \vee q)$
$=\quad \langle(3.47\text{b})\text{ De Morgan, }\neg(p \vee q) \equiv \neg p \wedge \neg q\rangle$
$\neg(p \vee q) \,\mathcal{U}\, (p \vee q)$
$=\quad \langle(192)\text{ Rule of }\mathcal{U}\text{ with } q := p \vee q\rangle$
$\Diamond(p \vee q)\quad \blacksquare$

(253) **$\mathcal{W}$ ordering:** $\quad \neg p \,\mathcal{W}\, q \vee \neg q \,\mathcal{W}\, p$

Proof:
$\neg p \,\mathcal{W}\, q \vee \neg q \,\mathcal{W}\, p$
$=\quad \langle(195)\text{ Conjunction rule of }\mathcal{W}\text{, twice}\rangle$
$(\neg p \wedge \neg q) \,\mathcal{W}\, q \vee (\neg q \wedge \neg p) \,\mathcal{W}\, p$
$=\quad \langle(184)\text{ Left distributivity of }\mathcal{W}\text{ over }\vee\rangle$
$(\neg p \wedge \neg q) \,\mathcal{W}\, (p \vee q)$
$=\quad \langle(3.47\text{b})\text{ De Morgan, }\neg(p \vee q) \equiv \neg p \wedge \neg q\rangle$
$\neg(p \vee q) \,\mathcal{W}\, (p \vee q)$
$=\quad \langle(191)\text{ Rule of }\mathcal{W}\text{ with } q := p \vee q\rangle$
true$\quad \blacksquare$

Pepperdine Papers on LTL   119

(254) $\mathcal{W}$ **implication ordering:** $p\,\mathcal{W}\,q \wedge \neg q\,\mathcal{W}\,r \Rightarrow p\,\mathcal{W}\,r$

Proof: The proof is based on the following lemmas.

Lemma A: $p\,\mathcal{U}\,q \wedge \Box\neg q \equiv \textit{false}$

Proof:

$\quad p\,\mathcal{U}\,q \wedge \Box\neg q$
$= \quad \langle(171)\ \mathcal{U}\ \text{in terms of}\ \mathcal{W}\rangle$
$\quad p\,\mathcal{W}\,q \wedge \Diamond q \wedge \Box\neg q$
$= \quad \langle(92)\ \Diamond\ \text{contradiction}\rangle$
$\quad p\,\mathcal{W}\,q \wedge \textit{false}$
$= \quad \langle(3.40)\ \text{Zero of}\ \wedge,\ p \wedge \textit{false} \equiv \textit{false}\rangle$
$\quad \textit{false}\quad\blacksquare$

Lemma B: $\Box p \wedge \Box\neg q \Rightarrow p\,\mathcal{W}\,r$

Proof:

$\quad \Box p \wedge \Box\neg q$
$\Rightarrow \quad \langle(3.76\text{b})\ \text{Strengthening},\ p \wedge q \Rightarrow p\rangle$
$\quad \Box p$
$\Rightarrow \quad \langle(179)\ \text{Perpetuity}\rangle$
$\quad p\,\mathcal{W}\,r\quad\blacksquare$

Lemma C: $\Box p \wedge \neg q\,\mathcal{U}\,r \Rightarrow p\,\mathcal{W}\,r$

Proof:

$\quad \Box p \wedge \neg q\,\mathcal{U}\,r$
$\Rightarrow \quad \langle(3.76\text{b})\ \text{Strengthening},\ p \wedge q \Rightarrow p\rangle$
$\quad \Box p$
$\Rightarrow \quad \langle(179)\ \text{Perpetuity}\rangle$
$\quad p\,\mathcal{W}\,r\quad\blacksquare$

And now,

$\quad p\,\mathcal{W}\,q \wedge \neg q\,\mathcal{W}\,r$
$= \quad \langle(169)\ \text{Definition of}\ \mathcal{W},\ \text{twice}\rangle$
$\quad (\Box p \vee p\,\mathcal{U}\,q) \wedge (\Box\neg q \vee \neg q\,\mathcal{U}\,r)$
$= \quad \langle(3.46)\ \text{Distributivity of}\ \wedge\ \text{over}\ \vee,\ p \wedge (q \vee r) \equiv (p \wedge q) \vee (p \wedge r)\rangle$
$\quad (\Box p \wedge \Box\neg q) \vee (\Box p \wedge \neg q\,\mathcal{U}\,r) \vee (p\,\mathcal{U}\,q \wedge \Box\neg q) \vee (p\,\mathcal{U}\,q \wedge \neg q\,\mathcal{U}\,r)$
$= \quad \langle\text{Lemma A}\rangle$

$$(\Box p \wedge \Box \neg q) \vee (\Box p \wedge \neg q\, \mathcal{U}\, r) \vee \mathit{false} \vee (p\, \mathcal{U}\, q \wedge \neg q\, \mathcal{U}\, r)$$
$= \quad \langle (3.30)\text{ Identity of } \vee,\ p \vee \mathit{false} \equiv p \rangle$
$$(\Box p \wedge \Box \neg q) \vee (\Box p \wedge \neg q\, \mathcal{U}\, r) \vee (p\, \mathcal{U}\, q \wedge \neg q\, \mathcal{U}\, r)$$
$\Rightarrow \quad \langle \text{Lemma B, Lemma C, and (4.2) Monotonicity of } \vee \rangle$
$$p\, \mathcal{W}\, r \vee p\, \mathcal{W}\, r \vee (p\, \mathcal{U}\, q \wedge \neg q\, \mathcal{U}\, r)$$
$\Rightarrow \quad \langle (16)\ \mathcal{U} \text{ implication ordering and (4.2) Monotonicity of } \vee \rangle$
$$p\, \mathcal{W}\, r \vee p\, \mathcal{W}\, r \vee p\, \mathcal{U}\, r$$
$\Rightarrow \quad \langle (174)\ \mathcal{U} \text{ implication and (4.2) Monotonicity of } \vee \rangle$
$$p\, \mathcal{W}\, r \vee p\, \mathcal{W}\, r \vee p\, \mathcal{W}\, r$$
$= \quad \langle (3.26) \text{ Idempotency of } \vee,\ p \vee p \equiv p \rangle$
$$p\, \mathcal{W}\, r \quad \blacksquare$$

(255) **Lemmon formula:** $\Box(\Box p \Rightarrow q) \vee \Box(\Box q \Rightarrow p)$

Proof: The proof is by (4.7.1) Truth implication.

$$\Box(\Box p \Rightarrow q) \vee \Box(\Box q \Rightarrow p)$$
$= \quad \langle (3.59) \text{ Implication, } p \Rightarrow q \equiv \neg p \vee q,\ \text{twice} \rangle$
$$\Box(\neg \Box p \vee q) \vee \Box(\neg \Box q \vee p)$$
$\Leftarrow \quad \langle (206) \text{ twice, (120) Monotonicity of } \Box,\ \text{and (4.2) Monotonicity of } \vee \rangle$
$$\Box(\neg \Box p\, \mathcal{W}\, q) \vee \Box(\neg \Box q\, \mathcal{W}\, p)$$
$\Leftarrow \quad \langle (177)\ \mathcal{W}\, \Box \text{ implication and (4.2) Monotonicity of } \vee \rangle$
$$\neg \Box p\, \mathcal{W}\, \Box q \vee \neg \Box q\, \mathcal{W}\, \Box p$$
$= \quad \langle (253)\ \mathcal{W} \text{ ordering with } p, q := \Box p, \Box q \rangle$
$$\mathit{true} \quad \blacksquare$$

The ten axioms that define the behavior of the *until* operator are (9), (10), (11), (12), (13), (14), (15), (16), (17), and (18). The corresponding theorems for the *wait* operator are (180), (181), (224), (184), (185), (186), (187), (254), (250), and (251). Of these ten theorems, nine are identical, with the substitution of $\mathcal{W}$ for $\mathcal{U}$, to the corresponding axioms that define the *until* operator. In addition, (170), the axiom that describes the distributivity of $\neg$ over $\mathcal{W}$, is identical to (173) with the interchange of $\mathcal{W}$ and $\mathcal{U}$. The one theorem that distinguishes $\mathcal{W}$ from the defining axioms of $\mathcal{U}$ is

(11) **Axiom, Right zero of $\mathcal{U}$:** $p\, \mathcal{U}\, \mathit{false} \equiv \mathit{false}$

for the *until* operator versus

(224) **$\Box$ to $\mathcal{W}$ law:** $\Box p \equiv p\, \mathcal{W}\, \mathit{false}$

for the *wait* operator.

4 Comparison with Previous Work

Linear temporal logic is just one of many different modal logic systems. This section shows how linear temporal logic fits in general systems of modal logic, how calculational logic applies to modal logic, and compares this system with other LTL axiomatizations.

4.1 Modal Logic Systems

In the terminology of modal logic, each state of an anchored sequence is called a *world*. In LTL, each world represents the state of a computation at each discrete point of time, and so one world is related to another world as one point in time is related to another point in time, *i.e.*, as occurring before or after it. In general, worlds need not have such an interpretation. Rather, a specific modal logic system is defined by a nonempty set of worlds W and a relation ρ over W. A frame is the ordered pair $\langle W, \rho \rangle$, and different modal systems are specified by different frames.

The most general modal system, known as K, extends propositional calculus by adding the unary operator $\Box$ together with the axiom $\Box(p \Rightarrow q) \Rightarrow (\Box p \Rightarrow \Box q)$, which is our theorem (120) Monotonicity of $\Box$. The Hilbert style inference rules of Uniform Substitution and Modus Ponens are extended by adding the rule of Necessitation: If P is a theorem then so is $\Box P$, which is Case 1 of our (136) Metatheorem. The operator $\Diamond p$ is defined to be an abbreviation for $\neg \Box \neg p$ as in (59). With these extensions, there is no restriction on relation ρ in the frame $\langle W, \rho \rangle$ that defines K.

A stronger modal system, known as T, extends K by adding the axiom $\Box p \Rightarrow p$, sometimes called the Axiom of Necessity, which is our theorem (76). Every theorem in K is also a theorem in T, but not every theorem in T is a theorem in K. It has been shown that theorems in T are valid on every frame $\langle W, \rho \rangle$ in which ρ is reflexive, *i.e.*, in which for all $w \in W$, $w \rho w$. [18]

A stronger yet modal system, known as S4, extends T by adding the single extra axiom $\Box p \Rightarrow \Box \Box p$, which is part of our theorem (72) Absorption of $\Box$, $\Box \Box p \equiv \Box p$ by virtue of mutual implication. Theorems in S4 are valid on every frame $\langle W, \rho \rangle$ in which ρ is both reflexive and transitive, *i.e.*, in which for all $w, x, y \in W$, $w \rho x \wedge x \rho y \Rightarrow w \rho y$.

Linear temporal systems model time as discrete points on the number line. Each world is on a time line and is either contemporaneous with, or earlier than, itself and all worlds that come after it. This "earlier than" relation is defined formally as a connected relation, in which for all $w, x, y \in W$, $w \rho x \wedge w \rho y \Rightarrow x \rho y \vee y \rho x$. In other words, if w is contemporaneous with or earlier than x, and w is contemporaneous with or earlier than y, then either x is contemporaneous with or earlier than y, or y is contemporaneous with or earlier than x. The temporal modal system, known as S4.3 [8], extends S4 by the axiom $\Box(\Box p \Rightarrow q) \vee \Box(\Box q \Rightarrow p)$, which is our theorem (255) Lemmon formula. Theorems in S4.3 are valid on every frame $\langle W, \rho \rangle$ in which ρ is reflexive, transitive, and connected. [18]

The Lemmon formula imposes linearity of the time line. The Dummett formula

$$\Box(\Box(p \Rightarrow \Box p) \Rightarrow \Box p) \Rightarrow (\Diamond \Box p \Rightarrow \Box p)$$

imposes discreteness of the time line. [9, 22] The modal system S4.3 with the addition of the Dummett formula is known as S4.3*Dum*. Our LTL system contains S4.3, because it includes as theorems the axioms of S4.3. The LTL systems we surveyed are close to S4.3*Dum* but do not contain the Dummett formula as an axiom. Instead, discreteness of the time line is implicit in the definition of the *next* operator $\bigcirc$. The binary operators *until* $\mathcal{U}$ and *wait* $\mathcal{W}$ are absent in classical modal logic systems.

4.2 Calculational Modal Logic Systems

Gries and Schneider [17, 14] describe the benefits of the calculational system for propositional logic and its application to discrete math. Gries and Schneider [16] and Tourlakis [25] demonstrate the soundness and completeness of such systems. The calculational method is the subject of a special issue in [1].

Gries and Schneider [15] extend calculational logic to the Carnapian modal system known as C. Theorems in C are valid on every frame $\langle W, \rho \rangle$ in which ρ is the universal accessibility relation, *i.e.*, each state is related to all states. Because C is not a linear temporal logic, the operator $\Box$ has a different interpretation from that in LTL. Denoting the Carnapian operator as $\Box_c$ the interpretation of $\Box_c P$, where P is a propositional logic expression, is "P is true in all states," or, equivalently, "P is valid." Gries and Schneider point out that this logic "... can be used for proving theorems that could otherwise be handled only at the meta-level, and most likely informally." As an example, (2.3b) Metatheorem Duality, $P \equiv Q$ is valid iff $P_D \equiv Q_D$ is valid, can be written as $\Box_c(P \equiv Q) \equiv \Box_c(P_D \equiv Q_D)$. In such a system, metatheorems become formulas in the logic and are thus directly available for use in calculational reasoning.

Schneider's graduate text *On Concurrent Programming* [24] extends the calculational logic system to linear temporal logic. It claims, and we concur, that "... proofs in Temporal Logic are often easier to read and to construct when an equational format is available." As justification for the extension, it gives a temporal logic analogue of (I1) Leibniz, TL Substitution of Equals. For proof steps that assert $p \Rightarrow q$, it gives two additional temporal logic rules – TL Monotonicity Rule and TL Antimonotonicity Rule. These rules are the basis for the justification of temporal logic calculational proofs.

4.3 Survey of LTL Deductive Systems

As a check on comprehensiveness, we inventoried the linear temporal logic theorems and inference rules in Ben-Ari [3], Emerson [10], Kröger and Merz[20], Manna and Pnueli [21], and Schneider [24], and included them all in this survey. This section compares the axiomatization systems of those sources with this work.

Ben-Ari [3] takes the temporal operators $\square$, $\circ$ and $\mathcal{U}$ as basic and defines $\diamond$ to be an abbreviation for $\neg\square\neg$ as in (59). The $\mathcal{W}$ operator is not covered in his system for LTL.

The two stated rules of inference in the Ben-Ari deductive system correspond to (3.77) Modus ponens and the generalization part of (136) Metatheorem. The five axioms that define $\circ$ and $\square$ correspond to (120) Monotonicity of $\square$, (2) Distributivity of $\circ$ over $\Rightarrow$, (67) Expansion of $\square$, (57) $\square$ induction, and (3) Linearity. The two axioms that define $\mathcal{U}$ correspond to (10) Expansion of $\mathcal{U}$ and (42) Eventuality.

Emerson [10] uses the symbol X to denote the *next* operator, F to denote *eventually*, and G to denote *always*. He also writes $\overset{\infty}{F}$ to denote *always eventually* and $\overset{\infty}{G}$ to denote *eventually always*. With these conventions $\square p \equiv p \wedge \circ \square p$ is written as G $p \equiv p \wedge$ X G p, and $\neg \diamond \square p \equiv \square \diamond \neg p$ is written as $\neg \overset{\infty}{G} p \equiv \overset{\infty}{F} \neg p$. Emerson's notation for the *until* operator is U, which is sometimes written $U_\exists$ to distinguish it from the *wait* operator, which is written $U_\forall$. With these conventions $p \mathcal{U} q \equiv p \mathcal{W} q \wedge \diamond q$ is written as $p U_\exists q \equiv p U_\forall q \wedge F q$.

Emerson's deductive system is for Computational Tree Logic (CTL) which is a system of branching time logic. In contrast to linear time logics, which model time as an anchored sequence of states, branching time logics model time as a tree structure where at each point in time the computation may split into several possible future states. CTL extends LTL to multiple timelines (branches) through the addition of two additional quantifiers over these branches, A (for all futures) and E (for some futures). Consequently, Emerson's deductive system does not directly apply to LTL. The CTL inference rules and axioms without the additional quantifiers, however, do correspond to the LTL deductive system of our system as follows.

As with the LTL system of Ben-Ari, the two stated rules of inference correspond to (3.77) Modus ponens and the generalization part of (136) Metatheorem. The nine axioms correspond to (38) Definition of $\diamond$, (54) Definition of $\square$, (4) Distributivity of $\circ$ over $\vee$, (3) Linearity, (10) Expansion of $\mathcal{U}$, (7) Truth of $\circ$, (129) and (130) Induction rules $\square$ and $\mathcal{U}$, and (118) Monotonicity of $\circ$.

Kröger and Merz [20] use the symbol **unt** (an abbreviation for until) to denote the *until* operator $\mathcal{U}$ and the symbol **unl** (an abbreviation for unless) to denote the *wait* operator $\mathcal{W}$. They call these *non-strict* operators in contrast to variations of the operators termed *strict* operators.

Kröger and Merz first define a *propositional linear temporal logic* consisting of only the unary temporal operators $\circ$ and $\square$. They define $\diamond p$ to be an abbreviation for $\neg \square \neg p$ as in (59). The three inference rules correspond to (3.77) Modus ponens, (78) Strengthening of $\square$, and (129) Induction rule $\square$. The three axioms correspond to (1) Self-dual, (2) Distributivity of $\circ$ over $\Rightarrow$, and (66) Expansion of $\square$. They introduce the *until* operator and the *wait* operator as an extension to the propositional linear temporal logic. The four axioms for the extension correspond to (10) Expansion of $\mathcal{U}$, (42) Eventuality, (181) Expansion of $\mathcal{W}$, and (179) Perpetuity.

Manna and Pnueli [21] use the $\rightarrow$ symbol for implication, so that $p \Rightarrow q$ of our system is written as $p \rightarrow q$ in theirs. They introduce the symbol $\Rightarrow$ to stand for always implies, so that $\square\,(p \to q)$ is written as $p \Rightarrow q$ in their system. Similarly, $p \equiv q$ of our system is written as $p \leftrightarrow q$, in theirs and $\square\,(p \leftrightarrow q)$ is written as $p \Leftrightarrow q$.

Manna and Pnueli define a proof system that combines the future operators $\bigcirc$, $\Diamond$, $\square$, $\mathcal{U}$, and $\mathcal{W}$ with a set of corresponding past operators. For the future operators they take $\bigcirc$ and $\mathcal{W}$ as basic and define $\square\,p$ to be an abbreviation for $p\,\mathcal{W}\,\mathit{false}$ as in (224), $\Diamond\,p$ to be an abbreviation for $\neg\square\,\neg p$ as in (59), and $p\,\mathcal{U}\,q$ to be an abbreviation for $p\,\mathcal{W}\,q \wedge \Diamond\,q$ as in (171) $\mathcal{U}$ in terms of $\mathcal{W}$.

The four basic inference rules in the Manna and Pnueli deductive system are generalization and specialization, corresponding to (136) Metatheorem, instantiation, corresponding to inference rule Substitution (Section 2.1), and modus ponens, corresponding to (3.77). The derived rules include particularization, corresponding to (76) Strengthening of $\square$, entailment modus ponens, corresponding to $\square\,(p \Rightarrow q) \wedge \square\,p \Rightarrow \square\,q$, and entailment transitivity, corresponding to $\square\,(p \Rightarrow q) \wedge \square\,(q \Rightarrow r) \Rightarrow \square\,(p \Rightarrow r)$. The last two expressions are easy to prove from (120) Monotonicity of $\square$.

Their eight future axioms correspond to (76) Strengthening of $\square$, (1) Self-dual, (2) Distributivity of $\bigcirc$ over $\Rightarrow$, (120) Monotonicity of $\square$, (80) $\bigcirc$ generalization, (57) $\square$ induction, (181) Expansion of $\mathcal{W}$, and (179) Perpetuity.

Schneider [24] is the only treatment of LTL that is based on the calculational deductive system developed in Gries and Schneider's LADM. [13] He does not consider the *until* operator $\mathcal{U}$ (apart from an exercise for the student) but uses the symbol $\mathcal{U}$ for the *wait* operator. The expression $p\,\mathcal{W}\,q$ is written $p\,\mathcal{U}\,q$ and read "p unless q".

The rules of inference in Schneider's deductive system include TL substitution, corresponding to inference rule Substitution, TL Modus ponens, corresponding to (3.77) Modus ponens, and Temporal generalization rule, corresponding to the generalization part of (136) Metatheorem. There are also a set of derived inference rules corresponding to (123) Consequence rule of $\square$, (125) Catenation rule of $\square$, (122) Consequence rule of $\Diamond$, (124) Catenation rule of $\Diamond$, (121) Consequence rule of $\bigcirc$, (57) $\square$ induction, and (129) Induction rule $\square$.

Schneider's deductive system consists of two axioms for $\square$ corresponding to (76) Strengthening of $\square$ and (120) Monotonicity of $\square$. It defines $\Diamond$ as in theorem (59). It has five axioms for $\bigcirc$ corresponding to (1) Self-dual, (2) Distributivity of $\bigcirc$ over $\Rightarrow$, (78) Strengthening of $\square$, (79) Strengthening of $\square$, and (57) $\square$ induction. The two axioms for the *wait* operator correspond to (179) Perpetuity and (181) Expansion of $\mathcal{W}$.

A striking feature of this survey is how each author chooses to include a set of theorems that all other authors omit. Ben-Ari has 13 unique theorems, Emerson has 14, Kröger and Merz have 10, Manna and Pnueli have 18, and Schneider has 19. There are 19 theorems that are common to all five sources in the survey, namely (1), (2), (4), (5), (45), (46), (50), (51), (52), (66), (72), (73), (76), (78), (99), (119), (120), (151), and (152).

This system includes all the theorems from the survey. It has 254 theorems, 18 of which are axioms. Of the axioms, six are theorems that do not appear in other LTL systems, namely (11), (16), (17), (18), (55), and (56).

5 Conclusion

Dijkstra and Scholten [7], and Feijen [11] originally developed $\mathcal{E}$ as a logic system to prove program correctness based on a calculational style. Gries and Schneider [13] extend that system to a theory of sets, a theory of sequences, relations and functions, a theory of integers, recurrence relations, modern algebra, and a theory of graphs. Similarly, this system extends $\mathcal{E}$ to a theory of linear temporal logic. It takes unary operator *next* $\bigcirc$ and binary operator *until* $\mathcal{U}$ as primitives and defines *eventually* $\Diamond$, *always* $\square$, and *wait* $\mathcal{W}$ in terms of them.

The calculational deductive system $\mathcal{E}$ has several advantages over other logic systems. The primary advantage is that the calculational system has only four inference rules. Consequently, proofs of theorems are easier to understand and more intuitive to those schooled in that system. One goal of this basic introduction is to make linear temporal logic accessible at the undergraduate level.

Many users of LTL are concerned with the scalability of reasoning and automated theorem proving. In contrast, the goal of this project is to make manual proofs accessible to human users. It would be an interesting area for future research to determine if and/or how the calculational system might be applied to LTL synthesis.

In our judgement, the progress we were able to make in exploring the structure of linear temporal systems is directly attributable to our training in the calculational deductive system $\mathcal{E}$. We believe the advantages of $\mathcal{E}$ over other logic systems is so substantial that it should be the tool of choice for computer science theory. We hope that this extension of $\mathcal{E}$ to linear temporal logic will not only be of use in the temporal logic community, but will serve as an example to promote $\mathcal{E}$ in the broader computer science community.

6 Acknowledgements

The authors would like to thank David Gries, who provided a prepublication manuscript of LADM, which we immediately adopted and have used at our institution ever since. He has been a source of constant encouragement over the years. David Gries and Fred Schneider provided valuable comments on draft manuscripts of this paper. Mordechai Ben-Ari's text [4] inspired us to use LTL to teach concurrent programming principles. Fred Schneider's graduate-level text [24] inspired us to create this calculational system to be accessible at the undergraduate level. Ray McIntyre, Ravi Mohan, Michael Ortiz, Kyle Sundman and John Wiegley contributed to the proofs.

7 References

[1] Roland Backhouse et al. "Special issue on the calculational method". In: *Inf. Process. Lett.* 53.3 (1995). ISSN: 0020-0190.

[2]   Christel Baier and Joost-Pieter Katoen. *Principles of Model Checking*. Cambridge, MA: MIT Press, 2008.

[3]   Mordechai Ben-Ari. *Mathematical Logic for Computer Science*. Third. London: Springer-Verlag, 2012.

[4]   Mordechai Ben-Ari. *Principles of Concurrent and Distributed Programming*. 2nd. Harlow, England: Addison-Wesley Pearson, 2006.

[5]   Edward Cohen. *Programming in the 1990's: An Introduction to the Calculation of Programs*. New York: Springer-Verlag, 1990.

[6]   Edsger W. Dijkstra. "Letters to the Editor: Go to Statement Considered Harmful". In: *Commun. ACM* 11.3 (Mar. 1968), pp. 147–148. ISSN: 0001-0782. DOI: 10.1145/362929.362947. URL: http://doi.acm.org/10.1145/362929.362947.

[7]   Edsger W. Dijkstra and Carel S. Scholten. *Predicate Calculus and Program Semantics*. New York: Springer-Verlag, 1990.

[8]   M.A.E. Dummett and E.J. Lemmon. "Modal logics between S4 and S5". In: *Zeitschrift für mathematische Logik und Grundlagen der Mathematik* 5 (1959), pp. 250–64.

[9]   Michael A. E. Dummett. *Truth and Other Enigmas*. Cambridge: Harvard University Press, 1978.

[10] E. Allen Emerson. "Handbook of Theoretical Computer Science (Vol. B)". In: ed. by Jan van Leeuwen. Cambridge, MA, USA: MIT Press, 1990. Chap. Temporal and Modal Logic, pp. 995–1072. ISBN: 0-444-88074-7. URL: http://dl.acm.org/citation.cfm?id=114891.114907.

[11] Wim H. H. Feijen. "Exercises in formula manipulation". In: *Formal Development of Programs*. Ed. by E.W. Dijkstra. Menlo Park: Addison-Wesley, 1990, pp. 139–158.

[12] David Gries. "Monotonicity in Calculational Proofs". In: *Correct System Design, Recent Insight and Advances, (to Hans Langmaack on the Occasion of His Retirement from His Professorship at the University of Kiel)*. Berlin, Heidelberg: Springer-Verlag, 1999, pp. 79–85. ISBN: 3-540-66624-9. URL: http://dl.acm.org/citation.cfm?id=646005.673872.

[13] David Gries and Fred B. Schneider. *A Logical Approach to Discrete Math*. New York: Springer-Verlag, 1994.

[14] David Gries and Fred B. Schneider. "A New Approach to Teaching Discrete Mathematics". In: *PRIMUS* 5.2 (1995), pp. 113–138. DOI: 10.1080/10511979508965779. eprint: https://doi.org/10.1080/10511979508965779. URL: https://doi.org/10.1080/10511979508965779.

[15] David Gries and Fred B. Schneider. "Adding the Everywhere Operator to Propositional Logic". In: *Journal of Logic and Computation* 8 (Dec. 1998).

[16] David Gries and Fred B. Schneider. "Equational Propositional Logic". In: *Information Processing Letters* 53.3 (1995), pp. 145–152.

[17] David Gries and Fred B. Schneider. "Teaching Math More Effectively, Through Calculational Proofs". In: *The American Mathematical Monthly* 102.8 (1995), pp. 691–697. ISSN: 00029890, 19300972. URL: `http://www.jstor.org/stable/2974638`.

[18] G. E. Hughes and M. J. Cresswell. *A New Introduction to Modal Logic*. New York: Routledge, 1996.

[19] Anne Kaldewaij. *Programming: The Derivation of Algorithms*. Hertfordshire, UK: Prentice-Hall International (UK) Ltd., 1990.

[20] Fred Kröger and Stephan Merz. *Temporal Logic and State Systems*. First. Berlin: Springer-Verlag, 2008.

[21] Zohar Manna and Amir Pnueli. *The Temporal Logic of Reactive and Concurrent Systems: v. 1, Specification*. First. New York: Springer-Verlag, 1992.

[22] Arthur N. Prior. *Past, Present and Future*. Oxford University Press, 1967.

[23] Kenneth H. Rosen. *Discrete Mathematics and Its Applications*. 6th. New York: McGraw-Hill, 2007.

[24] Fred B. Schneider. *On Concurrent Programming*. First. New York: Springer-Verlag, 1997.

[25] George Tourlakis. "On the Soundness and Completeness of Equational Predicate Logics". In: *J. Logic Computation* 11.4 (2001), pp. 623–653.

Theorems from CDS4LTL

J. Stanley Warford
Computer Science Department
Pepperdine University
Malibu, CA 90263

David Vega *
The Aerospace Corporation
El Segundo, CA 90245

Scott M. Staley
Ford Motor Company Research Labs (retired)
Dearborn, MI 48124

Abstract

The first section of this document is a collection of the axioms and theorems of the propositional calculus in Gries and Schneider's book *A Logical Approach to Discrete Math*, Springer-Verlag, 1993 (LADM). The numbering is consistent with that text with the chapter number followed by the equation number separated by a period. Additional theorems, either not included in LADM or included but not numbered, are indicated by a three-part number with two period separators. The second section is a collection of the axioms and theorems of linear temporal logic in Warford, Vega, and Staley's paper *A Computational Deductive System for Linear Temporal Logic* (CDS4LTL), Pepperdine University Research Report, Natural Science Division, 2019.

Table of Precedences

$[x := e]$ (textual substitution)	Highest precedence
$\neg \quad \bigcirc \quad \Diamond \quad \Box$	
$\mathcal{U} \quad \mathcal{W}$	
$=$ (conjunctional)	
$\vee \quad \wedge$	
$\Rightarrow \quad \Leftarrow$	
$\equiv$ (associative)	Lowest precedence

---

*Research supported by Tooma Undergraduate Research Fellowship Program, Pepperdine University, Summer 2009 and academic year 2009-10.

Theorems of the Propositional Calculus

Equivalence and *true*

(3.1) **Axiom, Associativity of $\equiv$:** $((p \equiv q) \equiv r) \equiv (p \equiv (q \equiv r))$
(3.2) **Axiom, Symmetry of $\equiv$:** $p \equiv q \equiv q \equiv p$
(3.3) **Axiom, Identity of $\equiv$:** $true \equiv q \equiv q$
(3.4) $true$
(3.5) **Reflexivity of $\equiv$:** $p \equiv p$

Negation, inequivalence, and *false*

(3.8) **Definition of $false$:** $false \equiv \neg true$
(3.9) **Axiom, Distributivity of $\neg$ over $\equiv$:** $\neg(p \equiv q) \equiv \neg p \equiv q$
(3.10) **Definition of $\not\equiv$:** $(p \not\equiv q) \equiv \neg(p \equiv q)$
(3.11) $\neg p \equiv q \equiv p \equiv \neg q$
(3.12) **Double negation:** $\neg\neg p \equiv p$
(3.13) **Negation of $false$:** $\neg false \equiv true$
(3.14) $(p \not\equiv q) \equiv \neg p \equiv q$
(3.15) $\neg p \equiv p \equiv false$
(3.16) **Symmetry of $\not\equiv$:** $(p \not\equiv q) \equiv (q \not\equiv p)$
(3.17) **Associativity of $\not\equiv$:** $((p \not\equiv q) \not\equiv r) \equiv (p \not\equiv (q \not\equiv r))$
(3.18) **Mutual associativity:** $((p \not\equiv q) \equiv r) \equiv (p \not\equiv (q \equiv r))$
(3.19) **Mutual interchangeability:** $p \not\equiv q \equiv r \equiv p \equiv q \not\equiv r$
(3.19.1) $p \not\equiv p \not\equiv q \equiv q$

Disjunction

(3.24) **Axiom, Symmetry of $\vee$:** $p \vee q \equiv q \vee p$
(3.25) **Axiom, Associativity of $\vee$:** $(p \vee q) \vee r \equiv p \vee (q \vee r)$
(3.26) **Axiom, Idempotency of $\vee$:** $p \vee p \equiv p$
(3.27) **Axiom, Distributivity of $\vee$ over $\equiv$:** $p \vee (q \equiv r) \equiv p \vee q \equiv p \vee r$
(3.28) **Axiom, Excluded middle:** $p \vee \neg p$
(3.29) **Zero of $\vee$:** $p \vee true \equiv true$
(3.30) **Identity of $\vee$:** $p \vee false \equiv p$
(3.31) **Distributivity of $\vee$ over $\vee$:** $p \vee (q \vee r) \equiv (p \vee q) \vee (p \vee r)$
(3.32) $p \vee q \equiv p \vee \neg q \equiv p$

Conjunction

(3.35) **Axiom, Golden rule:** $p \wedge q \equiv p \equiv q \equiv p \vee q$
(3.36) **Symmetry of** $\wedge$: $p \wedge q \equiv q \wedge p$
(3.37) **Associativity of** $\wedge$: $(p \wedge q) \wedge r \equiv p \wedge (q \wedge r)$
(3.38) **Idempotency of** $\wedge$: $p \wedge p \equiv p$
(3.39) **Identity of** $\wedge$: $p \wedge true \equiv p$
(3.40) **Zero of** $\wedge$: $p \wedge false \equiv false$
(3.41) **Distributivity of** $\wedge$ **over** $\wedge$: $p \wedge (q \wedge r) \equiv (p \wedge q) \wedge (p \wedge r)$
(3.42) **Contradiction:** $p \wedge \neg p \equiv false$
(3.43) **Absorption:**
  (a) $p \wedge (p \vee q) \equiv p$
  (b) $p \vee (p \wedge q) \equiv p$
(3.44) **Absorption:**
  (a) $p \wedge (\neg p \vee q) \equiv p \wedge q$
  (b) $p \vee (\neg p \wedge q) \equiv p \vee q$
(3.45) **Distributivity of** $\vee$ **over** $\wedge$: $p \vee (q \wedge r) \equiv (p \vee q) \wedge (p \vee r)$
(3.46) **Distributivity of** $\wedge$ **over** $\vee$: $p \wedge (q \vee r) \equiv (p \wedge q) \vee (p \wedge r)$
(3.47) **De Morgan:**
  (a) $\neg (p \wedge q) \equiv \neg p \vee \neg q$
  (b) $\neg (p \vee q) \equiv \neg p \wedge \neg q$
(3.48) $p \wedge q \equiv p \wedge \neg q \equiv \neg p$
(3.49) $p \wedge (q \equiv r) \equiv p \wedge q \equiv p \wedge r \equiv p$
(3.50) $p \wedge (q \equiv p) \equiv p \wedge q$
(3.51) **Replacement:** $(p \equiv q) \wedge (r \equiv p) \equiv (p \equiv q) \wedge (r \equiv q)$
(3.52) **Equivalence:** $p \equiv q \equiv (p \wedge q) \vee (\neg p \wedge \neg q)$
(3.53) **Exclusive or:** $p \not\equiv q \equiv (\neg p \wedge q) \vee (p \wedge \neg q)$
(3.55) $(p \wedge q) \wedge r \equiv p \equiv q \equiv r \equiv p \vee q \equiv q \vee r \equiv r \vee p \equiv p \vee q \vee r$

Implication

(3.57) **Definition of implication:** $p \Rightarrow q \equiv p \vee q \equiv q$
(3.58) **Axiom, Consequence:** $p \Leftarrow q \equiv q \Rightarrow p$
(3.59) **Implication:** $p \Rightarrow q \equiv \neg p \vee q$
(3.60) **Implication:** $p \Rightarrow q \equiv p \wedge q \equiv p$
(3.61) **Contrapositive:** $p \Rightarrow q \equiv \neg q \Rightarrow \neg p$
(3.62) $p \Rightarrow (q \equiv r) \equiv p \wedge q \equiv p \wedge r$

(3.63) **Distributivity of $\Rightarrow$ over $\equiv$:** $p \Rightarrow (q \equiv r) \equiv (p \Rightarrow q) \equiv (p \Rightarrow r)$
(3.63.1) **Distributivity of $\Rightarrow$ over $\wedge$:** $p \Rightarrow q \wedge r \equiv (p \Rightarrow q) \wedge (p \Rightarrow r)$
(3.63.2) **Distributivity of $\Rightarrow$ over $\vee$:** $p \Rightarrow q \vee r \equiv (p \Rightarrow q) \vee (p \Rightarrow r)$
(3.64) $p \Rightarrow (q \Rightarrow r) \equiv (p \Rightarrow q) \Rightarrow (p \Rightarrow r)$
(3.65) **Shunting:** $p \wedge q \Rightarrow r \equiv p \Rightarrow (q \Rightarrow r)$
(3.66) $p \wedge (p \Rightarrow q) \equiv p \wedge q$
(3.67) $p \wedge (q \Rightarrow p) \equiv p$
(3.68) $p \vee (p \Rightarrow q) \equiv true$
(3.69) $p \vee (q \Rightarrow p) \equiv q \Rightarrow p$
(3.70) $p \vee q \Rightarrow p \wedge q \equiv p \equiv q$
(3.71) **Reflexivity of $\Rightarrow$:** $p \Rightarrow p$
(3.72) **Right zero of $\Rightarrow$:** $p \Rightarrow true \equiv true$
(3.73) **Left identity of $\Rightarrow$:** $true \Rightarrow p \equiv p$
(3.74) $p \Rightarrow false \equiv \neg p$
(3.74.1) $\neg p \Rightarrow false \equiv p$
(3.75) $false \Rightarrow p \equiv true$
(3.76) **Weakening/strengthening:**
    (a) $p \Rightarrow p \vee q$ (Weakening the consequent)
    (b) $p \wedge q \Rightarrow p$ (Strengthening the antecedent)
    (c) $p \wedge q \Rightarrow p \vee q$ (Weakening/strengthening)
    (d) $p \vee (q \wedge r) \Rightarrow p \vee q$
    (e) $p \wedge q \Rightarrow p \wedge (q \vee r)$
(3.76.1) $p \wedge q \Rightarrow p \vee r$ (Weakening/strengthening)
(3.76.2) $(p \Rightarrow q) \Rightarrow ((q \Rightarrow r) \Rightarrow (p \Rightarrow r))$
(3.76.3) $(p \vee q) \wedge (q \Rightarrow r) \Rightarrow p \vee r$
(3.77) **Modus ponens:** $p \wedge (p \Rightarrow q) \Rightarrow q$
(3.77.1) **Modus tollens:** $(p \Rightarrow q) \wedge \neg q \Rightarrow \neg p$
(3.77.2) $((p \Rightarrow q) \Rightarrow (r \Rightarrow s)) \wedge (s \Rightarrow t) \Rightarrow ((p \Rightarrow q) \Rightarrow (r \Rightarrow t))$
(3.77.3) $((p \Rightarrow (q \Rightarrow r)) \wedge (r \Rightarrow s)) \Rightarrow (p \Rightarrow (q \Rightarrow s))$
(3.78) $(p \Rightarrow r) \wedge (q \Rightarrow r) \equiv p \vee q \Rightarrow r$
(3.78.1) $(p \Rightarrow r) \vee (q \Rightarrow r) \equiv p \wedge q \Rightarrow r$
(3.79) $(p \Rightarrow r) \wedge (\neg p \Rightarrow r) \equiv r$
(3.80) **Mutual implication:** $(p \Rightarrow q) \wedge (q \Rightarrow p) \equiv (p \equiv q)$
(3.81) **Antisymmetry:** $(p \Rightarrow q) \wedge (q \Rightarrow p) \Rightarrow (p \equiv q)$
(3.82) **Transitivity:**
    (a) $(p \Rightarrow q) \wedge (q \Rightarrow r) \Rightarrow (p \Rightarrow r)$
    (b) $(p \equiv q) \wedge (q \Rightarrow r) \Rightarrow (p \Rightarrow r)$

(c) $(p \Rightarrow q) \wedge (q \equiv r) \Rightarrow (p \Rightarrow r)$
(3.82.1) **Transitivity of** $\equiv$: $(p \equiv q) \wedge (q \equiv r) \Rightarrow (p \equiv r)$
(3.82.2) $(p \equiv q) \Rightarrow (p \Rightarrow q)$

Leibniz as an axiom

This section uses the following notation: E_X^z means $E[z := X]$.

(3.83) **Axiom, Leibniz:** $e = f \Rightarrow E_e^z = E_f^z$

(3.84) **Substitution:**
(a) $(e = f) \wedge E_e^z \equiv (e = f) \wedge E_f^z$
(b) $(e = f) \Rightarrow E_e^z \equiv (e = f) \Rightarrow E_f^z$
(c) $q \wedge (e = f) \Rightarrow E_e^z \equiv q \wedge (e = f) \Rightarrow E_f^z$

(3.85) **Replace by** *true*:
(a) $p \Rightarrow E_p^z \equiv p \Rightarrow E_{true}^z$
(b) $q \wedge p \Rightarrow E_p^z \equiv q \wedge p \Rightarrow E_{true}^z$

(3.86) **Replace by** *false*:
(a) $E_p^z \Rightarrow p \equiv E_{false}^z \Rightarrow p$
(b) $E_p^z \Rightarrow p \vee q \equiv E_{false}^z \Rightarrow p \vee q$

(3.87) **Replace by** *true*: $p \wedge E_p^z \equiv p \wedge E_{true}^z$

(3.88) **Replace by** *false*: $p \vee E_p^z \equiv p \vee E_{false}^z$

(3.89) **Shannon:** $E_p^z \equiv (p \wedge E_{true}^z) \vee (\neg p \wedge E_{false}^z)$

(3.89.1) $E_{true}^z \wedge E_{false}^z \Rightarrow E_p^z$

Additional theorems concerning implication

(4.1) $p \Rightarrow (q \Rightarrow p)$

(4.2) **Monotonicity of** $\vee$: $(p \Rightarrow q) \Rightarrow (p \vee r \Rightarrow q \vee r)$

(4.3) **Monotonicity of** $\wedge$: $(p \Rightarrow q) \Rightarrow (p \wedge r \Rightarrow q \wedge r)$

Proof technique metatheorems

(4.4) **Deduction (assume conjuncts of antecedent):**
To prove $P_1 \wedge P_2 \Rightarrow Q$, assume P_1 and P_2, and prove Q
You cannot use textual substitution in P_1 or P_2.

(4.7) **Mutual implication:** To prove $P \equiv Q$, prove $P \Rightarrow Q$ and $Q \Rightarrow P$
(4.7.1) **Truth implication:** To prove P, prove $true \Rightarrow P$
(4.9) **Proof by contradiction:** To prove P, prove $\neg P \Rightarrow false$
(4.12) **Proof by contrapositive:** To prove $P \Rightarrow Q$, prove $\neg Q \Rightarrow \neg P$

Theorems of Linear Temporal Logic

Next $\circ$

(1) **Axiom, Self-dual:** $\circ \neg p \equiv \neg \circ p$

(2) **Axiom, Distributivity of $\circ$ over $\Rightarrow$:** $\circ (p \Rightarrow q) \equiv \circ p \Rightarrow \circ q$

(3) **Linearity:** $\circ p \equiv \neg \circ \neg p$

(4) **Distributivity of $\circ$ over $\vee$:** $\circ (p \vee q) \equiv \circ p \vee \circ q$

(5) **Distributivity of $\circ$ over $\wedge$:** $\circ (p \wedge q) \equiv \circ p \wedge \circ q$

(6) **Distributivity of $\circ$ over $\equiv$:** $\circ (p \equiv q) \equiv \circ p \equiv \circ q$

(7) **Truth of $\circ$:** $\circ\, true \equiv true$

(8) **Falsehood of $\circ$:** $\circ\, false \equiv false$

Until $\mathcal{U}$

(9) **Axiom, Distributivity of $\circ$ over $\mathcal{U}$:** $\circ (p\, \mathcal{U}\, q) \equiv \circ p\, \mathcal{U}\, \circ q$

(10) **Axiom, Expansion of $\mathcal{U}$:** $p\, \mathcal{U}\, q \equiv q \vee (p \wedge \circ (p\, \mathcal{U}\, q))$

(11) **Axiom, Right zero of $\mathcal{U}$:** $p\, \mathcal{U}\, false \equiv false$

(12) **Axiom, Left distributivity of $\mathcal{U}$ over $\vee$:** $p\, \mathcal{U}\, (q \vee r) \equiv p\, \mathcal{U}\, q \vee p\, \mathcal{U}\, r$

(13) **Axiom, Right distributivity of $\mathcal{U}$ over $\vee$:** $p\, \mathcal{U}\, r \vee q\, \mathcal{U}\, r \Rightarrow (p \vee q)\, \mathcal{U}\, r$

(14) **Axiom, Left distributivity of $\mathcal{U}$ over $\wedge$:** $p\,\mathcal{U}\,(q \wedge r) \Rightarrow p\,\mathcal{U}\,q \wedge p\,\mathcal{U}\,r$

(15) **Axiom, Right distributivity of $\mathcal{U}$ over $\wedge$:** $(p \wedge q)\,\mathcal{U}\,r \equiv p\,\mathcal{U}\,r \wedge q\,\mathcal{U}\,r$

(16) **Axiom, $\mathcal{U}$ implication ordering:** $p\,\mathcal{U}\,q \wedge \neg q\,\mathcal{U}\,r \Rightarrow p\,\mathcal{U}\,r$

(17) **Axiom, Right $\mathcal{U} \vee$ ordering:** $p\,\mathcal{U}\,(q\,\mathcal{U}\,r) \Rightarrow (p \vee q)\,\mathcal{U}\,r$

(18) **Axiom, Right $\wedge\,\mathcal{U}$ ordering:** $p\,\mathcal{U}\,(q \wedge r) \Rightarrow (p\,\mathcal{U}\,q)\,\mathcal{U}\,r$

(19) **Right distributivity of $\mathcal{U}$ over $\Rightarrow$:** $(p \Rightarrow q)\,\mathcal{U}\,r \Rightarrow (p\,\mathcal{U}\,r \Rightarrow q\,\mathcal{U}\,r)$

(20) **Right zero of $\mathcal{U}$:** $p\,\mathcal{U}\,\mathit{true} \equiv \mathit{true}$

(21) **Left identity of $\mathcal{U}$:** $\mathit{false}\,\mathcal{U}\,q \equiv q$

(22) **Idempotency of $\mathcal{U}$:** $p\,\mathcal{U}\,p \equiv p$

(23) **$\mathcal{U}$ excluded middle:** $p\,\mathcal{U}\,q \vee p\,\mathcal{U}\,\neg q$

(24) $\neg p\,\mathcal{U}\,(q\,\mathcal{U}\,r) \wedge p\,\mathcal{U}\,r \Rightarrow q\,\mathcal{U}\,r$

(25) $p\,\mathcal{U}\,(\neg q\,\mathcal{U}\,r) \wedge q\,\mathcal{U}\,r \Rightarrow p\,\mathcal{U}\,r$

(26) $p\,\mathcal{U}\,q \wedge \neg q\,\mathcal{U}\,p \Rightarrow p$

(27) $p \wedge \neg p\,\mathcal{U}\,q \Rightarrow q$

(28) $p\,\mathcal{U}\,q \Rightarrow p \vee q$

(29) **$\mathcal{U}$ insertion:** $q \Rightarrow p\,\mathcal{U}\,q$

(30) $p \wedge q \Rightarrow p\,\mathcal{U}\,q$

(31) **Absorption:** $p \vee p\,\mathcal{U}\,q \equiv p \vee q$

(32) **Absorption:** $p\,\mathcal{U}\,q \vee q \equiv p\,\mathcal{U}\,q$

(33) **Absorption:** $p\,\mathcal{U}\,q \wedge q \equiv q$

(34) **Absorption:** $p\,\mathcal{U}\,q \vee (p \wedge q) \equiv p\,\mathcal{U}\,q$

(35) **Absorption:** $p\,\mathcal{U}\,q \wedge (p \vee q) \equiv p\,\mathcal{U}\,q$

(36) **Left absorption of $\mathcal{U}$:** $p\,\mathcal{U}\,(p\,\mathcal{U}\,q) \equiv p\,\mathcal{U}\,q$

(37) **Right absorption of $\mathcal{U}$:** $(p\,\mathcal{U}\,q)\,\mathcal{U}\,q \equiv p\,\mathcal{U}\,q$

Eventually $\Diamond$

(38) **Definition of** $\Diamond$: $\quad \Diamond q \equiv \text{true} \: \mathcal{U} \: q$

(39) **Absorption of** $\Diamond$ **into** $\mathcal{U}$: $\quad p \: \mathcal{U} \: q \wedge \Diamond q \equiv p \: \mathcal{U} \: q$

(40) **Absorption of** $\mathcal{U}$ **into** $\Diamond$: $\quad p \: \mathcal{U} \: q \vee \Diamond q \equiv \Diamond q$

(41) **Absorption of** $\mathcal{U}$ **into** $\Diamond$: $\quad p \: \mathcal{U} \Diamond q \equiv \Diamond q$

(42) **Eventuality:** $\quad p \: \mathcal{U} \: q \Rightarrow \Diamond q$

(43) **Truth of** $\Diamond$: $\quad \Diamond \text{true} \equiv \text{true}$

(44) **Falsehood of** $\Diamond$: $\quad \Diamond \text{false} \equiv \text{false}$

(45) **Expansion of** $\Diamond$: $\quad \Diamond p \equiv p \vee \circ \Diamond p$

(46) **Weakening of** $\Diamond$: $\quad p \Rightarrow \Diamond p$

(47) **Weakening of** $\Diamond$: $\quad \circ p \Rightarrow \Diamond p$

(48) **Absorption of** $\vee$ **into** $\Diamond$: $\quad p \vee \Diamond p \equiv \Diamond p$

(49) **Absorption of** $\Diamond$ **into** $\wedge$: $\quad \Diamond p \wedge p \equiv p$

(50) **Absorption of** $\Diamond$: $\quad \Diamond \Diamond p \equiv \Diamond p$

(51) **Exchange of** $\circ$ **and** $\Diamond$: $\quad \circ \Diamond p \equiv \Diamond \circ p$

(52) **Distributivity of** $\Diamond$ **over** $\vee$: $\quad \Diamond (p \vee q) \equiv \Diamond p \vee \Diamond q$

(53) **Distributivity of** $\Diamond$ **over** $\wedge$: $\quad \Diamond (p \wedge q) \Rightarrow \Diamond p \wedge \Diamond q$

Always $\square$

(54) **Definition of $\square$:** $\square p \equiv \neg \Diamond \neg p$

(55) **Axiom, $\mathcal{U}$ induction:** $\square(p \Rightarrow (\circ p \wedge q) \vee r) \Rightarrow (p \Rightarrow \square q \vee q \, \mathcal{U} \, r)$

(56) **Axiom, $\mathcal{U}$ induction:** $\square(p \Rightarrow \circ(p \vee q)) \Rightarrow (p \Rightarrow \square p \vee p \, \mathcal{U} \, q)$

(57) **$\square$ induction:** $\square(p \Rightarrow \circ p) \Rightarrow (p \Rightarrow \square p)$

(58) **$\Diamond$ induction:** $\square(\circ p \Rightarrow p) \Rightarrow (\Diamond p \Rightarrow p)$

(59) $\Diamond p \equiv \neg \square \neg p$

(60) **Dual of $\square$:** $\neg \square p \equiv \Diamond \neg p$

(61) **Dual of $\Diamond$:** $\neg \Diamond p \equiv \square \neg p$

(62) **Dual of $\Diamond \square$:** $\neg \Diamond \square p \equiv \square \Diamond \neg p$

(63) **Dual of $\square \Diamond$:** $\neg \square \Diamond p \equiv \Diamond \square \neg p$

(64) **Truth of $\square$:** $\square \mathit{true} \equiv \mathit{true}$

(65) **Falsehood of $\square$:** $\square \mathit{false} \equiv \mathit{false}$

(66) **Expansion of $\square$:** $\square p \equiv p \wedge \circ \square p$

(67) **Expansion of $\square$:** $\square p \equiv p \wedge \circ p \wedge \circ \square p$

(68) **Absorption of $\wedge$ into $\square$:** $p \wedge \square p \equiv \square p$

(69) **Absorption of $\square$ into $\vee$:** $\square p \vee p \equiv p$

(70) **Absorption of $\Diamond$ into $\square$:** $\Diamond p \wedge \square p \equiv \square p$

(71) **Absorption of $\square$ into $\Diamond$:** $\square p \vee \Diamond p \equiv \Diamond p$

(72) **Absorption of $\square$:** $\square \square p \equiv \square p$

(73) **Exchange of $\circ$ and $\square$:** $\circ \square p \equiv \square \circ p$

(74) $p \Rightarrow \square p \equiv p \Rightarrow \circ \square p$

(75) $p \wedge \Diamond \neg p \Rightarrow \Diamond (p \wedge \circ \neg p)$

(76) **Strengthening of $\square$:** $\square p \Rightarrow p$

(77) **Strengthening of $\square$:** $\square p \Rightarrow \Diamond p$

(78) **Strengthening of $\square$:** $\square p \Rightarrow \circ p$

(79) **Strengthening of $\square$:** $\square p \Rightarrow \circ \square p$

(80) **$\circ$ generalization:** $\square p \Rightarrow \square \circ p$

(81) $\square p \Rightarrow \neg(q \, \mathcal{U} \, \neg p)$

Temporal Deduction

(82) **Temporal deduction:**

To prove $\Box P_1 \wedge \Box P_2 \Rightarrow Q$, assume P_1 and P_2, and prove Q.
You cannot use textual substitution in P_1 or P_2.

Always, continued

(83) **Distributivity of $\wedge$ over $\mathcal{U}$:** $\Box p \wedge q \,\mathcal{U}\, r \Rightarrow (p \wedge q) \,\mathcal{U}\, (p \wedge r)$

(84) **$\mathcal{U}$ implication:** $\Box p \wedge \Diamond q \Rightarrow p \,\mathcal{U}\, q$

(85) **Right monotonicity of $\mathcal{U}$:** $\Box(p \Rightarrow q) \Rightarrow (r \,\mathcal{U}\, p \Rightarrow r \,\mathcal{U}\, q)$

(86) **Left monotonicity of $\mathcal{U}$:** $\Box(p \Rightarrow q) \Rightarrow (p \,\mathcal{U}\, r \Rightarrow q \,\mathcal{U}\, r)$

(87) **Distributivity of $\neg$ over $\Box$:** $\Box \neg p \Rightarrow \neg \Box p$

(88) **Distributivity of $\Diamond$ over $\wedge$:** $\Box p \wedge \Diamond q \Rightarrow \Diamond(p \wedge q)$

(89) **$\Diamond$ excluded middle:** $\Diamond p \vee \Box \neg p$

(90) **$\Box$ excluded middle:** $\Box p \vee \Diamond \neg p$

(91) **Temporal excluded middle:** $\Diamond p \vee \Diamond \neg p$

(92) **$\Diamond$ contradiction:** $\Diamond p \wedge \Box \neg p \equiv false$

(93) **$\Box$ contradiction:** $\Box p \wedge \Diamond \neg p \equiv false$

(94) **Temporal contradiction:** $\Box p \wedge \Box \neg p \equiv false$

(95) **$\Box \Diamond$ excluded middle:** $\Box \Diamond p \vee \Diamond \Box \neg p$

(96) **$\Diamond \Box$ excluded middle:** $\Diamond \Box p \vee \Box \Diamond \neg p$

(97) **□◇ contradiction:** $\Box\Diamond p \land \Diamond\Box\neg p \equiv \mathit{false}$
(98) **◇□ contradiction:** $\Diamond\Box p \land \Box\Diamond\neg p \equiv \mathit{false}$
(99) **Distributivity of □ over ∧:** $\Box(p \land q) \equiv \Box p \land \Box q$
(100) **Distributivity of □ over ∨:** $\Box p \lor \Box q \Rightarrow \Box(p \lor q)$
(101) **Logical equivalence law of ○:** $\Box(p \equiv q) \Rightarrow (\bigcirc p \equiv \bigcirc q)$
(102) **Logical equivalence law of ◇:** $\Box(p \equiv q) \Rightarrow (\Diamond p \equiv \Diamond q)$
(103) **Logical equivalence law of □:** $\Box(p \equiv q) \Rightarrow (\Box p \equiv \Box q)$
(104) **Distributivity of ◇ over ⇒:** $\Diamond(p \Rightarrow q) \equiv (\Box p \Rightarrow \Diamond q)$
(105) **Distributivity of ◇ over ⇒:** $(\Diamond p \Rightarrow \Diamond q) \Rightarrow \Diamond(p \Rightarrow q)$
(106) **∧ frame law of ○:** $\Box p \Rightarrow (\bigcirc q \Rightarrow \bigcirc(p \land q))$
(107) **∧ frame law of ◇:** $\Box p \Rightarrow (\Diamond q \Rightarrow \Diamond(p \land q))$
(108) **∧ frame law of □:** $\Box p \Rightarrow (\Box q \Rightarrow \Box(p \land q))$
(109) **∨ frame law of ○:** $\Box p \Rightarrow (\bigcirc q \Rightarrow \bigcirc(p \lor q))$
(110) **∨ frame law of ◇:** $\Box p \Rightarrow (\Diamond q \Rightarrow \Diamond(p \lor q))$
(111) **∨ frame law of □:** $\Box p \Rightarrow (\Box q \Rightarrow \Box(p \lor q))$
(112) **⇒ frame law of ○:** $\Box p \Rightarrow (\bigcirc q \Rightarrow \bigcirc(p \Rightarrow q))$
(113) **⇒ frame law of ◇:** $\Box p \Rightarrow (\Diamond q \Rightarrow \Diamond(p \Rightarrow q))$
(114) **⇒ frame law of □:** $\Box p \Rightarrow (\Box q \Rightarrow \Box(p \Rightarrow q))$
(115) **≡ frame law of ○:** $\Box p \Rightarrow (\bigcirc q \Rightarrow \bigcirc(p \equiv q))$
(116) **≡ frame law of ◇:** $\Box p \Rightarrow (\Diamond q \Rightarrow \Diamond(p \equiv q))$
(117) **≡ frame law of □:** $\Box p \Rightarrow (\Box q \Rightarrow \Box(p \equiv q))$
(118) **Monotonicity of ○:** $\Box(p \Rightarrow q) \Rightarrow (\bigcirc p \Rightarrow \bigcirc q)$
(119) **Monotonicity of ◇:** $\Box(p \Rightarrow q) \Rightarrow (\Diamond p \Rightarrow \Diamond q)$
(120) **Monotonicity of □:** $\Box(p \Rightarrow q) \Rightarrow (\Box p \Rightarrow \Box q)$
(121) **Consequence rule of ○:** $\Box((p \Rightarrow q) \land (q \Rightarrow \bigcirc r) \land (r \Rightarrow s)) \Rightarrow (p \Rightarrow \bigcirc s)$
(122) **Consequence rule of ◇:** $\Box((p \Rightarrow q) \land (q \Rightarrow \Diamond r) \land (r \Rightarrow s)) \Rightarrow (p \Rightarrow \Diamond s)$
(123) **Consequence rule of □:** $\Box((p \Rightarrow q) \land (q \Rightarrow \Box r) \land (r \Rightarrow s)) \Rightarrow (p \Rightarrow \Box s)$
(124) **Catenation rule of ◇:** $\Box((p \Rightarrow \Diamond q) \land (q \Rightarrow \Diamond r)) \Rightarrow (p \Rightarrow \Diamond r)$
(125) **Catenation rule of □:** $\Box((p \Rightarrow \Box q) \land (q \Rightarrow \Box r)) \Rightarrow (p \Rightarrow \Box r)$
(126) **Catenation rule of $\mathcal{U}$:** $\Box((p \Rightarrow q\,\mathcal{U}\,r) \land (r \Rightarrow q\,\mathcal{U}\,s)) \Rightarrow (p \Rightarrow q\,\mathcal{U}\,s)$
(127) **$\mathcal{U}$ strengthening rule:** $\Box((p \Rightarrow r) \land (q \Rightarrow s)) \Rightarrow (p\,\mathcal{U}\,q \Rightarrow r\,\mathcal{U}\,s)$
(128) **Induction rule ◇:** $\Box(p \lor \bigcirc q \Rightarrow q) \Rightarrow (\Diamond p \Rightarrow q)$
(129) **Induction rule □:** $\Box(p \Rightarrow q \land \bigcirc p) \Rightarrow (p \Rightarrow \Box q)$
(130) **Induction rule $\mathcal{U}$:** $\Box(p \Rightarrow \neg q \land \bigcirc p) \Rightarrow (p \Rightarrow \neg(r\,\mathcal{U}\,q))$
(131) **◇ confluence:** $\Box((p \Rightarrow \Diamond(q \lor r)) \land (q \Rightarrow \Diamond t) \land (r \Rightarrow \Diamond t)) \Rightarrow (p \Rightarrow \Diamond t)$
(132) **Temporal generalization law:** $\Box(\Box p \Rightarrow q) \Rightarrow (\Box p \Rightarrow \Box q)$
(133) **Temporal particularization law:** $\Box(p \Rightarrow \Diamond q) \Rightarrow (\Diamond p \Rightarrow \Diamond q)$
(134) $\Box(p \Rightarrow \bigcirc q) \Rightarrow (p \Rightarrow \Diamond q)$
(135) $\Box(p \Rightarrow \bigcirc \neg p) \Rightarrow (p \Rightarrow \neg \Box p)$

Proof Metatheorems

(136) **Metatheorem:** P is a theorem iff $\Box P$ is a theorem.
(137) **Metatheorem** $\circ$: If $P \Rightarrow Q$ is a theorem then $\circ P \Rightarrow \circ Q$ is a theorem.
(138) **Metatheorem** $\Diamond$: If $P \Rightarrow Q$ is a theorem then $\Diamond P \Rightarrow \Diamond Q$ is a theorem.
(139) **Metatheorem** $\Box$: If $P \Rightarrow Q$ is a theorem then $\Box P \Rightarrow \Box Q$ is a theorem.

Always, continued

(140) $\mathcal{U}\Box$ **implication:** $p\,\mathcal{U}\,\Box q \Rightarrow \Box(p\,\mathcal{U}\,q)$
(141) **Absorption of** $\mathcal{U}$ **into** $\Box$: $p\,\mathcal{U}\,\Box p \equiv \Box p$
(142) **Right** $\wedge\,\mathcal{U}$ **strengthening:** $p\,\mathcal{U}\,(q \wedge r) \Rightarrow p\,\mathcal{U}\,(q\,\mathcal{U}\,r)$
(143) **Left** $\wedge\,\mathcal{U}$ **strengthening:** $(p \wedge q)\,\mathcal{U}\,r \Rightarrow (p\,\mathcal{U}\,q)\,\mathcal{U}\,r$
(144) **Left** $\wedge\,\mathcal{U}$ **ordering:** $(p \wedge q)\,\mathcal{U}\,r \Rightarrow p\,\mathcal{U}\,(q\,\mathcal{U}\,r)$
(145) $\Diamond\Box$ **implication:** $\Diamond\Box p \Rightarrow \Box\Diamond p$
(146) $\Box\Diamond$ **excluded middle:** $\Box\Diamond p \vee \Box\Diamond \neg p$
(147) $\Diamond\Box$ **contradiction:** $\Diamond\Box p \wedge \Diamond\Box \neg p \equiv false$
(148) $\mathcal{U}$ **frame law of** $\circ$: $\Box p \Rightarrow (\circ q \Rightarrow \circ(p\,\mathcal{U}\,q))$
(149) $\mathcal{U}$ **frame law of** $\Diamond$: $\Box p \Rightarrow (\Diamond q \Rightarrow \Diamond(p\,\mathcal{U}\,q))$
(150) $\mathcal{U}$ **frame law of** $\Box$: $\Box p \Rightarrow (\Box q \Rightarrow \Box(p\,\mathcal{U}\,q))$
(151) **Absorption of** $\Diamond$ **into** $\Box\Diamond$: $\Diamond\Box\Diamond p \equiv \Box\Diamond p$
(152) **Absorption of** $\Box$ **into** $\Diamond\Box$: $\Box\Diamond\Box p \equiv \Diamond\Box p$
(153) **Absorption of** $\Box\Diamond$: $\Box\Diamond\Box\Diamond p \equiv \Box\Diamond p$
(154) **Absorption of** $\Diamond\Box$: $\Diamond\Box\Diamond\Box p \equiv \Diamond\Box p$
(155) **Absorption of** $\circ$ **into** $\Box\Diamond$: $\circ\Box\Diamond p \equiv \Box\Diamond p$
(156) **Absorption of** $\circ$ **into** $\Diamond\Box$: $\circ\Diamond\Box p \equiv \Diamond\Box p$
(157) **Monotonicity of** $\Box\Diamond$: $\Box(p \Rightarrow q) \Rightarrow (\Box\Diamond p \Rightarrow \Box\Diamond q)$
(158) **Monotonicity of** $\Diamond\Box$: $\Box(p \Rightarrow q) \Rightarrow (\Diamond\Box p \Rightarrow \Diamond\Box q)$
(159) **Distributivity of** $\Box\Diamond$ **over** $\wedge$: $\Box\Diamond(p \wedge q) \Rightarrow \Box\Diamond p \wedge \Box\Diamond q$
(160) **Distributivity of** $\Diamond\Box$ **over** $\vee$: $\Diamond\Box p \vee \Diamond\Box q \Rightarrow \Diamond\Box(p \vee q)$
(161) **Distributivity of** $\Box\Diamond$ **over** $\vee$: $\Box\Diamond(p \vee q) \equiv \Box\Diamond p \vee \Box\Diamond q$
(162) **Distributivity of** $\Diamond\Box$ **over** $\wedge$: $\Diamond\Box(p \wedge q) \equiv \Diamond\Box p \wedge \Diamond\Box q$
(163) **Eventual latching:** $\Diamond\Box(p \Rightarrow \Box q) \equiv \Diamond\Box \neg p \vee \Diamond\Box q$
(164) $\Box(\Box\Diamond p \Rightarrow \Diamond q) \equiv \Diamond\Box \neg p \vee \Box\Diamond q$
(165) $\Box((p \vee \Box q) \wedge (\Box p \vee q)) \equiv \Box p \vee \Box q$
(166) $\Diamond\Box p \wedge \Box\Diamond q \Rightarrow \Box\Diamond(p \wedge q)$
(167) $\Box((\Box p \Rightarrow \Diamond q) \wedge (q \Rightarrow \circ r)) \Rightarrow (\Box p \Rightarrow \circ\Box\Diamond r)$
(168) **Progress proof rule:** $\Diamond\Box p \wedge \Box(\Box p \Rightarrow \Diamond q) \Rightarrow \Diamond q$

Wait $\mathcal{W}$

(169) **Definition of** $\mathcal{W}$: $p\,\mathcal{W}\,q \equiv \Box p \vee p\,\mathcal{U}\,q$

(170) **Axiom, Distributivity of** $\neg$ **over** $\mathcal{W}$: $\neg(p\,\mathcal{W}\,q) \equiv \neg q\,\mathcal{U}\,(\neg p \wedge \neg q)$

(171) $\mathcal{U}$ **in terms of** $\mathcal{W}$: $p\,\mathcal{U}\,q \equiv p\,\mathcal{W}\,q \wedge \Diamond q$

(172) $p\,\mathcal{W}\,q \equiv \Box(p \wedge \neg q) \vee p\,\mathcal{U}\,q$

(173) **Distributivity of** $\neg$ **over** $\mathcal{U}$: $\neg(p\,\mathcal{U}\,q) \equiv \neg q\,\mathcal{W}\,(\neg p \wedge \neg q)$

(174) $\mathcal{U}$ **implication:** $p\,\mathcal{U}\,q \Rightarrow p\,\mathcal{W}\,q$

(175) **Distributivity of** $\wedge$ **over** $\mathcal{W}$: $\Box p \wedge q\,\mathcal{W}\,r \Rightarrow (p \wedge q)\,\mathcal{W}\,(p \wedge r)$

(176) $\mathcal{W}\Diamond$ **equivalence:** $p\,\mathcal{W}\,\Diamond q \equiv \Box p \vee \Diamond q$

(177) $\mathcal{W}\Box$ **implication:** $p\,\mathcal{W}\,\Box q \Rightarrow \Box(p\,\mathcal{W}\,q)$

(178) **Absorption of** $\mathcal{W}$ **into** $\Box$: $p\,\mathcal{W}\,\Box p \equiv \Box p$

(179) **Perpetuity:** $\Box p \Rightarrow p\,\mathcal{W}\,q$

(180) **Distributivity of** $\circ$ **over** $\mathcal{W}$: $\circ(p\,\mathcal{W}\,q) \equiv \circ p\,\mathcal{W}\,\circ q$

(181) **Expansion of** $\mathcal{W}$: $p\,\mathcal{W}\,q \equiv q \vee (p \wedge \circ(p\,\mathcal{W}\,q))$

(182) $\mathcal{W}$ **excluded middle:** $p\,\mathcal{W}\,q \vee p\,\mathcal{W}\,\neg q$

(183) **Left zero of** $\mathcal{W}$: $true\,\mathcal{W}\,q \equiv true$

(184) **Left distributivity of** $\mathcal{W}$ **over** $\vee$: $p\,\mathcal{W}\,(q \vee r) \equiv p\,\mathcal{W}\,q \vee p\,\mathcal{W}\,r$

(185) **Right distributivity of** $\mathcal{W}$ **over** $\vee$: $p\,\mathcal{W}\,r \vee q\,\mathcal{W}\,r \Rightarrow (p \vee q)\,\mathcal{W}\,r$

(186) **Left distributivity of** $\mathcal{W}$ **over** $\wedge$: $p\,\mathcal{W}\,(q \wedge r) \Rightarrow p\,\mathcal{W}\,q \wedge p\,\mathcal{W}\,r$

(187) **Right distributivity of** $\mathcal{W}$ **over** $\wedge$: $(p \wedge q) \mathcal{W} r \equiv p \mathcal{W} r \wedge q \mathcal{W} r$

(188) **Right distributivity of** $\mathcal{W}$ **over** $\Rightarrow$: $(p \Rightarrow q) \mathcal{W} r \Rightarrow (p \mathcal{W} r \Rightarrow q \mathcal{W} r)$

(189) **Disjunction rule of** $\mathcal{W}$: $p \mathcal{W} q \equiv (p \vee q) \mathcal{W} q$

(190) **Disjunction rule of** $\mathcal{U}$: $p \mathcal{U} q \equiv (p \vee q) \mathcal{U} q$

(191) **Rule of** $\mathcal{W}$: $\neg q \mathcal{W} q$

(192) **Rule of** $\mathcal{U}$: $\neg q \mathcal{U} q \equiv \Diamond q$

(193) $(p \Rightarrow q) \mathcal{W} p$

(194) $\Diamond p \Rightarrow (p \Rightarrow q) \mathcal{U} p$

(195) **Conjunction rule of** $\mathcal{W}$: $p \mathcal{W} q \equiv (p \wedge \neg q) \mathcal{W} q$

(196) **Conjunction rule of** $\mathcal{U}$: $p \mathcal{U} q \equiv (p \wedge \neg q) \mathcal{U} q$

(197) **Distributivity of** $\neg$ **over** $\mathcal{W}$: $\neg(p \mathcal{W} q) \equiv (p \wedge \neg q) \mathcal{U} (\neg p \wedge \neg q)$

(198) **Distributivity of** $\neg$ **over** $\mathcal{U}$: $\neg(p \mathcal{U} q) \equiv (p \wedge \neg q) \mathcal{W} (\neg p \wedge \neg q)$

(199) **Dual of** $\mathcal{U}$: $\neg(\neg p \mathcal{U} \neg q) \equiv q \mathcal{W} (p \wedge q)$

(200) **Dual of** $\mathcal{U}$: $\neg(\neg p \mathcal{U} \neg q) \equiv (\neg p \wedge q) \mathcal{W} (p \wedge q)$

(201) **Dual of** $\mathcal{W}$: $\neg(\neg p \mathcal{W} \neg q) \equiv q \mathcal{U} (p \wedge q)$

(202) **Dual of** $\mathcal{W}$: $\neg(\neg p \mathcal{W} \neg q) \equiv (\neg p \wedge q) \mathcal{U} (p \wedge q)$

(203) **Idempotency of** $\mathcal{W}$: $p \mathcal{W} p \equiv p$

(204) **Right zero of** $\mathcal{W}$: $p \mathcal{W} \text{true} \equiv \text{true}$

(205) **Left identity of** $\mathcal{W}$: $\mathit{false}\,\mathcal{W}\,q \equiv q$

(206) $p\,\mathcal{W}\,q \Rightarrow p \vee q$

(207) $\Box(p \vee q) \Rightarrow p\,\mathcal{W}\,q$

(208) $\Box(\neg q \Rightarrow p) \Rightarrow p\,\mathcal{W}\,q$

(209) $\mathcal{W}$ **insertion:** $q \Rightarrow p\,\mathcal{W}\,q$

(210) $\mathcal{W}$ **frame law of** $\circ$: $\Box p \Rightarrow (\circ q \Rightarrow \circ(p\,\mathcal{W}\,q))$

(211) $\mathcal{W}$ **frame law of** $\Diamond$: $\Box p \Rightarrow (\Diamond q \Rightarrow \Diamond(p\,\mathcal{W}\,q))$

(212) $\mathcal{W}$ **frame law of** $\Box$: $\Box p \Rightarrow (\Box q \Rightarrow \Box(p\,\mathcal{W}\,q))$

(213) $\mathcal{W}$ **induction:** $\Box(p \Rightarrow (\circ p \wedge q) \vee r) \Rightarrow (p \Rightarrow q\,\mathcal{W}\,r)$

(214) $\mathcal{W}$ **induction:** $\Box(p \Rightarrow \circ(p \vee q)) \Rightarrow (p \Rightarrow p\,\mathcal{W}\,q)$

(215) $\mathcal{W}$ **induction:** $\Box(p \Rightarrow \circ p) \Rightarrow (p \Rightarrow p\,\mathcal{W}\,q)$

(216) $\mathcal{W}$ **induction:** $\Box(p \Rightarrow q \wedge \circ p) \Rightarrow (p \Rightarrow p\,\mathcal{W}\,q)$

(217) **Absorption:** $p \vee p\,\mathcal{W}\,q \equiv p \vee q$

(218) **Absorption:** $p\,\mathcal{W}\,q \vee q \equiv p\,\mathcal{W}\,q$

(219) **Absorption:** $p\,\mathcal{W}\,q \wedge q \equiv q$

(220) **Absorption:** $p\,\mathcal{W}\,q \wedge (p \vee q) \equiv p\,\mathcal{W}\,q$

(221) **Absorption:** $p\,\mathcal{W}\,q \vee (p \wedge q) \equiv p\,\mathcal{W}\,q$

(222) **Left absorption of** $\mathcal{W}$: $p\,\mathcal{W}\,(p\,\mathcal{W}\,q) \equiv p\,\mathcal{W}\,q$

(223) **Right absorption of** $\mathcal{W}$: $(p\,\mathcal{W}\,q)\,\mathcal{W}\,q \equiv p\,\mathcal{W}\,q$
(224) $\square$ **to** $\mathcal{W}$ **law:** $\square p \equiv p\,\mathcal{W}\,\mathit{false}$
(225) $\diamond$ **to** $\mathcal{W}$ **law:** $\diamond p \equiv \neg(\neg p\,\mathcal{W}\,\mathit{false})$
(226) $\mathcal{W}$ **implication:** $p\,\mathcal{W}\,q \Rightarrow \square p \vee \diamond q$
(227) **Absorption:** $p\,\mathcal{W}\,(p\,\mathcal{U}\,q) \equiv p\,\mathcal{W}\,q$
(228) **Absorption:** $(p\,\mathcal{U}\,q)\,\mathcal{W}\,q \equiv p\,\mathcal{U}\,q$
(229) **Absorption:** $p\,\mathcal{U}\,(p\,\mathcal{W}\,q) \equiv p\,\mathcal{W}\,q$
(230) **Absorption:** $(p\,\mathcal{W}\,q)\,\mathcal{U}\,q \equiv p\,\mathcal{U}\,q$
(231) **Absorption of** $\mathcal{W}$ **into** $\diamond$: $\diamond q\,\mathcal{W}\,q \equiv \diamond q$
(232) **Absorption of** $\mathcal{W}$ **into** $\square$: $\square p \wedge p\,\mathcal{W}\,q \equiv \square p$
(233) **Absorption of** $\square$ **into** $\mathcal{W}$: $\square p \vee p\,\mathcal{W}\,q \equiv p\,\mathcal{W}\,q$
(234) $p\,\mathcal{W}\,q \wedge \square\neg q \Rightarrow \square p$
(235) $\square p \Rightarrow p\,\mathcal{U}\,q \vee \square\neg q$
(236) $\neg\square p \wedge p\,\mathcal{W}\,q \Rightarrow \diamond q$
(237) $\diamond q \Rightarrow \neg\square p \vee p\,\mathcal{U}\,q$
(238) $\square\neg p \wedge p\,\mathcal{U}\,q \Rightarrow q$
(239) **Left monotonicity of** $\mathcal{W}$: $\square(p \Rightarrow q) \Rightarrow (p\,\mathcal{W}\,r \Rightarrow q\,\mathcal{W}\,r)$
(240) **Right monotonicity of** $\mathcal{W}$: $\square(p \Rightarrow q) \Rightarrow (r\,\mathcal{W}\,p \Rightarrow r\,\mathcal{W}\,q)$
(241) $\mathcal{W}$ **strengthening rule:** $\square((p \Rightarrow r) \wedge (q \Rightarrow s)) \Rightarrow (p\,\mathcal{W}\,q \Rightarrow r\,\mathcal{W}\,s)$
(242) $\mathcal{W}$ **catenation rule:** $\square((p \Rightarrow q\,\mathcal{W}\,r) \wedge (r \Rightarrow q\,\mathcal{W}\,s)) \Rightarrow (p \Rightarrow q\,\mathcal{W}\,s)$
(243) **Left** $\mathcal{U}\,\mathcal{W}$ **implication:** $(p\,\mathcal{U}\,q)\,\mathcal{W}\,r \Rightarrow (p\,\mathcal{W}\,q)\,\mathcal{W}\,r$
(244) **Right** $\mathcal{W}\,\mathcal{U}$ **implication:** $p\,\mathcal{W}\,(q\,\mathcal{U}\,r) \Rightarrow p\,\mathcal{W}\,(q\,\mathcal{W}\,r)$
(245) **Right** $\mathcal{U}\,\mathcal{U}$ **implication:** $p\,\mathcal{U}\,(q\,\mathcal{U}\,r) \Rightarrow p\,\mathcal{U}\,(q\,\mathcal{W}\,r)$
(246) **Left** $\mathcal{U}\,\mathcal{U}$ **implication:** $(p\,\mathcal{U}\,q)\,\mathcal{U}\,r \Rightarrow (p\,\mathcal{W}\,q)\,\mathcal{U}\,r$
(247) **Left** $\mathcal{U}\,\vee$ **strengthening:** $(p\,\mathcal{U}\,q)\,\mathcal{U}\,r \Rightarrow (p \vee q)\,\mathcal{U}\,r$
(248) **Left** $\mathcal{W}\,\vee$ **strengthening:** $(p\,\mathcal{W}\,q)\,\mathcal{W}\,r \Rightarrow (p \vee q)\,\mathcal{W}\,r$
(249) **Right** $\mathcal{W}\,\vee$ **strengthening:** $p\,\mathcal{W}\,(q\,\mathcal{W}\,r) \Rightarrow p\,\mathcal{W}\,(q \vee r)$
(250) **Right** $\mathcal{W}\,\vee$ **ordering:** $p\,\mathcal{W}\,(q\,\mathcal{W}\,r) \Rightarrow (p \vee q)\,\mathcal{W}\,r$
(251) **Right** $\wedge\,\mathcal{W}$ **ordering:** $p\,\mathcal{W}\,(q \wedge r) \Rightarrow (p\,\mathcal{W}\,q)\,\mathcal{W}\,r$
(252) $\mathcal{U}$ **ordering:** $\neg p\,\mathcal{U}\,q \vee \neg q\,\mathcal{U}\,p \equiv \diamond(p \vee q)$
(253) $\mathcal{W}$ **ordering:** $\neg p\,\mathcal{W}\,q \vee \neg q\,\mathcal{W}\,p$
(254) $\mathcal{W}$ **implication ordering:** $p\,\mathcal{W}\,q \wedge \neg q\,\mathcal{W}\,r \Rightarrow p\,\mathcal{W}\,r$
(255) **Lemmon formula:** $\square(\square p \Rightarrow q) \vee \square(\square q \Rightarrow p)$

A Calculational Deductive System for Linear Temporal Logic: Additional Theorems

Scott M. Staley

Ford Motor Company Research Labs (retired)

Seattle, WA 98198

Abstract

This paper presents additional theorems that are not contained in *A Calculational Deductive System for Linear Temporal Logic* by Warford, Vega and Staley. The theorems derived herein are proved sequentially, assuming all the theorems of CDS4LTL are available for proof hints. Duality is extended to include the release operators. Counterexamples of several expressions thought to be theorems are presented.

1 Preliminaries

This paper presents additional theorems proved subsequently to the publication of *A Calculational Deductive System for Linear Temporal Logic* (CDS4LTL) by Warford, Vega and Staley. [16] It is not intended to be tutorial of linear temporal logic (LTL), which is the stated goal of CDS4LTL. Some of the theorems are lemmas that can help shorten more significant proofs. The theorems of this paper are collected together with the previous theorems of CDS4LTL in a supporting document [17].

Section 2 presents the additional theorems of LTL. Section 3 presents counterexamples of two expressions previously thought to be LTL theorems. Section 4 presents a collection of left-over lemmas. Finally, Section 5 gives a brief summary.

2 Additional Theorems

The theorem numbering system used in this paper is compatible with the numbering system in CDS4LTL. A numbered reference enclosed in parentheses, prefixed with an S is a reference to an axiom or a previously-proved theorem in this paper. A numbered reference enclosed in parentheses *without* a period is a reference to an axiom or a previously-proved theorem in the CDS4LTL paper. A numbered reference enclosed in parentheses *with* a period is a reference to an axiom or a theorem from the propositional calculus in LADM [9]. Additional propositional and predicate calculus theorems, either not included in LADM or included but not numbered, are indicated by a three-part number with two period separators. So, for example (S22) $\Box p \land \Box q \Rightarrow \bigcirc (\Box p \land \Box q)$ is a theorem from this paper. (104) Distributivity of $\Diamond$ over $\Rightarrow$ is a theorem from CDS4LTL and (3.65) Shunting, $p \land q \Rightarrow r \equiv p \Rightarrow (q \Rightarrow r)$ is a theorem from LADM.

2.1 Temporal Modus Ponens and Modus Tollens

The first theorem (S1) is a simple non-temporal lemma that is useful in proofs involving Duality (see section 2.6). It is also used in the proof of (S7) Prior Formula.

(S1) **Negation of $\Rightarrow$:** $\quad \neg(p \Rightarrow q) \equiv p \wedge \neg q$

Proof:

$$\begin{aligned}
& \neg(p \Rightarrow q) \\
= \quad & \langle (3.59) \text{ Implication}, p \Rightarrow q \equiv \neg p \vee q \rangle \\
& \neg(\neg p \vee q) \\
= \quad & \langle (3.47\text{b}) \text{ De Morgan}, \neg(p \vee q) \equiv \neg p \wedge \neg q \rangle \\
& p \wedge \neg q \quad \blacksquare
\end{aligned}$$

The next five theorems (S2), (S3), (S4), (S5) and (S6) are temporal extensions of (3.77) Modus ponens, $p \wedge (p \Rightarrow q) \Rightarrow q$. They each have 1-step proofs. They are just a (3.65) Shunting step away from the monotonicity theorems (118), (119), (120), (157) and (158), so they are very simple lemmas. Nevertheless, they can sometimes be useful in a proof step where they appear, as is the case, for example, in (S7) Prior Formula [3] and (S30) $\square \diamond$ induction. Notice that (S7) Prior formula is equivalent to (S3) Temporal modus ponens of $\diamond$.

(S2) **Temporal modus ponens of $\circ$:** $\quad \square(p \Rightarrow q) \wedge \circ p \Rightarrow \circ q$

Proof:

$$\begin{aligned}
& \square(p \Rightarrow q) \wedge \circ p \Rightarrow \circ q \\
= \quad & \langle (3.65) \text{ Shunting}, p \wedge q \Rightarrow r \equiv p \Rightarrow (q \Rightarrow r) \rangle \\
& \square(p \Rightarrow q) \Rightarrow (\circ p \Rightarrow \circ q) \quad -(118) \text{ Monotonicity of } \circ \quad \blacksquare
\end{aligned}$$

(S3) **Temporal modus ponens of $\diamond$:** $\quad \square(p \Rightarrow q) \wedge \diamond p \Rightarrow \diamond q$

Proof:

$$\begin{aligned}
& \square(p \Rightarrow q) \wedge \diamond p \Rightarrow \diamond q \\
= \quad & \langle (3.65) \text{ Shunting}, p \wedge q \Rightarrow r \equiv p \Rightarrow (q \Rightarrow r) \rangle \\
& \square(p \Rightarrow q) \Rightarrow (\diamond p \Rightarrow \diamond q) \quad -(119) \text{ Monotonicity of } \diamond \quad \blacksquare
\end{aligned}$$

(S4) **Temporal modus ponens of $\square$:** $\quad \square(p \Rightarrow q) \wedge \square p \Rightarrow \square q$

Proof:

Pepperdine Papers on LTL

$$\square(p \Rightarrow q) \wedge \square p \Rightarrow \square q$$
$$= \quad \langle(3.65) \text{ Shunting}, p \wedge q \Rightarrow r \equiv p \Rightarrow (q \Rightarrow r)\rangle$$
$$\square(p \Rightarrow q) \Rightarrow (\square p \Rightarrow \square q) \quad -(120) \text{ Monotonicity of } \square \quad \blacksquare$$

(S5) **Temporal modus ponens of $\square\diamond$:** $\quad \square(p \Rightarrow q) \wedge \square\diamond p \Rightarrow \square\diamond q$

Proof:

$$\square(p \Rightarrow q) \wedge \square\diamond p \Rightarrow \square\diamond q$$
$$= \quad \langle(3.65) \text{ Shunting}, p \wedge q \Rightarrow r \equiv p \Rightarrow (q \Rightarrow r)\rangle$$
$$\square(p \Rightarrow q) \Rightarrow (\square\diamond p \Rightarrow \square\diamond q) \quad -(157) \text{ Monotonicity of } \square\diamond \quad \blacksquare$$

(S6) **Temporal modus ponens of $\diamond\square$:** $\quad \square(p \Rightarrow q) \wedge \diamond\square p \Rightarrow \diamond\square q$

Proof:

$$\square(p \Rightarrow q) \wedge \diamond\square p \Rightarrow \diamond\square q$$
$$= \quad \langle(3.65) \text{ Shunting}, p \wedge q \Rightarrow r \equiv p \Rightarrow (q \Rightarrow r)\rangle$$
$$\square(p \Rightarrow q) \Rightarrow (\diamond\square p \Rightarrow \diamond\square q) \quad -(158) \text{ Monotonicity of } \diamond\square \quad \blacksquare$$

(S7) **Prior formula:** $\quad \neg\diamond q \Rightarrow (\square(p \Rightarrow q) \Rightarrow \neg\diamond p)$

Proof:

$$\neg\diamond q \Rightarrow (\square(p \Rightarrow q) \Rightarrow \neg\diamond p)$$
$$= \quad \langle(3.61) \text{ Contrapositive}, p \Rightarrow q \equiv \neg q \Rightarrow \neg p\rangle$$
$$\neg(\square(p \Rightarrow q) \Rightarrow \neg\diamond p) \Rightarrow \diamond q$$
$$= \quad \langle(S1) \text{ Negation of } \Rightarrow\rangle$$
$$\square(p \Rightarrow q) \wedge \diamond p \Rightarrow \diamond q \quad -(S3) \text{ Temporal modus ponens of } \diamond \quad \blacksquare$$

The next six theorems (S8), (S9), (S10), (S11), (S12) and (S13) are temporal extensions of (3.77.1) Modus tollens, $(p \Rightarrow q) \wedge \neg q \Rightarrow \neg p$.

(S8) **Temporal modus tollens of $\square$:** $\quad \square(p \Rightarrow q) \wedge \neg\square q \Rightarrow \neg\square p$

Proof:

$$\square(p \Rightarrow q) \wedge \neg\square q$$
$$\Rightarrow \quad \langle(120) \text{ Monotonicity of } \square\rangle$$
$$(\square p \Rightarrow \square q) \wedge \neg\square q$$
$$\Rightarrow \quad \langle(3.77.1) \text{ Modus tollens}, (p \Rightarrow q) \wedge \neg q \Rightarrow \neg p \text{ with } p,q := \square p, \square q\rangle$$
$$\neg\square p \quad \blacksquare$$

(S9) **Temporal modus tollens of $\Box$:** $\Box(p \Rightarrow q) \wedge \Box \neg q \Rightarrow \Box \neg p$

Proof:

$\quad\quad \Box(p \Rightarrow q) \wedge \Box \neg q$
$= \quad \langle\text{(99) Distributivity of } \Box \text{ over } \wedge\rangle$
$\quad\quad \Box((p \Rightarrow q) \wedge \neg q)$
$\Rightarrow \quad \langle\text{(3.77.1) Modus tollens, } (p \Rightarrow q) \wedge \neg q \Rightarrow \neg p$
$\quad\quad\quad \text{with } p, q := \Box p, \Box q \text{ and (120) Monotonicity of } \Box\rangle$
$\quad\quad \Box \neg p \quad \blacksquare$

(S10) **Temporal modus tollens of $\Diamond$:** $\Box(p \Rightarrow q) \wedge \neg \Diamond q \Rightarrow \neg \Diamond p$

Proof:

$\quad\quad \Box(p \Rightarrow q) \wedge \neg \Diamond q \Rightarrow \neg \Diamond p$
$= \quad \langle\text{(61) Dual of } \Diamond, \text{twice}\rangle$
$\quad\quad \Box(p \Rightarrow q) \wedge \Box \neg q \Rightarrow \Box \neg p \quad -\text{(S9) Temporal modus tollens of } \Box \quad \blacksquare$

(S11) **Temporal modus tollens of $\Diamond$:** $\Box(p \Rightarrow q) \wedge \Diamond \neg q \Rightarrow \Diamond \neg p$

Proof:

$\quad\quad \Box(p \Rightarrow q) \wedge \Diamond \neg q \Rightarrow \Diamond \neg p$
$= \quad \langle\text{(60) Dual of } \Box, \text{twice}\rangle$
$\quad\quad \Box(p \Rightarrow q) \wedge \neg \Box q \Rightarrow \neg \Box p \quad -\text{(S8) Temporal modus tollens of } \Box \quad \blacksquare$

(S12) **Temporal modus tollens of $\circ$:** $\Box(p \Rightarrow q) \wedge \neg \circ q \Rightarrow \neg \circ p$

Proof:

$\quad\quad \Box(p \Rightarrow q) \wedge \neg \circ q$
$\Rightarrow \quad \langle\text{(118) Monotonicity of } \circ \text{ and (4.3) Monotonicity of } \wedge\rangle$
$\quad\quad (\circ p \Rightarrow \circ q) \wedge \neg \circ q$
$\Rightarrow \quad \langle\text{(3.77.1) Modus tollens, } (p \Rightarrow q) \wedge \neg q \Rightarrow \neg p \text{ with } p, q := \circ p, \circ q\rangle$
$\quad\quad \neg \circ p \quad \blacksquare$

(S13) **Temporal modus tollens of $\circ$:** $\Box(p \Rightarrow q) \wedge \circ \neg q \Rightarrow \circ \neg p$

Proof:

$\quad\quad \Box(p \Rightarrow q) \wedge \circ \neg q \Rightarrow \circ \neg p$
$= \quad \langle\text{(1) Self-dual, twice}\rangle$
$\quad\quad \Box(p \Rightarrow q) \wedge \neg \circ q \Rightarrow \neg \circ p \quad -\text{(S12) Temporal modus tollens of } \circ \quad \blacksquare$

2.2 Theorems from StackExchange and Spot

Spot is a C++14 library for LTL, ω-automata manipulation, and model checking [5]. Documentation on the application of Spot is accessible from the website [15]. Spot's online tools can be used to check the validity of LTL formulas which can be helpful in trying to prove LTL formulas. Most of these theorems came from the on-line document [4].

Stackexchange.com [2] is a popular website for exchanging information on a wide range of topics. It is a particularly good place to look for information on mathematics or computer science questions. The next five theorems are from the Stackexchange website which often discusses topics related to LTL.

Theorem (S14) appears several times in forum discussions. Theorem (S14) and theorem (104) Distributivity of $\Diamond$ over $\Rightarrow$, can be used with (3.51) Replacement: $(p \equiv q) \land (r \equiv p) \equiv (p \equiv q) \land (r \equiv q)$ to produce a new theorem (S15) $\Diamond(p \Rightarrow q) \equiv p \,\mathcal{U}\, (q \lor \neg p)$. An alternate proof for this theorem is shown below.

(S14) $\quad \Box p \Rightarrow \Diamond q \equiv p \,\mathcal{U}\, (q \lor \neg p)$

Proof: The proof is by (4.7) Mutual implication.

$$\begin{aligned} & \Box p \Rightarrow \Diamond q \equiv p \,\mathcal{U}\, (q \lor \neg p) \\ = \quad & \langle (3.59) \text{ Implication}, p \Rightarrow q \equiv \neg p \lor q \rangle \\ & \Box p \Rightarrow \Diamond q \equiv p \,\mathcal{U}\, (p \Rightarrow q) \end{aligned}$$

And now, the proof in the first direction follows.

$$\begin{aligned} & \Box p \Rightarrow \Diamond q \\ = \quad & \langle (104) \text{ Distributivity of } \Diamond \text{ over } \Rightarrow \rangle \\ & \Diamond(p \Rightarrow q) \\ = \quad & \langle (192) \text{ Rule of } \mathcal{U} \rangle \\ & \neg(p \Rightarrow q) \,\mathcal{U}\, (p \Rightarrow q) \\ = \quad & \langle (3.59) \text{ Implication}, p \Rightarrow q \equiv \neg p \lor q, \text{ twice} \rangle \\ & \neg(\neg p \lor q) \,\mathcal{U}\, (\neg p \lor q) \\ = \quad & \langle (3.47b) \text{ De Morgan}, \neg(p \lor q) \equiv \neg p \land \neg q \rangle \\ & (p \land \neg q) \,\mathcal{U}\, (\neg p \lor q) \\ = \quad & \langle (15) \text{ Right distributivity of } \mathcal{U} \text{ over } \land \rangle \\ & p \,\mathcal{U}\, (q \lor \neg p) \land \neg q \,\mathcal{U}\, (q \lor \neg p) \\ \Rightarrow \quad & \langle (3.76b) \text{ Strengthening}, p \land q \Rightarrow p \rangle \\ & p \,\mathcal{U}\, (q \lor \neg p) \\ = \quad & \langle (3.59) \text{ Implication}, p \Rightarrow q \equiv \neg p \lor q \rangle \\ & p \,\mathcal{U}\, (p \Rightarrow q) \end{aligned}$$

The proof in the second direction follows.

$$p \,\mathcal{U}\, (p \Rightarrow q)$$
$= \quad \langle(3.59) \text{ Implication, } p \Rightarrow q \equiv \neg p \vee q\rangle$
$$p \,\mathcal{U}\, (q \vee \neg p)$$
$\Rightarrow \quad \langle(42) \text{ Eventuality}\rangle$
$$\Diamond(q \vee \neg p)$$
$= \quad \langle(52) \text{ Distributivity of } \Diamond \text{ over } \vee\rangle$
$$\Diamond q \vee \Diamond \neg p$$
$= \quad \langle(60) \text{ Dual of } \Box\rangle$
$$\Diamond q \vee \neg \Box p$$
$= \quad \langle(3.59) \text{ Implication, } p \Rightarrow q \equiv \neg p \vee q\rangle$
$$\Box p \Rightarrow \Diamond q \quad \blacksquare$$

(S15) $\Diamond(p \Rightarrow q) \equiv p \,\mathcal{U}\, (q \vee \neg p)$

Proof:

$$p \,\mathcal{U}\, (q \vee \neg p)$$
$= \quad \langle(190) \text{ Disjunction rule of } \mathcal{U}\rangle$
$$(p \vee (q \vee \neg p)) \,\mathcal{U}\, (q \vee \neg p)$$
$= \quad \langle(3.31) \text{ Distributivity of } \vee \text{ over } \vee, p \vee (q \vee r) \equiv (p \vee q) \vee (p \vee r)\rangle$
$$((p \vee q) \vee (p \vee \neg p)) \,\mathcal{U}\, (q \vee \neg p)$$
$= \quad \langle(3.28) \text{ Excluded middle, } p \vee \neg p\rangle$
$$((p \vee q) \vee (true)) \,\mathcal{U}\, (q \vee \neg p)$$
$= \quad \langle(3.29) \text{ Zero of } \vee, p \vee true \equiv true\rangle$
$$true \,\mathcal{U}\, (q \vee \neg p)$$
$= \quad \langle(3.59) \text{ Implication, } p \Rightarrow q \equiv \neg p \vee q\rangle$
$$true \,\mathcal{U}\, (p \Rightarrow q)$$
$= \quad \langle(38) \text{ Definition of } \Diamond\rangle$
$$\Diamond(p \Rightarrow q) \quad \blacksquare$$

(S16) $p \,\mathcal{U}\, \neg p \equiv \neg \Box p$

Proof:

$$true$$
$= \quad \langle(192) \text{ Rule of } \mathcal{U} \text{ with } q := \neg p\rangle$
$$p \,\mathcal{U}\, \neg p \equiv \Diamond \neg p$$
$= \quad \langle(60) \text{ Dual of } \Box\rangle$
$$p \,\mathcal{U}\, \neg p \equiv \neg \Box p \quad \blacksquare$$

Pepperdine Papers on LTL

(S17) $\quad p \,\mathcal{U}\, (q \vee \neg p) \equiv \Box p \Rightarrow p \,\mathcal{U}\, q$

Proof:

$\quad\quad p \,\mathcal{U}\, (q \vee \neg p)$
$= \quad \langle (12) \text{ Left distributivity of } \mathcal{U} \text{ over } \vee \rangle$
$\quad\quad p \,\mathcal{U}\, q \vee p \,\mathcal{U}\, \neg p$
$= \quad \langle (S16) \; p \,\mathcal{U}\, \neg p \equiv \neg \Box p \rangle$
$\quad\quad p \,\mathcal{U}\, q \vee \neg \Box p$
$= \quad \langle (3.59) \text{ Implication, } p \Rightarrow q \equiv \neg p \vee q \rangle$
$\quad\quad \Box p \Rightarrow p \,\mathcal{U}\, q \quad \blacksquare$

Another theorem from StackExchange.com is (S18), which is proved with a supporting lemma typical of those found when using monotonicity theorems. There is also a simpler proof shown using temporal deduction.

(S18) $\quad \Box p \Rightarrow (p \,\mathcal{U}\, q) \,\mathcal{U}\, p$

Proof:

$\quad\quad (p \,\mathcal{U}\, q) \,\mathcal{U}\, p$
$\Leftarrow \quad \langle \text{Lemma A: } (p \,\mathcal{U}\, q) \,\mathcal{U}\, \Box p \Rightarrow (p \,\mathcal{U}\, q) \,\mathcal{U}\, p \rangle$
$\quad\quad (p \,\mathcal{U}\, q) \,\mathcal{U}\, \Box p$
$\Leftarrow \quad \langle (29) \; \mathcal{U} \text{ insertion with } p, q := p \,\mathcal{U}\, q, \Box p \rangle$
$\quad\quad \Box p \quad \blacksquare$

Lemma A: $(p \,\mathcal{U}\, q) \,\mathcal{U}\, \Box p \Rightarrow (p \,\mathcal{U}\, q) \,\mathcal{U}\, p$

Proof: The proof is by (4.7.1) Truth implication.

$\quad\quad true$
$= \quad \langle (76) \text{ Strengthening of } \Box \rangle$
$\quad\quad \Box p \Rightarrow p$
$= \quad \langle (136) \text{ Metatheorem with the above theorem} \rangle$
$\quad\quad \Box (\Box p \Rightarrow p)$
$\Rightarrow \quad \langle (85) \text{ Right monotonicity of } \mathcal{U} \text{ with } p, q, r := \Box p, p, p \,\mathcal{U}\, q \rangle$
$\quad\quad (p \,\mathcal{U}\, q) \,\mathcal{U}\, \Box p \Rightarrow (p \,\mathcal{U}\, q) \,\mathcal{U}\, p \quad \blacksquare$

Alternate Proof: The proof is by (82) Temporal deduction.

$\quad\quad (p \,\mathcal{U}\, q) \,\mathcal{U}\, p$
$= \quad \langle \text{Assume } p \rangle$
$\quad\quad (true \,\mathcal{U}\, q) \,\mathcal{U}\, true$
$= \quad \langle (20) \text{ Right zero of } \mathcal{U} \text{ with } p := (true \,\mathcal{U}\, q) \rangle$
$\quad\quad true \quad \blacksquare$

The next two theorems, (S19) and (S20), are from the Spot documentation [4]. Together, these two theorems show that $p \,\mathcal{U}\, (q \,\mathcal{U}\, p) \equiv (p \,\mathcal{U}\, q) \,\mathcal{U}\, p$. Similarly, theorems (36) and (37) show that $p \,\mathcal{U}\, (p \,\mathcal{U}\, q) \equiv (p \,\mathcal{U}\, q) \,\mathcal{U}\, q$. In general $\mathcal{U}$ is not associative, i.e. $p \,\mathcal{U}\, (q \,\mathcal{U}\, r) \not\equiv (p \,\mathcal{U}\, q) \,\mathcal{U}\, r$. Apparently, the $\mathcal{U}$ operator is associative for two propositional variables, but not for three.

(S19) $\quad p \,\mathcal{U}\, (q \,\mathcal{U}\, p) \equiv q \,\mathcal{U}\, p$

Proof: The proof is by (4.7) Mutual implication.
The proof in the first direction follows.

$$\begin{aligned}
& p \,\mathcal{U}\, (q \,\mathcal{U}\, p) \\
\Rightarrow \quad & \langle (28) \text{ with } q := q \,\mathcal{U}\, p \rangle \\
& p \vee q \,\mathcal{U}\, p \\
= \quad & \langle (32) \text{ Absorption with } p, q := q, p \rangle \\
& q \,\mathcal{U}\, p
\end{aligned}$$

The proof in the second direction follows.

$$\begin{aligned}
& q \,\mathcal{U}\, p \\
\Rightarrow \quad & \langle (29) \,\, \mathcal{U} \text{ insertion with } q := q \,\mathcal{U}\, p \rangle \\
& p \,\mathcal{U}\, (q \,\mathcal{U}\, p) \quad \blacksquare
\end{aligned}$$

(S20) $\quad (p \,\mathcal{U}\, q) \,\mathcal{U}\, p \equiv q \,\mathcal{U}\, p$

Proof: The proof is by (4.7) Mutual implication.
The proof in the first direction follows.

$$\begin{aligned}
& (p \,\mathcal{U}\, q) \,\mathcal{U}\, p \\
= \quad & \langle (190) \text{ Disjunction rule of } \mathcal{U} \rangle \\
& ((p \vee q) \,\mathcal{U}\, q) \,\mathcal{U}\, p \\
\Rightarrow \quad & \langle \text{Lemma A: } ((p \vee q) \,\mathcal{U}\, q) \,\mathcal{U}\, p \Rightarrow ((p \vee q) \,\mathcal{U}\, (p \vee q)) \,\mathcal{U}\, p \rangle \\
& ((p \vee q) \,\mathcal{U}\, (p \vee q)) \,\mathcal{U}\, p \\
= \quad & \langle (22) \text{ Idempotency of } \mathcal{U} \rangle \\
& (p \vee q) \,\mathcal{U}\, p \\
= \quad & \langle (190) \text{ Disjunction rule of } \mathcal{U} \rangle \\
& q \,\mathcal{U}\, p
\end{aligned}$$

The proof in the second direction follows.

$$\begin{aligned}
& q \,\mathcal{U}\, p \\
\Rightarrow \quad & \langle \text{Lemma B: } (q \,\mathcal{U}\, p) \Rightarrow (p \,\mathcal{U}\, q) \,\mathcal{U}\, p \rangle \\
& p \,\mathcal{U}\, (q \,\mathcal{U}\, p) \quad \blacksquare
\end{aligned}$$

Lemma A: $((p \vee q) \, \mathcal{U} \, q) \, \mathcal{U} \, p \Rightarrow ((p \vee q) \, \mathcal{U} \, (p \vee q)) \, \mathcal{U} \, p$

Proof: The proof is by (4.7.1) Truth implication.

$\quad$ *true*
$= \quad \langle$(3.76a) Weakening, $p \Rightarrow p \vee q$ with $p,q := q,p$ and
$\quad\quad$ (3.24) Symmetry of $\vee$, $p \vee q \equiv q \vee p\rangle$
$\quad q \Rightarrow p \vee q$
$= \quad \langle$(136) Metatheorem with the above theorem$\rangle$
$\quad \Box(q \Rightarrow p \vee q)$
$\Rightarrow \quad \langle$(85) Right monotonicity of $\mathcal{U}$ with $p,q,r := q, p \vee q, p \vee q\rangle$
$\quad (p \vee q) \, \mathcal{U} \, q \Rightarrow (p \vee q) \, \mathcal{U} \, (p \vee q)$
$= \quad \langle$(136) Metatheorem with the above theorem$\rangle$
$\quad \Box((p \vee q) \, \mathcal{U} \, q \Rightarrow (p \vee q) \, \mathcal{U} \, (p \vee q))$
$\Rightarrow \quad \langle$(86) Left monotonicity of $\mathcal{U}$ with $p,q,r := (p \vee q) \, \mathcal{U} \, q, (p \vee q) \, \mathcal{U} \, (p \vee q), p\rangle$
$\quad ((p \vee q) \, \mathcal{U} \, q) \, \mathcal{U} \, p \Rightarrow ((p \vee q) \, \mathcal{U} \, (p \vee q)) \, \mathcal{U} \, p \quad \blacksquare$

Lemma B: $(q \, \mathcal{U} \, p) \Rightarrow (p \, \mathcal{U} \, q) \, \mathcal{U} \, p$

Proof: The proof is by (4.7.1) Truth implication.

$\quad$ *true*
$\Rightarrow \quad \langle$(29) $\mathcal{U}$ insertion$\rangle$
$\quad q \Rightarrow p \, \mathcal{U} \, q$
$= \quad \langle$(136) Metatheorem with the above theorem$\rangle$
$\quad \Box(q \Rightarrow p \, \mathcal{U} \, q)$
$\Rightarrow \quad \langle$(86) Left monotonicity of $\mathcal{U}$ with $p,q,r := q, p \, \mathcal{U} \, q, p\rangle$
$\quad q \, \mathcal{U} \, p \Rightarrow (p \, \mathcal{U} \, q) \, \mathcal{U} \, p \quad \blacksquare$

2.3 Induction

This section presents some theorems related to induction. Theorem (165) includes proof of Lemma F which is used in the induction proof of theorem (165). A similar proof can be used to establish theorem (S22). Because Lemma F was not numbered in the CDS4LTL system, but just supported the proof of theorem (165), it is promoted to a numbered theorem here as (S21) so it can be used in other proofs.

(S21) (165) Lemma F: $\Box p \vee \Box q \Rightarrow \circ (\Box p \vee \Box q)$

Proof:

$\quad\quad \Box p \lor \Box q$
$=\quad \langle\text{(66) Expansion of } \Box, \text{twice}\rangle$
$\quad\quad (p \land \circ \Box p) \lor (q \land \circ \Box q)$
$\Rightarrow\quad \langle\text{(3.76b) Strengthening, } p \land q \Rightarrow p \text{ and (4.2) Monotonicity of } \lor, \text{twice}\rangle$
$\quad\quad \circ \Box p \lor \circ \Box q$
$=\quad \langle\text{(4) Distributivity of } \circ \text{ over } \lor\rangle$
$\quad\quad \circ(\Box p \lor \Box q) \quad \blacksquare$

Theorems (S22), (S23) and (S24) prove useful theorems about conjunction and disjunction of $\Box$ and $\Diamond$ in induction proofs.

(S22) $\quad \Box p \land \Box q \Rightarrow \circ(\Box p \land \Box q)$

Proof:

$\quad\quad \Box p \land \Box q$
$=\quad \langle\text{(66) Expansion of } \Box, \text{twice}\rangle$
$\quad\quad (p \land \circ \Box p) \land (q \land \circ \Box q)$
$=\quad \langle\text{(3.36) Symmetry of } \land, p \land q \equiv q \land p\rangle$
$\quad\quad (\circ \Box p \land \circ \Box q) \land (p \land q)$
$\Rightarrow\quad \langle\text{(3.76b) Strengthening, } p \land q \Rightarrow p \text{ with } p,q := (\circ \Box p \land \circ \Box q),(p \land q)\rangle$
$\quad\quad \circ \Box p \land \circ \Box q$
$=\quad \langle\text{(5) Distributivity of } \circ \text{ over } \land\rangle$
$\quad\quad \circ(\Box p \land \Box q) \quad \blacksquare$

(S23) $\quad \circ(\Diamond p \lor \Diamond q) \Rightarrow (\Diamond p \lor \Diamond q)$

Proof:

$\quad\quad \Diamond p \lor \Diamond q$
$=\quad \langle\text{(45) Expansion of } \Diamond, \text{twice}\rangle$
$\quad\quad (p \lor \circ \Diamond p) \lor (q \lor \circ \Diamond q)$
$=\quad \langle\text{(3.24) Symmetry of } \lor, p \lor q \equiv q \lor p\rangle$
$\quad\quad (\circ \Diamond p \lor \circ \Diamond q) \lor (p \lor q)$
$\Leftarrow\quad \langle\text{(3.76a) Weakening, } p \Rightarrow p \lor q \text{ and (4.2) Monotonicity of } \lor\rangle$
$\quad\quad (\circ \Diamond p \lor \circ \Diamond q)$
$=\quad \langle\text{(4) Distributivity of } \circ \text{ over } \lor\rangle$
$\quad\quad \circ(\Diamond p \lor \Diamond q) \quad \blacksquare$

(S24) $\quad \circ(\Diamond p \land \Diamond q) \Rightarrow (\Diamond p \land \Diamond q)$

Proof:

Pepperdine Papers on LTL

$\qquad \diamond p \wedge \diamond q$
$= \quad \langle (45) \text{ Expansion of } \diamond, \text{ twice} \rangle$
$\qquad (p \vee \circ \diamond p) \wedge (q \vee \circ \diamond q)$
$\Leftarrow \quad \langle (3.76a) \text{ Weakening}, p \Rightarrow p \vee q \text{ and } (4.3) \text{ Monotonicity of } \wedge, \text{ twice} \rangle$
$\qquad (\circ \diamond p \wedge \circ \diamond q)$
$= \quad \langle (5) \text{ Distributivity of } \circ \text{ over } \wedge \rangle$
$\qquad \circ (\diamond p \wedge \diamond q) \quad \blacksquare$

Theorems (S25) through (S28) are being used in the development of a proof of the Dummett theorem (see section 2.13).

(S25) $\Box (p \Rightarrow \circ p) \Rightarrow \Box (p \Rightarrow \Box p)$

Proof: The proof is by (4.7.1) Truth implication.

$\qquad true$
$= \quad \langle (57) \Box \text{ induction} \rangle$
$\qquad \Box (p \Rightarrow \circ p) \Rightarrow (p \Rightarrow \Box p)$
$= \quad \langle (139) \text{ Metatheorem } \Box \text{ with the above theorem} \rangle$
$\qquad \Box \Box (p \Rightarrow \circ p) \Rightarrow \Box (p \Rightarrow \Box p)$
$= \quad \langle (72) \text{ Absorption of } \Box \rangle$
$\qquad \Box (p \Rightarrow \circ p) \Rightarrow \Box (p \Rightarrow \Box p) \quad \blacksquare$

(S26) $\Box p \Rightarrow (p \Rightarrow \Box p)$

Proof: The proof is by (4.7.1) Truth implication.

$\qquad true$
$= \quad \langle (4.1)\ p \Rightarrow (q \Rightarrow p) \text{ with } p, q := \Box p, p \rangle$
$\qquad \Box p \Rightarrow (p \Rightarrow \Box p) \quad \blacksquare$

(S27) $\diamond \Box p \Rightarrow \diamond (p \Rightarrow \Box p)$

Proof: The proof is by (4.7.1) Truth implication.

$\qquad true$
$= \quad \langle (S26)\ \Box p \Rightarrow (p \Rightarrow \Box p) \rangle$
$\qquad \Box p \Rightarrow (p \Rightarrow \Box p)$
$= \quad \langle (136) \text{ Metatheorem with the above theorem} \rangle$
$\qquad \Box (\Box p \Rightarrow (p \Rightarrow \Box p))$
$\Rightarrow \quad \langle (119) \text{ Monotonicity of } \diamond \text{ with } p, q := \Box p, (p \Rightarrow \Box p) \rangle$
$\qquad \diamond \Box p \Rightarrow \diamond (p \Rightarrow \Box p) \quad \blacksquare$

Alternate Proof:

$$\diamond(p \Rightarrow \Box p)$$
$$= \quad \langle(3.59) \text{ Implication}, p \Rightarrow q \equiv \neg p \lor q\rangle$$
$$\diamond(\neg p \lor \Box p)$$
$$= \quad \langle(52) \text{ Distributivity of } \diamond \text{ over } \lor\rangle$$
$$\diamond \neg p \lor \diamond \Box p)$$
$$\Leftarrow \quad \langle(3.76a) \text{ Weakening}, p \Rightarrow p \lor q\rangle$$
$$\diamond \Box p \quad \blacksquare$$

(S28) $\quad \Box p \Rightarrow \diamond(p \Rightarrow \Box p)$

Proof:

$$\Box p$$
$$= \quad \langle(72) \text{ Absorption of } \Box\rangle$$
$$\Box \Box p$$
$$\Rightarrow \quad \langle(77) \text{ Strengthening of } \Box \text{ with } p := \Box p\rangle$$
$$\diamond \Box p$$
$$\Rightarrow \quad \langle(S27) \diamond \Box p \Rightarrow \diamond(p \Rightarrow \Box p)\rangle$$
$$\diamond(p \Rightarrow \Box p) \quad \blacksquare$$

Alternate Proof:

$$\Box p \Rightarrow \diamond(p \Rightarrow \Box p)$$
$$= \quad \langle(104) \text{ Distributivity of } \diamond \text{ over } \Rightarrow\rangle$$
$$\Box p \Rightarrow (\Box p \Rightarrow \diamond \Box p)$$
$$= \quad \langle(3.65) \text{ Shunting}, p \land q \Rightarrow r \equiv p \Rightarrow (q \Rightarrow r)\rangle$$
$$\Box p \land \Box p \Rightarrow \diamond \Box p)$$
$$= \quad \langle(3.38) \text{ Idempotency of } \land, p \land p \equiv p\rangle$$
$$\Box p \Rightarrow \diamond \Box p \quad -(46) \text{ Weakening of } \diamond \quad \blacksquare$$

The next four theorems (S29) through (S32) are new induction theorems. They do not appear in the LTL literature.

(S29) $\quad \Box \diamond$ **induction:** $\quad \Box((p \Rightarrow q \land \circ p) \land (q \Rightarrow \diamond p)) \Rightarrow (p \Rightarrow \Box \diamond p)$

Proof:

$$\Box((p \Rightarrow q \land \circ p) \land (q \Rightarrow \diamond p)) \Rightarrow (p \Rightarrow \Box \diamond p)$$
$$= \quad \langle(99) \text{ Distributivity of } \Box \text{ over } \land\rangle$$
$$\Box(p \Rightarrow q \land \circ p) \land \Box(q \Rightarrow \diamond p) \Rightarrow (p \Rightarrow \Box \diamond p)$$
$$= \quad \langle(3.65) \text{ Shunting}, p \land q \Rightarrow r \equiv p \Rightarrow (q \Rightarrow r)\rangle$$
$$\Box(p \Rightarrow q \land \circ p) \land \Box(q \Rightarrow \diamond p) \land p \Rightarrow \Box \diamond p$$

And now,

$$\Box(p \Rightarrow q \land \bigcirc p) \land \Box(q \Rightarrow \Diamond p) \land p$$
$\Rightarrow \quad \langle(129)\text{ Induction rule }\Box\text{ and }(4.3)\text{ Monotonicity of }\land\rangle$
$$(p \Rightarrow \Box q) \land \Box(q \Rightarrow \Diamond p) \land p$$
$\Rightarrow \quad \langle(120)\text{ Monotonicity of }\Box\text{ and }(4.3)\text{ Monotonicity of }\land\rangle$
$$(p \Rightarrow \Box q) \land (\Box q \Rightarrow \Box \Diamond p) \land p$$
$\Rightarrow \quad \langle(3.77)\text{ Modus ponens, }p \land (p \Rightarrow q) \Rightarrow q\text{ and }(4.3)\text{ Monotonicity of }\land\text{, twice}\rangle$
$$\Box \Diamond p \quad \blacksquare$$

(S30) **$\Box \Diamond$ induction:** $\quad \Box((p \Rightarrow \bigcirc p) \land (p \Rightarrow \Diamond p)) \Rightarrow (p \Rightarrow \Box \Diamond p)$

Proof:

$$\Box((p \Rightarrow \bigcirc p) \land (p \Rightarrow \Diamond p)) \Rightarrow (p \Rightarrow \Box \Diamond p)$$
$= \quad \langle(99)\text{ Distributivity of }\Box\text{ over }\land\rangle$
$$\Box(p \Rightarrow \bigcirc p) \land \Box(p \Rightarrow \Diamond p) \Rightarrow (p \Rightarrow \Box \Diamond p)$$
$= \quad \langle(3.65)\text{ Shunting, }p \land q \Rightarrow r \equiv p \Rightarrow (q \Rightarrow r)\rangle$
$$\Box(p \Rightarrow \bigcirc p) \land \Box(p \Rightarrow \Diamond p) \land p \Rightarrow \Box \Diamond p$$

And now,

$$\Box(p \Rightarrow \bigcirc p) \land \Box(p \Rightarrow \Diamond p) \land p$$
$\Rightarrow \quad \langle(57)\ \Box\text{ induction and }(4.3)\text{ Monotonicity of }\land\rangle$
$$(p \Rightarrow \Box p) \land \Box(p \Rightarrow \Diamond p) \land p$$
$\Rightarrow \quad \langle(3.77)\text{ Modus ponens, }p \land (p \Rightarrow q) \Rightarrow q\text{ and }(4.3)\text{ Monotonicity of }\land\rangle$
$$\Box p \land \Box(p \Rightarrow \Diamond p)$$
$\Rightarrow \quad \langle(S4)\text{ Temporal modus ponens of }\Box\rangle$
$$\Box \Diamond p \quad \blacksquare$$

This next theorem (S31) is related to (176) $\mathcal{W} \Diamond$ equivalence.

(S31) **Obligation induction:** $\quad \Box(p \Rightarrow \bigcirc p) \lor \Box(p \Rightarrow \bigcirc q) \Rightarrow (p \Rightarrow p \mathcal{W} \Diamond q)$

Proof:

$$\Box(p \Rightarrow \bigcirc p) \lor \Box(p \Rightarrow \bigcirc q)$$
$\Rightarrow \quad \langle(134)\ \Box(p \Rightarrow \bigcirc q) \Rightarrow (p \Rightarrow \Diamond q)\text{ and }(4.2)\text{ Monotonicity of }\lor\rangle$
$$\Box(p \Rightarrow \bigcirc p) \lor (p \Rightarrow \Diamond q)$$
$\Rightarrow \quad \langle(57)\ \Box\text{ induction and }(4.2)\text{ Monotonicity of }\lor\rangle$
$$(p \Rightarrow \Box p) \lor (p \Rightarrow \Diamond q)$$
$= \quad \langle(3.63.2)\text{ Distributivity of }\Rightarrow\text{ over }\lor, p \Rightarrow q \lor r \equiv (p \Rightarrow q) \lor (p \Rightarrow r)\rangle$
$$p \Rightarrow (\Box p \lor \Diamond q)$$
$= \quad \langle(176)\ \mathcal{W} \Diamond\text{ equivalence}\rangle$
$$p \Rightarrow p \mathcal{W} \Diamond q \quad \blacksquare$$

(S32) **Induction rule** $\mathcal{U}$: $\quad \Box(p \Rightarrow \bigcirc(p \land q)) \Rightarrow (p \Rightarrow p\,\mathcal{U}\,q)$

Proof:

$\quad\quad \Box(p \Rightarrow \bigcirc(p \land q))$
$= \quad \langle (5) \text{ Distributivity of } \bigcirc \text{ over } \land \rangle$
$\quad\quad \Box(p \Rightarrow \bigcirc p \land \bigcirc q))$
$= \quad \langle (3.63.1) \text{ Distributivity of } \Rightarrow \text{ over } \land, p \Rightarrow q \land r \equiv (p \Rightarrow q) \land (p \Rightarrow r) \rangle$
$\quad\quad \Box((p \Rightarrow \bigcirc p) \land (p \Rightarrow \bigcirc q))$
$= \quad \langle (99) \text{ Distributivity of } \Box \text{ over } \land \rangle$
$\quad\quad \Box(p \Rightarrow \bigcirc p) \land \Box(p \Rightarrow \bigcirc q))$
$\Rightarrow \quad \langle (57) \Box \text{ induction and } (4.3) \text{ Monotonicity of } \land \rangle$
$\quad\quad (p \Rightarrow \Box p) \land \Box(p \Rightarrow \bigcirc q))$
$\Rightarrow \quad \langle (134) \Box(p \Rightarrow \bigcirc q) \Rightarrow (p \Rightarrow \Diamond q) \text{ and } (4.3) \text{ Monotonicity of } \land \rangle$
$\quad\quad (p \Rightarrow \Box p) \land (p \Rightarrow \Diamond q))$
$= \quad \langle (3.63.1) \text{ Distributivity of } \Rightarrow \text{ over } \land, p \Rightarrow q \land r \equiv (p \Rightarrow q) \land (p \Rightarrow r) \rangle$
$\quad\quad p \Rightarrow \Box p \land \Diamond q$
$\Rightarrow \quad \langle (84) \; \mathcal{U} \text{ implication and } (3.82a) \text{ Transitivity} \rangle$
$\quad\quad p \Rightarrow p\,\mathcal{U}\,q \quad \blacksquare$

(S33) **Dummett induction:** $\quad \Box(p \equiv \bigcirc p) \Rightarrow (\Diamond \Box p \Rightarrow \Box p)$

Proof:

$\quad\quad \Box(p \equiv \bigcirc p) \Rightarrow (\Diamond \Box p \Rightarrow \Box p)$
$= \quad \langle (3.80) \text{ Mutual implication}, (p \Rightarrow q) \land (q \Rightarrow p) \equiv (p \equiv q) \rangle$
$\quad\quad \Box((p \Rightarrow \bigcirc p) \land (\bigcirc p \Rightarrow p)) \Rightarrow (\Diamond \Box p \Rightarrow \Box p)$
$= \quad \langle (99) \text{ Distributivity of } \Box \text{ over } \land \rangle$
$\quad\quad \Box(p \Rightarrow \bigcirc p) \land \Box(\bigcirc p \Rightarrow p) \Rightarrow (\Diamond \Box p \Rightarrow \Box p)$
$= \quad \langle (3.65) \text{ Shunting}, p \land q \Rightarrow r \equiv p \Rightarrow (q \Rightarrow r) \rangle$
$\quad\quad \Box(p \Rightarrow \bigcirc p) \land \Box(\bigcirc p \Rightarrow p) \land \Diamond \Box p \Rightarrow \Box p$

And now,

$\quad\quad \Box(p \Rightarrow \bigcirc p) \land \Box(\bigcirc p \Rightarrow p) \land \Diamond \Box p$
$\Rightarrow \quad \langle (57) \Box \text{ induction and } (4.3) \text{ Monotonicity of } \land \rangle$
$\quad\quad (p \Rightarrow \Box p) \land \Box(\bigcirc p \Rightarrow p) \land \Diamond \Box p$
$\Rightarrow \quad \langle (58) \Diamond \text{ induction and } (4.3) \text{ Monotonicity of } \land \rangle$
$\quad\quad (p \Rightarrow \Box p) \land (\Diamond p \Rightarrow p) \land \Diamond \Box p$
$\Rightarrow \quad \langle (3.82a) \text{ Transitivity}, (p \Rightarrow q) \land (q \Rightarrow r) \Rightarrow (p \Rightarrow r) \text{ and } (4.3) \text{ Monotonicity of } \land \rangle$

Pepperdine Papers on LTL 159

$$(\Diamond p \Rightarrow \Box p) \wedge \Diamond \Box p$$
$\Rightarrow \quad \langle (76) \text{ Strengthening of } \Box, (119) \text{ Monotonicity of } \Diamond$
$\quad\quad \text{ and } (4.3) \text{ Monotonicity of } \wedge \rangle$
$$(\Diamond p \Rightarrow \Box p) \wedge \Diamond p$$
$\Rightarrow \quad \langle (3.77) \text{ Modus ponens}, p \wedge (p \Rightarrow q) \Rightarrow q \rangle$
$$\Box p \quad \blacksquare$$

2.4 Absorption

Absorption theorems are always nice to have as they generally re-write a formula to a shorter, simpler form. For example, theorems (155) Absorption of $\bigcirc$ into $\Box \Diamond$ and (156) Absorption of $\bigcirc$ into $\Diamond \Box$ from CDS4LTL can be used to eliminate the $\bigcirc$ operator from a formula. Theorem (156) was used in the proof of (S77).

(S34) Absorption of $\Diamond$ into $\neg \Box$: $\quad \Diamond \neg \Box p \equiv \neg \Box p$

Proof:

$$\Diamond \neg \Box p$$
$= \quad \langle (60) \text{ Dual of } \Box \rangle$
$$\neg \Box \Box p$$
$= \quad \langle (72) \text{ Absorption of } \Box \rangle$
$$\neg \Box p \quad \blacksquare$$

(S35) Absorption of $\Box$ into $\neg \Diamond$: $\quad \Box \neg \Diamond p \equiv \neg \Diamond p$

Proof:

$$\Box \neg \Diamond p$$
$= \quad \langle (61) \text{ Dual of } \Diamond \rangle$
$$\neg \Diamond \Diamond p$$
$= \quad \langle (50) \text{ Absorption of } \Diamond \rangle$
$$\neg \Diamond p \quad \blacksquare$$

Theorem (S36) can be used to remove a $\mathcal{W}$ operator from this particular formula.

(S36) $p \mathcal{W} q \vee \neg q \mathcal{W} p \equiv q \vee \neg q \mathcal{W} p$

Proof:

$$p \,\mathcal{W}\, q \vee \neg q \,\mathcal{W}\, p$$
$= \quad \langle(3.39) \text{ Identity of } \wedge, p \wedge true \equiv p\rangle$
$$(p \,\mathcal{W}\, q \vee \neg q \,\mathcal{W}\, p) \wedge True$$
$= \quad \langle(253) \,\mathcal{W} \text{ ordering}\rangle$
$$(p \,\mathcal{W}\, q \vee \neg q \,\mathcal{W}\, p) \wedge (\neg p \,\mathcal{W}\, q \vee \neg q \,\mathcal{W}\, p)$$
$= \quad \langle(3.46) \text{ Distributivity of } \wedge \text{ over } \vee, p \wedge (q \vee r) \equiv (p \wedge q) \vee (p \wedge r)\rangle$
$$((p \,\mathcal{W}\, q \vee \neg q \,\mathcal{W}\, p) \wedge \neg p \,\mathcal{W}\, q) \vee ((p \,\mathcal{W}\, q \vee \neg q \,\mathcal{W}\, p) \wedge \neg q \,\mathcal{W}\, p)$$
$= \quad \langle(3.46) \text{ Distributivity of } \wedge \text{ over } \vee, p \wedge (q \vee r) \equiv (p \wedge q) \vee (p \wedge r)\rangle$
$$(p \,\mathcal{W}\, q \wedge \neg p \,\mathcal{W}\, q) \vee (\neg q \,\mathcal{W}\, p \wedge \neg p \,\mathcal{W}\, q) \vee (p \,\mathcal{W}\, q \wedge \neg q \,\mathcal{W}\, p) \vee (\neg q \,\mathcal{W}\, p \wedge \neg q \,\mathcal{W}\, p)$$
$= \quad \langle(187) \text{ Right distributivity of } \mathcal{W} \text{ over } \wedge,$
$\qquad \text{and (3.38) Idempotency of } \wedge, p \wedge p \equiv p\rangle$
$$(p \wedge \neg p) \,\mathcal{W}\, q \vee (\neg q \,\mathcal{W}\, p \wedge \neg p \,\mathcal{W}\, q) \vee (p \,\mathcal{W}\, q \wedge \neg q \,\mathcal{W}\, p) \vee \neg q \,\mathcal{W}\, p$$
$= \quad \langle(3.42) \text{ Contradiction}, p \wedge \neg p \equiv false$
$\qquad \text{and (205) Left identity of } \mathcal{W}\rangle$
$$q \vee (\neg q \,\mathcal{W}\, p \wedge \neg p \,\mathcal{W}\, q) \vee (p \,\mathcal{W}\, q \wedge \neg q \,\mathcal{W}\, p) \vee \neg q \,\mathcal{W}\, p$$
$= \quad \langle(3.43b) \text{ Absorption}, p \vee (p \wedge q) \equiv p, \text{ twice}\rangle$
$$q \vee \neg q \,\mathcal{W}\, p \quad \blacksquare$$

2.5 Frame Rules

The first theorem in this section comes from the work of Manna and Pneuli in [12]. Theorem (S37) is the only frame rule shown in [12]. The proof of theorem (S37), is supported by the proof of two lemmas.

(S37) Manna-Pnueli frame rule $\Diamond$: $\quad \Box(p \Rightarrow \Diamond q) \Rightarrow (\Box w \wedge p \Rightarrow \Diamond(\Box w \wedge q))$

Proof:
$$\Box(p \Rightarrow \Diamond q) \Rightarrow (\Box w \wedge p \Rightarrow \Diamond(\Box w \wedge q))$$
$= \quad \langle(3.65) \text{ Shunting}, p \wedge q \Rightarrow r \equiv p \Rightarrow (q \Rightarrow r)\rangle$
$$\Box(p \Rightarrow \Diamond q) \wedge \Box w \wedge p \Rightarrow \Diamond(\Box w \wedge q)$$

And now,
$$\Box(p \Rightarrow \Diamond q) \wedge \Box w \wedge p$$
$\Rightarrow \quad \langle(46) \text{ Weakening of } \Diamond \text{ and (4.3) Monotonicity of } \wedge\rangle$
$$\Box(p \Rightarrow \Diamond q) \wedge \Box w \wedge \Diamond p$$
$\Rightarrow \quad \langle\text{Lemma A: } \Box(p \Rightarrow \Diamond q) \wedge \Diamond p \Rightarrow \Diamond q \text{ and (4.3) Monotonicity of } \wedge\rangle$
$$\Box w \wedge \Diamond q$$
$\Rightarrow \quad \langle\text{Lemma B: } \Box w \wedge \Diamond q \Rightarrow \Diamond(\Box w \wedge q)\rangle$
$$\Diamond(\Box w \wedge q) \quad \blacksquare$$

Lemma A: $\Box(p \Rightarrow \Diamond q) \land \Diamond p \Rightarrow \Diamond q$

Proof:

$$\begin{aligned}
& \quad \textit{true} \\
& = \quad \langle\text{(S3) Temporal modus ponens of } \Diamond \text{ with } q := \Diamond q\rangle \\
& \quad \Box(p \Rightarrow \Diamond q) \land \Diamond p \Rightarrow \Diamond \Diamond q \\
& = \quad \langle\text{(50) Absorption of } \Diamond\rangle \\
& \quad \Box(p \Rightarrow \Diamond q) \land \Diamond p \Rightarrow \Diamond q \quad \blacksquare
\end{aligned}$$

Lemma B: $\Box w \land \Diamond q \Rightarrow \Diamond(\Box w \land q)$

Proof:

$$\begin{aligned}
& \quad \textit{true} \\
& = \quad \langle\text{(107)} \land \text{ frame law of } \Diamond \text{ with } p := \Box w\rangle \\
& \quad \Box\Box w \Rightarrow (\Diamond q \Rightarrow \Diamond(\Box w \land q)) \\
& = \quad \langle\text{(72) Absorption of } \Box \text{ and (3.65) Shunting, } p \land q \Rightarrow r \equiv p \Rightarrow (q \Rightarrow r)\rangle \\
& \quad \Box w \land \Diamond q \Rightarrow \Diamond(\Box w \land q) \quad \blacksquare
\end{aligned}$$

Similar to our earlier work on the frame law theorems (106) through (117), theorems (148) to (150), and theorems (210) to (212) we were motivated by the work of Kröger and Merz [11] to add additional theorems here, building on the work of Manna and Pneuli.

(S38) Manna-Pnueli frame rule $\Box$: $\quad \Box(p \Rightarrow \Box q) \Rightarrow (\Box w \land p \Rightarrow \Box(\Box w \land q))$

Proof:

$$\begin{aligned}
& \quad \Box(p \Rightarrow \Box q) \Rightarrow (\Box w \land p \Rightarrow \Box(\Box w \land q)) \\
& = \quad \langle\text{(3.65) Shunting, } p \land q \Rightarrow r \equiv p \Rightarrow (q \Rightarrow r)\rangle \\
& \quad \Box(p \Rightarrow \Box q) \land \Box w \land p \Rightarrow \Box(\Box w \land q) \\
& = \quad \langle\text{(99) Distributivity of } \Box \text{ over } \land \text{ and (72) Absorption of } \Box\rangle \\
& \quad \Box(p \Rightarrow \Box q) \land \Box w \land p \Rightarrow \Box w \land \Box q
\end{aligned}$$

And now,

$$\begin{aligned}
& \quad \Box(p \Rightarrow \Box q) \land \Box w \land p \\
& \Rightarrow \quad \langle\text{(76) Strengthening of } \Box \text{ and (4.3) Monotonicity of } \land\rangle \\
& \quad (p \Rightarrow \Box q) \land \Box w \land p \\
& \Rightarrow \quad \langle\text{(3.77) Modus ponens, } p \land (p \Rightarrow q) \Rightarrow q \text{ and (4.3) Monotonicity of } \land\rangle \\
& \quad \Box q \land \Box w \\
& = \quad \langle\text{(3.36) Symmetry of } \land, p \land q \equiv q \land p\rangle \\
& \quad \Box w \land \Box q \quad \blacksquare
\end{aligned}$$

(S39) **Manna-Pnueli frame rule** $\circ$: $\quad \Box(p \Rightarrow \circ q) \Rightarrow (\Box w \wedge p \Rightarrow \circ (\Box w \wedge q))$

Proof:

$$\Box(p \Rightarrow \circ q) \Rightarrow (\Box w \wedge p \Rightarrow \circ(\Box w \wedge q))$$
$$= \quad \langle (3.65) \text{ Shunting}, p \wedge q \Rightarrow r \equiv p \Rightarrow (q \Rightarrow r) \rangle$$
$$\Box(p \Rightarrow \circ q) \wedge \Box w \wedge p \Rightarrow \circ(\Box w \wedge q)$$

And now,

$$\Box(p \Rightarrow \circ q) \wedge \Box w \wedge p$$
$$\Rightarrow \quad \langle (76) \text{ and } (79) \text{ Strengthening of } \Box \text{ and } (4.3) \text{ Monotonicity of } \wedge \rangle$$
$$(p \Rightarrow \circ q) \wedge \circ \Box w \wedge p$$
$$\Rightarrow \quad \langle (3.77) \text{ Modus ponens}, p \wedge (p \Rightarrow q) \Rightarrow q \text{ and } (4.3) \text{ Monotonicity of } \wedge \rangle$$
$$\circ q \wedge \circ \Box w$$
$$= \quad \langle (3.36) \text{ Symmetry of } \wedge, p \wedge q \equiv q \wedge p \text{ and } (5) \text{ Distributivity of } \circ \text{ over } \wedge \rangle$$
$$\circ(\Box w \wedge q) \quad \blacksquare$$

2.6 Duality

The proof of (58) $\diamond$ induction is an example of a proof by Metatheorem Duality. The CDS4LTL paper mentions the release operator $\mathcal{R}$ as the dual of the until operator $\mathcal{U}$. This section gives a more complete description of LTL duality by formally defining the following two release operators [4].

(S40) **Definition of weak release** $\mathcal{R}$: $\quad p \mathcal{R} q \equiv q \mathcal{W} (p \wedge q)$

(S41) **Definition of strong release** $\mathcal{M}$: $\quad p \mathcal{M} q \equiv q \mathcal{U} (p \wedge q)$

The *dual* P_D of a LTL expression P is constructed from P by interchanging occurrences of:

true	and	*false*
$\wedge$	and	$\vee$
$\equiv$	and	$\not\equiv$
$\Rightarrow$	and	$\not\Leftarrow$
$\Leftarrow$	and	$\not\Rightarrow$
$\circ$	and	$\circ$ (self dual)
$\Box$	and	$\diamond$
$\mathcal{U}$	and	$\mathcal{R}$
$\mathcal{W}$	and	$\mathcal{M}$

Pepperdine Papers on LTL

With this more complete definition of P_D we can use (2.3) Metatheorem Duality in Gries and Schneider [9] section (2.3) without change.

(2.3) Metatheorem Duality. (a) P is valid iff $\neg P_D$ is valid
(b) $P \equiv Q$ is valid iff $P_D \equiv Q_D$ is valid

As an example of using duality to generate a new $\mathcal{U}$ theorem from an existing $\mathcal{W}$ theorem, consider theorem (253) Ordering: $\neg p \, \mathcal{W} \, q \vee \neg q \, \mathcal{W} \, p$. Now, using (2.3a) Metatheorem Duality the following new theorem is proven.

(S42) $\neg(q \, \mathcal{U} \, (\neg p \wedge q)) \vee \neg(p \, \mathcal{U} \, (\neg q \wedge p))$

Proof: The proof uses (2.3a) Metatheorem Duality.

$\quad\quad true$
$= \quad \langle (253) \text{ Ordering is a theorem} \rangle$
$\quad\quad \neg p \, \mathcal{W} \, q \vee \neg q \, \mathcal{W} \, p$
$= \quad \langle (2.3\text{a}) \text{ Metatheorem Duality}, P \equiv \neg P_D \text{ with } P := \neg p \, \mathcal{W} \, q \vee \neg q \, \mathcal{W} \, p$
$\quad\quad\quad \text{and } P_D := \neg p \, \mathcal{M} \, q \wedge \neg q \, \mathcal{M} \, p \, \rangle$
$\quad\quad \neg(\neg p \, \mathcal{M} \, q \wedge \neg q \, \mathcal{M} \, p)$
$= \quad \langle (\text{S41}) \text{ Definition of strong release } \mathcal{M}\!: p \, \mathcal{M} \, q \equiv q \, \mathcal{U} \, (p \wedge q), \text{ twice} \rangle$
$\quad\quad \neg(q \, \mathcal{U} \, (\neg p \wedge q) \wedge p \, \mathcal{U} \, (\neg q \wedge p))$
$= \quad \langle (3.47\text{a}) \text{ De Morgan}, \neg(p \wedge q) \equiv \neg p \vee \neg q \rangle$
$\quad\quad \neg(q \, \mathcal{U} \, (\neg p \wedge q)) \vee \neg(p \, \mathcal{U} \, (\neg q \wedge p)) \quad \blacksquare$

Now we prove a new theorem found in [6].

$$\Diamond p \wedge \Diamond q \Rightarrow (\Diamond(p \wedge q) \vee \Diamond(p \wedge \Diamond q) \vee \Diamond(\Diamond p \wedge q))$$

Before proving this theorem we prove a lemma that uses an Always $\square$ theorem (165) to prove a new Eventually $\Diamond$ theorem (S43). Theorem (S43) does not appear in the LTL literature, but is useful in proving the above theorem and is also used in the proof of theorem (S66). The lemma we prove is:

(S43) $\Diamond((p \wedge \Diamond q) \vee (\Diamond p \wedge q)) \equiv \Diamond p \wedge \Diamond q$

Proof: The proof uses (2.3b) Metatheorem Duality.

$\quad\quad true$
$= \quad \langle (165) \text{ is a theorem} \rangle$
$\quad\quad \square((p \vee \square q) \wedge (\square p \vee q)) \equiv \square p \vee \square q$
$= \quad \langle (2.3\text{b}) \text{ Metatheorem Duality}, (P \equiv Q) \equiv (P_D \equiv Q_D) \text{ with } P_D := \Diamond((p \wedge \Diamond q) \vee (\Diamond p \wedge q))$
$\quad\quad\quad \text{and } Q_D := \Diamond p \wedge \Diamond q \, \rangle$
$\quad\quad \Diamond((p \wedge \Diamond q) \vee (\Diamond p \wedge q)) \equiv \Diamond p \wedge \Diamond q \quad \blacksquare$

Now we prove the main result.

(S44) $\Diamond p \wedge \Diamond q \Rightarrow (\Diamond(p \wedge q) \vee \Diamond(p \wedge \Diamond q) \vee \Diamond(\Diamond p \wedge q))$

Proof:

$\quad \Diamond p \wedge \Diamond q \Rightarrow (\Diamond(p \wedge q) \vee \Diamond(p \wedge \Diamond q) \vee \Diamond(\Diamond p \wedge q))$
$= \quad \langle(S43)\rangle$
$\quad \Diamond p \wedge \Diamond q \Rightarrow (\Diamond(p \wedge q)) \vee (\Diamond p \wedge \Diamond q)) \quad -(3.76a)$ Weakening, $p \Rightarrow p \vee q$ ∎

The following axioms, definitions and theorems are often helpful in proofs using duality. Axiom (3.58) Consequence: $p \Leftarrow q \equiv q \Rightarrow p$. The definition

(S45) $p \not\Leftarrow q \equiv \neg(p \Leftarrow q)$

which is similar in form to (3.10) Definition of $\not\equiv$: $(p \not\equiv q) \equiv \neg(p \equiv q)$. Theorem (S1) and the next two theorems (S46) and (S47).

(S46) $p \Leftarrow q \equiv \neg q \vee p$

Proof:

$\quad p \Leftarrow q$
$= \quad \langle(3.58)$ Consequence, $p \Leftarrow q \equiv q \Rightarrow p\rangle$
$\quad q \Rightarrow p$
$= \quad \langle(3.59)$ Implication, $p \Rightarrow q \equiv \neg p \vee q\rangle$
$\quad \neg q \vee p$ ∎

(S47) $p \not\Leftarrow q \equiv \neg(q \Rightarrow p)$

Proof:

$\quad p \not\Leftarrow q$
$= \quad \langle(S45)\ p \not\Leftarrow q \equiv \neg(p \Leftarrow q)\rangle$
$\quad \neg(p \Leftarrow q)$
$= \quad \langle(S46)\ p \Leftarrow q \equiv \neg q \vee p\rangle$
$\quad \neg(\neg q \vee p)$
$= \quad \langle(3.59)$ Implication, $p \Rightarrow q \equiv \neg p \vee q\rangle$
$\quad \neg(q \Rightarrow p)$ ∎

2.7 Next $\circ$

This section proves some theorems involving the $\circ$ operator. The theorems (S48) and (S49) have the same antecedent as (57) $\square$ induction.

(S48) $\square(p \Rightarrow \circ p) \Rightarrow \square(p \Rightarrow \circ(p \vee q))$

Proof: First prove $(p \Rightarrow \circ p) \Rightarrow (p \Rightarrow \circ(p \vee q))$, and then use (139) Metatheorem $\square$

$\quad\quad p \Rightarrow \circ p$
$= \quad \langle (3.39) \text{ Identity of } \wedge, p \wedge true \equiv p \rangle$
$\quad\quad (p \Rightarrow \circ p) \wedge true$
$= \quad \langle (3.76a) \text{ Weakening}, p \Rightarrow p \vee q \text{ with } p,q := \circ p, \circ q \text{ is a theorem} \rangle$
$\quad\quad (p \Rightarrow \circ p) \wedge (\circ p \Rightarrow \circ p \vee \circ q)$
$\Rightarrow \quad \langle (3.82a) \text{ Transitivity}, (p \Rightarrow q) \wedge (q \Rightarrow r) \Rightarrow (p \Rightarrow r) \rangle$
$\quad\quad p \Rightarrow \circ p \vee \circ q$
$= \quad \langle (4) \text{ Distributivity of } \circ \text{ over } \vee \rangle$
$\quad\quad p \Rightarrow \circ(p \vee q)$

And now, by (139) Metatheorem $\square$, and the above theorem

$\quad\quad \square(p \Rightarrow \circ p) \Rightarrow \square(p \Rightarrow \circ(p \vee q))$ ∎

(S49) $\square(p \Rightarrow \circ(p \Rightarrow q) \wedge \circ p) \equiv \square(p \Rightarrow \circ(p \wedge q))$

Proof:

$\quad\quad \square(p \Rightarrow \circ(p \Rightarrow q) \wedge \circ p)$
$= \quad \langle (3.59) \text{ Implication}, p \Rightarrow q \equiv \neg p \vee q \rangle$
$\quad\quad \square(p \Rightarrow \circ(\neg p \vee q) \wedge \circ p)$
$= \quad \langle (4) \text{ Distributivity of } \circ \text{ over } \vee \rangle$
$\quad\quad \square(p \Rightarrow (\circ \neg p \vee \circ q) \wedge \circ p)$
$= \quad \langle (1) \text{ Self-dual} \rangle$
$\quad\quad \square(p \Rightarrow (\neg \circ p \vee \circ q) \wedge \circ p)$
$= \quad \langle (3.44a) \text{ Absorption}, p \wedge (\neg p \vee q) \equiv p \wedge q \rangle$
$\quad\quad \square(p \Rightarrow \circ p \wedge \circ q)$
$= \quad \langle (5) \text{ Distributivity of } \circ \text{ over } \wedge \rangle$
$\quad\quad \square(p \Rightarrow \circ(p \wedge q))$ ∎

This next theorem (S50) comes from Spot documentation [4].

(S50) $p \wedge \circ p \equiv \neg(\circ p \Rightarrow \neg p)$

Proof:

$\qquad \neg(\bigcirc p \Rightarrow \neg p)$
$=\quad \langle(3.59) \text{ Implication}, p \Rightarrow q \equiv \neg p \vee q\rangle$
$\qquad \neg(\neg \bigcirc p \vee \neg p)$
$=\quad \langle(3.47b) \text{ De Morgan}, \neg(p \vee q) \equiv \neg p \wedge \neg q$
$\qquad$ and (3.12) Double negation, $\neg\neg p \equiv p\rangle$
$\qquad \bigcirc p \wedge p \quad \blacksquare$

2.8 Until $\mathcal{U}$

This section collects some interesting $\mathcal{U}$ theorems that were not included in the CDS4LTL paper. Often a $\mathcal{W}$ theorem will suggest a corresponding $\mathcal{U}$ theorem and vice versa. In the case of (S51) it is suggested by (207) $\Box(p \vee q) \Rightarrow p \mathcal{W} q$.

(S51) $\quad \Box(p \wedge q) \Rightarrow p \mathcal{U} q$

Proof:

$\qquad \Box(p \wedge q)$
$\Rightarrow \quad \langle(76) \text{ Strengthening of } \Box \rangle$
$\qquad p \wedge q$
$\Rightarrow \quad \langle(30) \; p \wedge q \Rightarrow p \mathcal{U} q\rangle$
$\qquad p \mathcal{U} q \quad \blacksquare$

This next theorem (S52) comes from Spot documentation [4].

(S52) $\quad p \mathcal{U} (q \vee \Diamond r) \equiv p \mathcal{U} q \vee \Diamond r$

Proof:

$\qquad p \mathcal{U} (q \vee \Diamond r)$
$=\quad \langle(12) \text{ Left distributivity of } \mathcal{U} \text{ over } \vee\rangle$
$\qquad p \mathcal{U} q \vee p \mathcal{U} \Diamond r$
$=\quad \langle(41) \text{ Absorption of } \mathcal{U} \text{ into } \Diamond \rangle$
$\qquad p \mathcal{U} q \vee \Diamond r \quad \blacksquare$

(S53) $\quad \Box p \mathcal{U} q \Rightarrow p \mathcal{U} q \vee \Box \neg q$

Proof:

$\qquad \Box p \mathcal{U} q$
$\Rightarrow \quad \langle \text{Lemma A: } \Box p \mathcal{U} q \Rightarrow (p \mathcal{U} q \vee \Box \neg q) \mathcal{U} q \rangle$
$\qquad (p \mathcal{U} q \vee \Box \neg q) \mathcal{U} q$

$\quad = \quad \langle(10)\text{ Expansion of } \mathcal{U}\rangle$
$\quad\quad q \vee ((p\,\mathcal{U}\,q \vee \Box\neg q) \wedge \circ ((p\,\mathcal{U}\,q \vee \Box\neg q)\,\mathcal{U}\,q))$
$\quad = \quad \langle(3.45)\text{ Distributivity of } \vee \text{ over } \wedge,\ p \vee (q \wedge r) \equiv (p \vee q) \wedge (p \vee r)\rangle$
$\quad\quad (q \vee p\,\mathcal{U}\,q \vee \Box\neg q) \wedge (q \vee \circ ((p\,\mathcal{U}\,q \vee \Box\neg q)\,\mathcal{U}\,q))$
$\quad = \quad \langle(32)\text{ Absorption}\rangle$
$\quad\quad (p\,\mathcal{U}\,q \vee \Box\neg q) \wedge (q \vee \circ ((p\,\mathcal{U}\,q \vee \Box\neg q)\,\mathcal{U}\,q))$
$\quad \Rightarrow \quad \langle(3.76\text{b})\text{ Strengthening the antecedent, } p \wedge q \Rightarrow p\rangle$
$\quad\quad p\,\mathcal{U}\,q \vee \Box\neg q \quad \blacksquare$

Lemma A: $\Box p\,\mathcal{U}\,q \Rightarrow (p\,\mathcal{U}\,q \vee \Box\neg q)\,\mathcal{U}\,q$

Proof: The proof is by (4.7.1) Truth implication.

$\quad\quad true$
$\quad = \quad \langle(235)\rangle$
$\quad\quad \Box p \Rightarrow p\,\mathcal{U}\,q \vee \Box\neg q$
$\quad = \quad \langle(136)\text{ Metatheorem with the above theorem}\rangle$
$\quad\quad \Box(\Box p \Rightarrow p\,\mathcal{U}\,q \vee \Box\neg q)$
$\quad \Rightarrow \quad \langle(86)\text{ Left monotonicity of } \mathcal{U} \text{ with } p,q,r := \Box p, (p\,\mathcal{U}\,q \vee \Box\neg q), q\rangle$
$\quad\quad \Box p\,\mathcal{U}\,q \Rightarrow (p\,\mathcal{U}\,q \vee \Box\neg q)\,\mathcal{U}\,q \quad \blacksquare$

(S54) $\neg(p\,\mathcal{U}\,q) \Rightarrow p\,\mathcal{U}\,\neg q$

Proof:

$\quad\quad \neg(p\,\mathcal{U}\,q)$
$\quad = \quad \langle(10)\text{ Expansion of } \mathcal{U}\rangle$
$\quad\quad \neg(q \vee (p \wedge \circ (p\,\mathcal{U}\,q)))$
$\quad = \quad \langle(3.47\text{b})\text{ De Morgan, } \neg(p \vee q) \equiv \neg p \wedge \neg q\rangle$
$\quad\quad \neg q \wedge \neg(p \wedge \circ (p\,\mathcal{U}\,q))$
$\quad \Rightarrow \quad \langle(3.76\text{b})\text{ Strengthening, } p \wedge q \Rightarrow p\rangle$
$\quad\quad \neg q$
$\quad \Rightarrow \quad \langle(29)\ \mathcal{U}\text{ insertion}\rangle$
$\quad\quad p\,\mathcal{U}\,\neg q \quad \blacksquare$

The CDS4LTL paper has four theorems that state equivalences to $\neg(p\,\mathcal{U}\,q)$ and $\neg(p\,\mathcal{W}\,q)$, namely (170), (173), (197) and (198). The following two theorems (S55) and (S56) add to this collection.

(S55) **Distributivity of $\neg$ over $\mathcal{W}$:** $\neg(p\,\mathcal{W}\,q) \equiv \neg q\,\mathcal{W}\,(\neg p \wedge \neg q) \wedge \neg\Box p$

Proof:

$\quad\quad \neg(p \, \mathcal{W} \, q)$
$=\quad \langle (169) \text{ Definition of } \mathcal{W} \rangle$
$\quad\quad \neg(\Box p \lor p \, \mathcal{U} \, q)$
$=\quad \langle (3.47b) \text{ De Morgan, } \neg(p \lor q) \equiv \neg p \land \neg q \rangle$
$\quad\quad \neg \Box p \land \neg(p \, \mathcal{U} \, q)$
$=\quad \langle (173) \text{ Distributivity of } \neg \text{ over } \mathcal{U} \rangle$
$\quad\quad \neg \Box p \land \neg q \, \mathcal{W} \, (\neg p \land \neg q)$
$=\quad \langle (3.36) \text{ Symmetry of } \land, p \land q \equiv q \land p \rangle$
$\quad\quad \neg q \, \mathcal{W} \, (\neg p \land \neg q) \land \neg \Box p \quad \blacksquare$

(S56) Distributivity of $\neg$ over $\mathcal{U}$: $\quad \neg(p \, \mathcal{U} \, q) \equiv \neg q \, \mathcal{U} \, (\neg p \land \neg q) \lor \Box \neg q$

Proof:

$\quad\quad \neg(p \, \mathcal{U} \, q)$
$=\quad \langle (173) \text{ Distributivity of } \neg \text{ over } \mathcal{U} \rangle$
$\quad\quad \neg q \, \mathcal{W} \, (\neg p \land \neg q)$
$=\quad \langle (169) \text{ Definition of } \mathcal{W} \rangle$
$\quad\quad \Box \neg q \lor \neg q \, \mathcal{U} \, (\neg p \land \neg q)$
$=\quad \langle (3.24) \text{ Symmetry of } \lor, p \lor q \equiv q \lor p \rangle$
$\quad\quad \neg q \, \mathcal{U} \, (\neg p \land \neg q) \lor \Box \neg q \quad \blacksquare$

(S57) Weak symmetry of $\mathcal{U}$: $\quad (p \lor q) \, \mathcal{U} \, q \Rightarrow q \, \mathcal{U} \, (p \lor q)$

Proof:

$\quad\quad (p \lor q) \, \mathcal{U} \, q$
$=\quad \langle (190) \text{ Disjunction rule of } \mathcal{U} \rangle$
$\quad\quad p \, \mathcal{U} \, q$
$\Rightarrow\quad \langle \text{Lemma A: } p \, \mathcal{U} \, q \Rightarrow q \, \mathcal{U} \, (p \lor q) \rangle$
$\quad\quad q \, \mathcal{U} \, (p \lor q) \quad \blacksquare$

Lemma A: $p \, \mathcal{U} \, q \Rightarrow q \, \mathcal{U} \, (p \lor q)$

Proof: The proof is by (4.7.1) Truth implication.

$\quad\quad true$
$=\quad \langle \text{Lemma B: } p \, \mathcal{U} \, (q \, \mathcal{U} \, r) \Rightarrow q \, \mathcal{U} \, (p \lor r) \text{ with } r := q \text{ is a theorem} \rangle$
$\quad\quad p \, \mathcal{U} \, (q \, \mathcal{U} \, q) \Rightarrow q \, \mathcal{U} \, (p \lor q)$
$=\quad \langle (22) \text{ Idempotency of } \mathcal{U} \rangle$
$\quad\quad p \, \mathcal{U} \, q \Rightarrow q \, \mathcal{U} \, (p \lor q) \quad \blacksquare$

Lemma B: $p \, \mathcal{U} \, (q \, \mathcal{U} \, r) \Rightarrow q \, \mathcal{U} \, (p \vee r)$

Proof:

$\qquad p \, \mathcal{U} \, (q \, \mathcal{U} \, r)$
$\Rightarrow \quad \langle (28) \text{ with } q := q \, \mathcal{U} \, r \rangle$
$\qquad p \vee q \, \mathcal{U} \, r$
$\Rightarrow \quad \langle (29) \; \mathcal{U} \text{ insertion with } q, p := p, q \text{ and } (4.2) \text{ Monotonicity of } \vee \rangle$
$\qquad q \, \mathcal{U} \, p \vee q \, \mathcal{U} \, r$
$= \quad \langle (12) \text{ Left distributivity of } \mathcal{U} \text{ over } \vee \rangle$
$\qquad q \, \mathcal{U} \, (p \vee r)$ ∎

The next theorem (S58) is a generalized version of the $\mathcal{U}$ excluded middle theorem (23) in CDS4LTL.

(S58) Generalized $\mathcal{U}$ excluded middle: $\quad p \, \mathcal{U} \, q \vee r \, \mathcal{U} \, \neg q$

Proof:

$\qquad p \, \mathcal{U} \, q \vee r \, \mathcal{U} \, \neg q$
$= \quad \langle (10) \text{ Expansion, twice} \rangle$
$\qquad q \vee (p \wedge \circ (p \, \mathcal{U} \, q)) \vee \neg q \vee (r \wedge \circ (r \, \mathcal{U} \, \neg q))$
$= \quad \langle (3.28) \text{ Excluded middle, } p \vee \neg p \rangle$
$\qquad true \vee (p \wedge \circ (p \, \mathcal{U} \, q)) \vee (r \wedge \circ (r \, \mathcal{U} \, \neg q))$
$= \quad \langle (3.29) \text{ Zero of } \vee, p \vee true \equiv true \rangle$
$\qquad true$ ∎

The next theorem (S59) is a Spot equivalence [4].

(S59) $((\Box p \vee \Diamond q) \equiv p \, \mathcal{W} \, \Diamond q) \equiv \Box p \vee \Diamond q \vee \Box \neg q \, \mathcal{U} \, (\neg p \wedge \Box \neg q)$

Proof: The LHS is theorem (176) $\mathcal{W} \Diamond$ equivalence, so we prove the RHS is also a theorem.

$\qquad \Box p \vee \Diamond q \vee \Box \neg q \, \mathcal{U} \, (\neg p \wedge \Box \neg q)$
$= \quad \langle (201) \text{ Dual of } \mathcal{W} \text{ with } p, q := \neg p, \Box \neg q \rangle$
$\qquad \Box p \vee \Diamond q \vee \neg (p \, \mathcal{W} \, \Diamond q)$
$= \quad \langle (176) \; \mathcal{W} \Diamond \text{ equivalence} \rangle$
$\qquad p \, \mathcal{W} \, \Diamond q \vee \neg (p \, \mathcal{W} \, \Diamond q) \quad -(3.28) \text{ Excluded middle, } p \vee \neg p \text{ with } p := p \, \mathcal{W} \, \Diamond q$ ∎

2.8.1 In-state Expansion of $\mathcal{U}$

In CDS4LTL, (10) Expansion of $\mathcal{U}$, defines $p\,\mathcal{U}\,q$ in terms of $\circ(p\,\mathcal{U}\,q)$. This is taken without proof as an axiom in CDS4LTL. We will call this the *next-state expansion* of $\mathcal{U}$. It is also the case that $p\,\mathcal{U}\,q$ can be shown equivalent to a similar expansion formula that does not contain the $\circ$ operator and is a theorem of CDS4LTL. This is theorem (S61). There is a similar theorem for in-state expansion of $\mathcal{W}$ (S84) in section 2.12.

(S60) **In-state next-state equivalence:** $q \vee (p \wedge p\,\mathcal{U}\,q) \equiv q \vee (p \wedge \circ(p\,\mathcal{U}\,q))$

Proof:

$\quad q \vee (p \wedge (p\,\mathcal{U}\,q))$
$= \quad \langle(3.45)\text{ Distributivity of }\vee\text{ over }\wedge, p \vee (q \wedge r) \equiv (p \vee q) \wedge (p \vee r)\rangle$
$\quad (q \vee p) \wedge (q \vee p\,\mathcal{U}\,q)$
$= \quad \langle(10)\text{ Expansion of }\mathcal{U}\rangle$
$\quad (q \vee p) \wedge (q \vee (q \vee (p \wedge \circ(p\,\mathcal{U}\,q))))$
$= \quad \langle(3.25)\text{ Associativity of }\vee, (p \vee q) \vee r \equiv p \vee (q \vee r)$
$\quad\quad\text{ and }(3.26)\text{ Idempotency of }\vee, p \vee p \equiv p\rangle$
$\quad (q \vee p) \wedge (q \vee (p \wedge \circ(p\,\mathcal{U}\,q)))$
$= \quad \langle(3.45)\text{ Distributivity of }\vee\text{ over }\wedge, p \vee (q \wedge r) \equiv (p \vee q) \wedge (p \vee r)\rangle$
$\quad (q \vee p) \wedge (q \vee p) \wedge (q \vee \circ(p\,\mathcal{U}\,q))$
$= \quad \langle(3.38)\text{ Idempotency of }\wedge, p \wedge p \equiv p\rangle$
$\quad (q \vee p) \wedge (q \vee \circ(p\,\mathcal{U}\,q))$
$= \quad \langle(3.45)\text{ Distributivity of }\vee\text{ over }\wedge, p \vee (q \wedge r) \equiv (p \vee q) \wedge (p \vee r)\rangle$
$\quad q \vee (p \wedge \circ(p\,\mathcal{U}\,q))$ ∎

(S61) **In-state expansion of $\mathcal{U}$:** $p\,\mathcal{U}\,q \equiv q \vee (p \wedge p\,\mathcal{U}\,q)$

Proof:

$\quad q \vee (p \wedge p\,\mathcal{U}\,q)$
$= \quad \langle(S60)\text{ In-state next-state equivalence}\rangle$
$\quad q \vee (p \wedge \circ(p\,\mathcal{U}\,q))$
$= \quad \langle(10)\text{ Expansion of }\mathcal{U}\rangle$
$\quad p\,\mathcal{U}\,q$ ∎

Alternate Proof: Requires (S61) come after (35) in the CDS4LTL theorem sequence.

$\quad q \vee (p \wedge p\,\mathcal{U}\,q)$
$= \quad \langle(3.45)\text{ Distributivity of }\vee\text{ over }\wedge, p \vee (q \wedge r) \equiv (p \vee q) \wedge (p \vee r)\rangle$
$\quad (q \vee p) \wedge (q \vee p\,\mathcal{U}\,q)$
$= \quad \langle(32)\text{ Absorption}\rangle$
$\quad (q \vee p) \wedge p\,\mathcal{U}\,q$
$= \quad \langle(35)\text{ Absorption}\rangle$
$\quad p\,\mathcal{U}\,q$ ∎

2.8.2 Nested Insertion

In this section some results from working with the theorems in Table (1) are presented.

(S62) **Nested insertion:** $r \Rightarrow p \,\mathcal{U}\, (q \,\mathcal{U}\, r)$

Proof:

$\quad r$
$\Rightarrow \quad \langle (29) \; \mathcal{U} \text{ insertion with } p, q := q, r \rangle$
$\quad q \,\mathcal{U}\, r$
$\Rightarrow \quad \langle (29) \; \mathcal{U} \text{ insertion with } q := (q \,\mathcal{U}\, r) \rangle$
$\quad p \,\mathcal{U}\, (q \,\mathcal{U}\, r)$ ∎

Theorem Number	Theorem Name	Formula
(17)	Axiom, Right $\mathcal{U} \vee$ ordering	$p \,\mathcal{U}\, (q \,\mathcal{U}\, r) \Rightarrow (p \vee q) \,\mathcal{U}\, r$
(18)	Axiom, Right $\wedge \mathcal{U}$ ordering	$p \,\mathcal{U}\, (q \wedge r) \Rightarrow (p \,\mathcal{U}\, q) \,\mathcal{U}\, r$
(247)	Left $\mathcal{U} \vee$ strengthening	$(p \,\mathcal{U}\, q) \,\mathcal{U}\, r \Rightarrow (p \vee q) \,\mathcal{U}\, r$
(29)	$\mathcal{U}$ insertion	$q \Rightarrow p \,\mathcal{U}\, q$

Table 1: Theorems for nested insertion.

(S63) **Nested insertion:** $r \Rightarrow (p \,\mathcal{U}\, q) \,\mathcal{U}\, r$

Proof:

$\quad r$
$\Rightarrow \quad \langle (29) \; \mathcal{U} \text{ insertion with } q, p := r, (p \,\mathcal{U}\, q) \rangle$
$\quad (p \,\mathcal{U}\, q) \,\mathcal{U}\, r$ ∎

Additionally, what is interesting is that this nesting can be extended indefinitely. Changing notation we write theorems (S62) and (S63) as $x_3 \Rightarrow x_1 \,\mathcal{U}\, (x_2 \,\mathcal{U}\, x_3)$ and $x_3 \Rightarrow (x_1 \,\mathcal{U}\, x_2) \,\mathcal{U}\, x_3$ respectively. Then we claim that the following are also theorems.

(S64) **Indefinite nested insertion:** $x_n \Rightarrow x_1 \,\mathcal{U}\, (x_2 \,\mathcal{U}\, (\ldots \,\mathcal{U}\, (x_{n-1} \,\mathcal{U}\, x_n \underbrace{)\ldots))}_{n-2 \text{ times}}$ for $n \geq 3$

Proof: By mathematical induction.

    Base Case: $n = 3$. Prove $x_3 \Rightarrow x_1 \,\mathcal{U}\, (x_2 \,\mathcal{U}\, x_3)$

Proof:
    $true$
$=$ $\langle$(S62) Nested insertion with $p, q, r := x_1, x_2, x_3\rangle$
    $x_3 \Rightarrow x_1 \,\mathcal{U}\, (x_2 \,\mathcal{U}\, x_3)$ ∎

Induction Step: Prove $x_n \Rightarrow x_1 \,\mathcal{U}\, (x_2 \,\mathcal{U}\, (\ldots \,\mathcal{U}\, (x_{n-1} \,\mathcal{U}\, x_n \underbrace{)\ldots))}_{n-2 \text{ times}}$

assuming $x_{n-1} \Rightarrow x_1 \,\mathcal{U}\, (x_2 \,\mathcal{U}\, (\ldots \,\mathcal{U}\, (x_{n-2} \,\mathcal{U}\, x_{n-1} \underbrace{)\ldots))}_{n-3 \text{ times}}$

as the induction hypothesis.

Proof: The proof is by (4.7.1) Truth implication.
    $true$
$=$ $\langle$Assume the Induction Hypothesis is $true\rangle$
    $x_{n-1} \Rightarrow x_1 \,\mathcal{U}\, (x_2 \,\mathcal{U}\, (\ldots \,\mathcal{U}\, (x_{n-2} \,\mathcal{U}\, x_{n-1} \underbrace{)\ldots))}_{n-3 \text{ times}}$

$=$ $\langle$The above theorem with $x_{n-1} := x_{n-1} \,\mathcal{U}\, x_n\rangle$
    $x_{n-1} \,\mathcal{U}\, x_n \Rightarrow x_1 \,\mathcal{U}\, (x_2 \,\mathcal{U}\, (\ldots \,\mathcal{U}\, (x_{n-2} \,\mathcal{U}\, (x_{n-1} \,\mathcal{U}\, x_n) \underbrace{)\ldots))}_{n-3 \text{ times}}$

$\Rightarrow$ $\langle$(29) $\mathcal{U}$ insertion with $q, p := x_n, x_{n-1}$
       and (3.82a) Transitivity, $(p \Rightarrow q) \land (q \Rightarrow r) \Rightarrow (p \Rightarrow r)\,\rangle$
    $x_n \Rightarrow x_1 \,\mathcal{U}\, (x_2 \,\mathcal{U}\, (\ldots \,\mathcal{U}\, (x_{n-1} \,\mathcal{U}\, x_n \underbrace{)\ldots))}_{n-2 \text{ times}}$ ∎

(S65) **Indefinite nested insertion:** $x_n \Rightarrow \underbrace{(\ldots(}_{n-2 \text{ times}} x_1 \,\mathcal{U}\, x_2) \,\mathcal{U}\, x_3) \ldots \,\mathcal{U}\, x_{n-1}) \,\mathcal{U}\, x_n$

Proof:

    x_n
$\Rightarrow$ $\langle$(29) $\mathcal{U}$ insertion with $q, p := x_n, \underbrace{(\ldots(}_{n-2 \text{ times}} x_1 \,\mathcal{U}\, x_2) \,\mathcal{U}\, x_3) \ldots \,\mathcal{U}\, x_{n-1}\rangle$
    $\underbrace{(\ldots(}_{n-2 \text{ times}} x_1 \,\mathcal{U}\, x_2) \,\mathcal{U}\, x_3) \ldots \,\mathcal{U}\, x_{n-1}) \,\mathcal{U}\, x_n$ ∎

Pepperdine Papers on LTL

2.9 Eventually $\diamond$

This section collects $\diamond$ theorems that were not included in the CDS4LTL paper. The first two theorems (S66), (S67) are from Figure 11 on page 143 in [13]. There they are labeled (lin1) and (4M) respectively.

(S66) $\diamond p \wedge \diamond q \Rightarrow \diamond(p \wedge \diamond q) \vee \diamond(\diamond p \wedge q)$

Proof: The proof is by (4.7.1) Truth implication.

$\qquad$ *true*
$\quad = \quad \langle$(S43) and (3.2) Symmetry of $\equiv$, $p \equiv q \equiv q \equiv p\rangle$
$\qquad \diamond p \wedge \diamond q \equiv \diamond((p \wedge \diamond q) \vee (\diamond p \wedge q))$
$\quad \Rightarrow \quad \langle$(3.82.2) $(p \equiv q) \Rightarrow (p \Rightarrow q)\rangle$
$\qquad \diamond p \wedge \diamond q \Rightarrow \diamond((p \wedge \diamond q) \vee (\diamond p \wedge q))$
$\quad = \quad \langle$(52) Distributivity of $\diamond$ over $\vee\rangle$
$\qquad \diamond p \wedge \diamond q \Rightarrow \diamond(p \wedge \diamond q) \vee \diamond(\diamond p \wedge q)$ ∎

(S67) $\square p \wedge \diamond q \Rightarrow \diamond(\square p \wedge q)$

Proof:

$\qquad$ *true*
$\quad = \quad \langle$(107) $\wedge$ frame law of $\diamond$ with $p := \square p$ is a theorem$\rangle$
$\qquad \square\square p \Rightarrow (\diamond q \Rightarrow \diamond(\square p \wedge q))$
$\quad = \quad \langle$(72) Absorption of $\square\rangle$
$\qquad \square p \Rightarrow (\diamond q \Rightarrow \diamond(\square p \wedge q))$
$\quad = \quad \langle$(3.65) Shunting, $p \wedge q \Rightarrow r \equiv p \Rightarrow (q \Rightarrow r)\rangle$
$\qquad \square p \wedge \diamond q \Rightarrow \diamond(\square p \wedge q))$ ∎

This next theorem (S68) is closely related to theorem (S77).

(S68) $\square(p \Rightarrow \circ(p \Rightarrow q)) \Rightarrow \diamond(p \Rightarrow q)$

Proof:

$\qquad$ *true*
$\quad = \quad \langle$(134) with $q := p \Rightarrow q$ is a theorem$\rangle$
$\qquad \square(p \Rightarrow \circ(p \Rightarrow q)) \Rightarrow (p \Rightarrow \diamond(p \Rightarrow q))$
$\quad = \quad \langle$(105) Distributivity of $\diamond$ over $\Rightarrow\rangle$
$\qquad \square(p \Rightarrow \circ(p \Rightarrow q)) \Rightarrow (p \Rightarrow (\square p \Rightarrow \diamond q))$
$\quad = \quad \langle$(3.65) Shunting, $p \wedge q \Rightarrow r \equiv p \Rightarrow (q \Rightarrow r)\rangle$
$\qquad \square(p \Rightarrow \circ(p \Rightarrow q)) \Rightarrow (p \wedge \square p \Rightarrow \diamond q)$
$\quad = \quad \langle$(68) Absorption of $\wedge$ into $\square\rangle$
$\qquad \square(p \Rightarrow \circ(p \Rightarrow q)) \Rightarrow (\square p \Rightarrow \diamond q)$
$\quad = \quad \langle$(105) Distributivity of $\diamond$ over $\Rightarrow\rangle$
$\qquad \square(p \Rightarrow \circ(p \Rightarrow q)) \Rightarrow \diamond(p \Rightarrow q)$ ∎

Theorem (S69) comes from Spot documentation [4].

(S69) $\Diamond(p\,\mathcal{U}\,q) \equiv \Diamond q$

Proof: The proof is by (4.7) Mutual implication.
The proof in the first direction follows.

$$\begin{aligned}
& \Diamond(p\,\mathcal{U}\,q) \\
= \quad & \langle (39) \text{ Absorption of } \Diamond \text{ into } \mathcal{U} \rangle \\
& \Diamond(p\,\mathcal{U}\,q \wedge \Diamond q) \\
\Rightarrow \quad & \langle (53) \text{ Distributivity of } \Diamond \text{ over } \wedge \rangle \\
& \Diamond(p\,\mathcal{U}\,q) \wedge \Diamond\Diamond q \\
= \quad & \langle (50) \text{ Absorption of } \Diamond \rangle \\
& \Diamond(p\,\mathcal{U}\,q) \wedge \Diamond q \\
\Rightarrow \quad & \langle (3.76b) \text{ Strengthening}, p \wedge q \Rightarrow p \rangle \\
& \Diamond q
\end{aligned}$$

The proof in the second direction follows.

$$\begin{aligned}
& \Diamond(p\,\mathcal{U}\,q) \\
= \quad & \langle (10) \text{ Expansion of } \mathcal{U} \rangle \\
& \Diamond(q \vee (p \wedge \bigcirc(p\,\mathcal{U}\,q))) \\
= \quad & \langle (52) \text{ Distributivity of } \Diamond \text{ over } \vee \rangle \\
& \Diamond q \vee \Diamond(p \wedge \bigcirc(p\,\mathcal{U}\,q)) \\
\Leftarrow \quad & \langle (3.76a) \text{ Weakening}, p \Rightarrow p \vee q \rangle \\
& \Diamond q \quad \blacksquare
\end{aligned}$$

Alternate Proof: The proof is by (4.7) Mutual implication.
The proof in the first direction follows.

$$\begin{aligned}
& \Diamond(p\,\mathcal{U}\,q) \\
\Rightarrow \quad & \langle (42) \text{ Eventuality and } (119) \text{ Monotonicity of } \Diamond \rangle \\
& \Diamond q
\end{aligned}$$

The proof in the second direction follows.

$$\begin{aligned}
& \Diamond(p\,\mathcal{U}\,q) \\
\Leftarrow \quad & \langle (29) \, \mathcal{U} \text{ insertion and } (119) \text{ Monotonicity of } \Diamond \rangle \\
& \Diamond q \quad \blacksquare
\end{aligned}$$

(S70) $p \, \mathcal{U} \, \Diamond q \equiv \Diamond (p \, \mathcal{U} \, q)$

Proof:

$\quad p \, \mathcal{U} \, \Diamond q$
$= \quad \langle (41) \text{ Absorption of } \mathcal{U} \text{ into } \Diamond \rangle$
$\quad \Diamond q$
$= \quad \langle (S69) \, \Diamond (p \, \mathcal{U} \, q) \equiv \Diamond q \rangle$
$\quad \Diamond (p \, \mathcal{U} \, q) \quad \blacksquare$

(S71) $\Diamond q \Rightarrow (p \, \mathcal{W} \, q \equiv p \, \mathcal{U} \, q)$

Proof:

$\quad true$
$= \quad \langle (3.5) \text{ Reflexivity of } \equiv, p \equiv p \rangle$
$\quad p \, \mathcal{U} \, q \equiv p \, \mathcal{U} \, q$
$= \quad \langle (171) \, \mathcal{U} \text{ in terms of } \mathcal{W} \rangle$
$\quad p \, \mathcal{U} \, q \equiv \Diamond q \wedge p \, \mathcal{W} \, q$
$= \quad \langle (39) \text{ Absorption of } \Diamond \text{ into } \mathcal{U} \rangle$
$\quad \Diamond q \wedge p \, \mathcal{U} \, q \equiv \Diamond q \wedge p \, \mathcal{W} \, q$
$= \quad \langle (3.49) \, p \wedge (q \equiv r) \equiv p \wedge q \equiv p \wedge r \equiv p$
$\quad \text{ with } p, q, r := \Diamond q, p \, \mathcal{U} \, q, p \, \mathcal{W} \, q \rangle$
$\quad \Diamond q \equiv \Diamond q \wedge (p \, \mathcal{U} \, q \equiv p \, \mathcal{W} \, q)$
$= \quad \langle (3.60) \text{ Implication}, p \Rightarrow q \equiv p \wedge q \equiv p \text{ and}$
$\quad (3.2) \text{ Symmetry of } \equiv, p \equiv q \equiv q \equiv p \rangle$
$\quad \Diamond q \Rightarrow (p \, \mathcal{W} \, q \equiv p \, \mathcal{U} \, q) \quad \blacksquare$

2.10 Always $\Box$

This section collects $\Box$ theorems that were not included in the CDS4LTL paper. The first two theorems (S72), (S73) are from Figure 11 on page 143 in [13]. There they are labeled (H) and (H+) respectively. They are also labeled (H) and (H+) in Goré [7, 8].

(S72) $\Box (p \vee q) \wedge \Box (\Box p \vee q) \wedge \Box (p \vee \Box q) \Rightarrow \Box p \vee \Box q$

Proof:

$\quad \Box (p \vee q) \wedge \Box (\Box p \vee q) \wedge \Box (p \vee \Box q)$
$= \quad \langle (165) \, \Box ((p \vee \Box q) \wedge (\Box p \vee q)) \equiv \Box p \vee \Box q \rangle$
$\quad \Box (p \vee q) \wedge (\Box p \vee \Box q)$
$\Rightarrow \quad \langle (3.76b) \text{ Strengthening}, p \wedge q \Rightarrow p \rangle$
$\quad \Box p \vee \Box q \quad \blacksquare$

(S73) $\Box(\Box p \vee q) \wedge \Box(p \vee \Box q) \Rightarrow \Box p \vee \Box q$

Proof: The proof is by (4.7.1) Truth implication.

$\quad\quad true$
$= \quad \langle(165) \text{ and } (99) \text{ Distributivity of } \Box \text{ over } \wedge \text{ is a theorem}\rangle$
$\quad\quad \Box(\Box p \vee q) \wedge \Box(p \vee \Box q) \equiv \Box p \vee \Box q$
$\Rightarrow \quad \langle(3.82.2) \ (p \equiv q) \Rightarrow (p \Rightarrow q)\rangle$
$\quad\quad \Box(\Box p \vee q) \wedge \Box(p \vee \Box q) \Rightarrow \Box p \vee \Box q \quad \blacksquare$

The next two theorems (S74) (S75) come from [13] where they are labeled (L) and (L++). They are both proved using the theorem (255) Lemmon formula, which is labeled (L+) in [13]. These formulas are also labeled (L), (L+) and (L++) in Goré [7, 8].

(S74) $\Box(p \wedge \Box p \Rightarrow q) \vee \Box(q \wedge \Box q \Rightarrow p)$

Proof:

$\quad\quad \Box(p \wedge \Box p \Rightarrow q) \vee \Box(q \wedge \Box q \Rightarrow p)$
$= \quad \langle(68) \text{ Absorption of } \wedge \text{ into } \Box, \text{ twice}\rangle$
$\quad\quad \Box(\Box p \Rightarrow q) \vee \Box(\Box q \Rightarrow p) \quad -(255) \text{ Lemmon formula} \quad \blacksquare$

(S75) $\Box(\Box p \Rightarrow \Box q) \vee \Box(\Box q \Rightarrow \Box p)$

Proof:

$\quad\quad true$
$= \quad \langle(255) \text{ Lemmon formula with } p, q := \Box p, \Box q \text{ is a theorem}\rangle$
$\quad\quad \Box(\Box \Box p \Rightarrow \Box q) \vee \Box(\Box \Box q \Rightarrow \Box p)$
$= \quad \langle(72) \text{ Absorption of } \Box, \text{ twice}\rangle$
$\quad\quad \Box(\Box p \Rightarrow \Box q) \vee \Box(\Box q \Rightarrow \Box p) \quad \blacksquare$

The next theorem (S76) shows another temporal formula that is equivalent to $\Box p$. First, two lemmas are proved. The first lemma is a propositional calculus theorem.

Lemma A: $((p \Rightarrow q) \Rightarrow q) \wedge ((q \Rightarrow p) \Rightarrow p) \equiv p \vee q$

Proof:

$\quad\quad ((p \Rightarrow q) \Rightarrow q) \wedge ((q \Rightarrow p) \Rightarrow p)$
$= \quad \langle(3.59) \text{ Implication, } p \Rightarrow q \equiv \neg p \vee q, \text{ four times}\rangle$
$\quad\quad (\neg(\neg p \vee q) \vee q) \wedge (\neg(\neg q \vee p) \vee p)$
$= \quad \langle(3.47b) \text{ De Morgan, } \neg(p \vee q) \equiv \neg p \wedge \neg q, \text{ twice}\rangle$
$\quad\quad ((p \wedge \neg q) \vee q) \wedge ((q \wedge \neg p) \vee p)$
$= \quad \langle(3.44b) \text{ Absorption, } p \vee (\neg p \wedge q) \equiv p \vee q, \text{ twice}\rangle$
$\quad\quad (p \vee q) \wedge (p \vee q)$
$= \quad \langle(3.38) \text{ Idempotency of } \wedge, p \wedge p \equiv p\rangle$
$\quad\quad p \vee q \quad \blacksquare$

Pepperdine Papers on LTL

The second lemma is a temporal logic theorem. Lemma B does not follow from lemma A, but it uses the same proof strategy until the last step.

Lemma B: $\Box(((p \Rightarrow \Box q) \Rightarrow \Box q) \land ((q \Rightarrow \Box p) \Rightarrow \Box p)) \equiv \Box p \lor \Box q$

Proof:

$\quad\quad \Box(((p \Rightarrow \Box q) \Rightarrow \Box q) \land ((q \Rightarrow \Box p) \Rightarrow \Box p))$
$= \quad \langle(3.59)\text{ Implication, } p \Rightarrow q \equiv \neg p \lor q, \text{four times}\rangle$
$\quad\quad \Box((\neg(\neg p \lor \Box q) \lor \Box q) \land (\neg(\neg q \lor \Box p) \lor \Box p))$
$= \quad \langle(3.47b)\text{ De Morgan, }\neg(p \lor q) \equiv \neg p \land \neg q, \text{twice}\rangle$
$\quad\quad \Box(((p \land \neg\Box q) \lor \Box q) \land ((q \land \neg\Box p) \lor \Box p))$
$= \quad \langle(3.44b)\text{ Absorption, } p \lor (\neg p \land q) \equiv p \lor q, \text{twice}\rangle$
$\quad\quad \Box((p \lor \Box q) \land (\Box p \lor q))$
$= \quad \langle(165)\ \Box((p \lor \Box q) \land (\Box p \lor q)) \equiv \Box p \lor \Box q\rangle$
$\quad\quad \Box p \lor \Box q$ ∎

Now a proof of the main result, Theorem (S76), can be constructed using lemma B as follows.

(S76) $\Box((p \Rightarrow \Box p) \Rightarrow \Box p) \equiv \Box p$

Proof:

$\quad\quad \textit{true}$
$= \quad \langle\text{Lemma B: } \Box(((p \Rightarrow \Box q) \Rightarrow \Box q) \land ((q \Rightarrow \Box p) \Rightarrow \Box p)) \equiv \Box p \lor \Box q \text{ with } q := p\rangle$
$\quad\quad \Box(((p \Rightarrow \Box p) \Rightarrow \Box p) \land ((p \Rightarrow \Box p) \Rightarrow \Box p)) \equiv \Box p \lor \Box p$
$= \quad \langle(3.38)\text{ Idempotency of } \land, p \land p \equiv p \text{ and } (3.26)\text{ Idempotency of } \lor, p \lor p \equiv p\rangle$
$\quad\quad \Box((p \Rightarrow \Box p) \Rightarrow \Box p) \equiv \Box p$ ∎

2.11 Always Eventually $\Box\Diamond$ and its Dual $\Diamond\Box$

This section collects $\Box\Diamond$ theorems and $\Diamond\Box$ theorems that did not make it into the CDS4LTL paper. Note that (S68) follows easily from (S77). And, as the alternate proof shows, (S77) follows easily from (S68).

(S77) $\Box(p \Rightarrow \circ(p \Rightarrow q)) \Rightarrow \Box\Diamond(p \Rightarrow q)$

Proof:

$\quad\quad \Box(p \Rightarrow \circ(p \Rightarrow q))$
$= \quad \langle(158)\text{ Monotonicity of } \Diamond\Box\rangle$
$\quad\quad \Diamond\Box p \Rightarrow \Diamond\Box\circ(p \Rightarrow q)$

$\quad = \quad \langle(51)$ Exchange of $\circ$ and $\diamond$ and (73) Exchange of $\circ$ and $\square\rangle$
$\qquad \diamond\square p \Rightarrow \circ\diamond\square(p \Rightarrow q)$
$\quad = \quad \langle(156)$ Absorption of $\circ$ into $\diamond\square\rangle$
$\qquad \diamond\square p \Rightarrow \diamond\square(p \Rightarrow q)$
$\quad = \quad \langle(3.59)$ Implication, $p \Rightarrow q \equiv \neg p \vee q\rangle$
$\qquad \diamond\square p \Rightarrow \diamond\square(\neg p \vee q)$
$\quad \Rightarrow \quad \langle(145) \diamond\square$ implication and $(3.82a)$ Transitivity$\rangle$
$\qquad \diamond\square p \Rightarrow \square\diamond(\neg p \vee q)$
$\quad = \quad \langle(161)$ Distributivity of $\square\diamond$ over $\vee\rangle$
$\qquad \diamond\square p \Rightarrow \square\diamond\neg p \vee \square\diamond q$
$\quad = \quad \langle(62)$ Dual of $\diamond\square\rangle$
$\qquad \diamond\square p \Rightarrow \neg\diamond\square p \vee \square\diamond q$
$\quad = \quad \langle(3.59)$ Implication, $p \Rightarrow q \equiv \neg p \vee q$ and (3.26) Idempotency of $\vee$, $p \vee p \equiv p\rangle$
$\qquad \neg\diamond\square p \vee \square\diamond q$
$\quad = \quad \langle(62)$ Dual of $\diamond\square\rangle$
$\qquad \square\diamond\neg p \vee \square\diamond q$
$\quad = \quad \langle(161)$ Distributivity of $\square\diamond$ over $\vee$ and (3.59) Implication, $p \Rightarrow q \equiv \neg p \vee q\rangle$
$\qquad \square\diamond(p \Rightarrow q) \quad \blacksquare$

Alternate Proof:

$\qquad true$
$\quad = \quad \langle(S68)$ is a theorem$\rangle$
$\qquad \square(p \Rightarrow \circ(p \Rightarrow q)) \Rightarrow \diamond(p \Rightarrow q)$
$\quad = \quad \langle(139)$ Metatheorem $\square$ with the above theorem$\rangle$
$\qquad \square\square(p \Rightarrow \circ(p \Rightarrow q)) \Rightarrow \square\diamond(p \Rightarrow q)$
$\quad = \quad \langle(72)$ Absorption of $\square\rangle$
$\qquad \square(p \Rightarrow \circ(p \Rightarrow q)) \Rightarrow \square\diamond(p \Rightarrow q) \quad \blacksquare$

The next 6 theorems (S78), (S79), (S80), (S81), (S82) and (S83) come from Spot [4].

(S78) $\quad \square\diamond(p \vee \circ q) \equiv \square\diamond(p \vee q)$

Proof:

$\qquad \square\diamond(p \vee \circ q)$
$\quad = \quad \langle(161)$ Distributivity of $\square\diamond$ over $\vee\rangle$
$\qquad \square\diamond p \vee \square\diamond\circ q$
$\quad = \quad \langle(51)$ Exchange of $\circ$ and $\diamond$ and (73) Exchange of $\circ$ and $\square\rangle$

$\square\diamond p \vee \circ \square\diamond q$
= ⟨(155) Absorption of $\circ$ into $\square\diamond$⟩
$\square\diamond p \vee \square\diamond q$
= ⟨(161) Distributivity of $\square\diamond$ over $\vee$⟩
$\square\diamond(p \vee q)$ ∎

(S79) $\diamond\square(p \wedge \circ q) \equiv \diamond\square(p \wedge q)$

Proof:

$\diamond\square(p \wedge \circ q)$
= ⟨(162) Distributivity of $\diamond\square$ over $\wedge$⟩
$\diamond\square p \wedge \diamond\square \circ q$
= ⟨(51) Exchange of $\circ$ and $\diamond$ and (73) Exchange of $\circ$ and $\square$⟩
$\diamond\square p \wedge \circ\diamond\square q$
= ⟨(156) Absorption of $\circ$ into $\diamond\square$⟩
$\diamond\square p \wedge \diamond\square q$
= ⟨(162) Distributivity of $\diamond\square$ over $\wedge$⟩
$\diamond\square(p \wedge q)$ ∎

(S80) $\square\diamond(p \vee \diamond q) \equiv \square\diamond(p \vee q)$

Proof:

$\square\diamond(p \vee \diamond q)$
= ⟨(161) Distributivity of $\square\diamond$ over $\vee$⟩
$\square\diamond p \vee \square\diamond\diamond q$
= ⟨(50) Absorption of $\diamond$⟩
$\square\diamond p \vee \square\diamond q$
= ⟨(161) Distributivity of $\square\diamond$ over $\vee$⟩
$\square\diamond(p \vee q)$ ∎

(S81) $\diamond\square(p \wedge \square q) \equiv \diamond\square(p \wedge q)$

Proof:

$\diamond\square(p \wedge \square q)$
= ⟨(162) Distributivity of $\diamond\square$ over $\wedge$⟩
$\diamond\square p \wedge \diamond\square\square q$
= ⟨(72) Absorption of $\square$⟩
$\diamond\square p \wedge \diamond\square q$
= ⟨(162) Distributivity of $\diamond\square$ over $\wedge$⟩
$\diamond\square(p \wedge q)$ ∎

180   *CDS4LTL: Additional Theorems*

(S82) $\circ p \vee \square \diamond q \equiv \circ (p \vee \square \diamond q)$

Proof:

$\quad\quad \circ p \vee \square \diamond q$
$=\quad \langle(155) \text{ Absorption of } \circ \text{ into } \square \diamond \rangle$
$\quad\quad \circ p \vee \circ \square \diamond q$
$=\quad \langle(4) \text{ Distributivty of } \circ \text{ over } \vee \rangle$
$\quad\quad \circ (p \vee \square \diamond q)$ ∎

(S83) $\circ p \wedge \diamond \square q \equiv \circ (p \wedge \diamond \square q)$

Proof:

$\quad\quad \circ p \wedge \diamond \square q$
$=\quad \langle(156) \text{ Absorption of } \circ \text{ into } \diamond \square \rangle$
$\quad\quad \circ p \wedge \circ \diamond \square q$
$=\quad \langle(5) \text{ Distributivity of } \circ \text{ over } \wedge \rangle$
$\quad\quad \circ (p \wedge \diamond \square q)$ ∎

2.12 Wait $\mathcal{W}$

This section collects $\mathcal{W}$ theorems that were not included in the CDS4LTL paper. Similar to (S61) $\mathcal{W}$ has an in-state expansion. The proof of this makes use of the theorem (S61) In-state expansion of $\mathcal{U}$.

(S84) **In-state expansion of** $\mathcal{W}$: $p \mathcal{W} q \equiv q \vee (p \wedge p \mathcal{W} q)$

Proof:

$\quad\quad q \vee (p \wedge p \mathcal{W} q)$
$=\quad \langle(169) \text{ Definition of } \mathcal{W} \rangle$
$\quad\quad q \vee (p \wedge (\square p \vee p \mathcal{U} q))$
$=\quad \langle(3.46) \text{ Distributivity of } \wedge \text{ over } \vee, p \wedge (q \vee r) \equiv (p \wedge q) \vee (p \wedge r)\rangle$
$\quad\quad q \vee (p \wedge \square p) \vee (p \wedge p \mathcal{U} q)$
$=\quad \langle(68) \text{ Absorption of } \wedge \text{ into } \square \rangle$
$\quad\quad q \vee \square p \vee (p \wedge p \mathcal{U} q)$
$=\quad \langle(S61) \text{ In-state expansion of } \mathcal{U} \text{ and } (3.24) \text{ Symmetry of } \vee, p \vee q \equiv q \vee p \rangle$
$\quad\quad \square p \vee p \mathcal{U} q$
$=\quad \langle(169) \text{ Definition of } \mathcal{W} \rangle$
$\quad\quad p \mathcal{W} q$ ∎

Alternate Proof: Requires (S84) come after (220) in the CDS4LTL theorem sequence.

$$q \vee (p \wedge p \mathcal{W} q)$$
$= \quad \langle(3.45) \text{ Distributivity of } \vee \text{ over } \wedge, p \vee (q \wedge r) \equiv (p \vee q) \wedge (p \vee r)\rangle$
$$(q \vee p) \wedge (q \vee p \mathcal{W} q)$$
$= \quad \langle(218) \text{ Absorption}\rangle$
$$(q \vee p) \wedge p \mathcal{W} q$$
$= \quad \langle(220) \text{ Absorption}\rangle$
$$p \mathcal{W} q \quad \blacksquare$$

$\mathcal{W}$ Theorem (S85) is inspired by $\mathcal{U}$ Theorem (S54). While the formulas have identical forms, with $\mathcal{W}$ substituted for $\mathcal{U}$, the proofs are quite different.

(S85) $\neg(p \mathcal{W} q) \Rightarrow p \mathcal{W} \neg q$

Proof:

$$true$$
$= \quad \langle(182) \, \mathcal{W} \text{ excluded middle}\rangle$
$$p \mathcal{W} q \vee p \mathcal{W} \neg q$$
$= \quad \langle(3.12) \text{ Double negation}, \neg\neg p \equiv p\rangle$
$$\neg\neg(p \mathcal{W} q) \vee p \mathcal{W} \neg q$$
$= \quad \langle(3.59) \text{ Implication}, p \Rightarrow q \equiv \neg p \vee q \text{ with } p, q := \neg(p \mathcal{W} q), p \mathcal{W} \neg q\rangle$
$$\neg(p \mathcal{W} q) \Rightarrow p \mathcal{W} \neg q \quad \blacksquare$$

(S86) $p \mathcal{W} \Box q \wedge \Diamond \Box q \Rightarrow \Box (p \mathcal{W} q)$

Proof: The proof is by (4.7.1) Truth implication.

$$true$$
$= \quad \langle(140) \, \mathcal{U} \Box \text{ implication is a theorem}\rangle$
$$p \, \mathcal{U} \Box q \Rightarrow \Box (p \, \mathcal{U} q)$$
$= \quad \langle(171) \, \mathcal{U} \text{ in terms of } \mathcal{W}, \text{twice}\rangle$
$$p \mathcal{W} \Box q \wedge \Diamond \Box q \Rightarrow \Box (p \mathcal{W} q \wedge \Diamond q)$$
$= \quad \langle(99) \text{ Distributivity of } \Box \text{ over } \wedge\rangle$
$$p \mathcal{W} \Box q \wedge \Diamond \Box q \Rightarrow \Box (p \mathcal{W} q) \wedge \Box \Diamond q)$$
$\Rightarrow \quad \langle(3.76b) \text{ Strengthening}, p \wedge q \Rightarrow p \text{ and } (3.82a) \text{ Transitivity}\rangle$
$$p \mathcal{W} \Box q \wedge \Diamond \Box q \Rightarrow \Box (p \mathcal{W} q) \quad \blacksquare$$

The next theorem (S87) is a generalized version of the $\mathcal{W}$ excluded middle theorem (182) in CDS4LTL.

(S87) **Generalized $\mathcal{W}$ excluded middle:** $p \mathcal{W} q \vee r \mathcal{W} \neg q$

Proof:

$\qquad p \mathcal{W} q \vee r \mathcal{W} \neg q$
$= \quad \langle (181) \text{ Expansion, twice} \rangle$
$\qquad q \vee (p \wedge \bigcirc (p \mathcal{W} q)) \vee \neg q \vee (r \wedge \bigcirc (r \mathcal{W} \neg q))$
$= \quad \langle (3.28) \text{ Excluded middle}, p \vee \neg p \rangle$
$\qquad true \vee (p \wedge \bigcirc (p \mathcal{W} q)) \vee (r \wedge \bigcirc (r \mathcal{W} \neg q))$
$= \quad \langle (3.29) \text{ Zero of } \vee, p \vee true \equiv true \rangle$
$\qquad true \quad \blacksquare$

(S88) $p \mathcal{W} q \equiv \Diamond \neg p \Rightarrow p \mathcal{U} q$

Proof:

$\qquad \Diamond \neg p \Rightarrow p \mathcal{U} q$
$= \quad \langle (60) \text{ Dual of } \Box \rangle$
$\qquad \neg \Box p \Rightarrow p \mathcal{U} q$
$= \quad \langle (3.59) \text{ Implication}, p \Rightarrow q \equiv \neg p \vee q \text{ with } p,q := \neg \Box p, p \mathcal{U} q \rangle$
$\qquad \neg \neg \Box p \vee p \mathcal{U} q$
$= \quad \langle (3.12) \text{ Double negation}, \neg \neg p \equiv p \rangle$
$\qquad \Box p \vee p \mathcal{U} q$
$= \quad \langle (169) \text{ Definition of } \mathcal{W} \rangle$
$\qquad p \mathcal{W} q \quad \blacksquare$

The next theorem (S89) comes from Spot documentation [4].

(S89) $p \mathcal{W} q \equiv p \mathcal{U} (q \vee \Box p)$

Proof:

$\qquad p \mathcal{U} (q \vee \Box p)$
$= \quad \langle (12) \text{ Left distributivity of } \mathcal{U} \text{ over } \vee \rangle$
$\qquad p \mathcal{U} q \vee p \mathcal{U} \Box p$
$= \quad \langle (141) \text{ Absorption of } \mathcal{U} \text{ into } \Box \rangle$
$\qquad p \mathcal{U} q \vee \Box p$
$= \quad \langle (169) \text{ Definition of } \mathcal{W} \rangle$
$\qquad p \mathcal{W} q \quad \blacksquare$

(S90) $q \mathcal{W} \Box \neg p \Rightarrow (\Diamond p \Rightarrow q)$

Proof:

$\quad q \mathcal{W} \Box \neg p$
$\Rightarrow \quad \langle (206)\ p \mathcal{W} q \Rightarrow p \vee q \rangle$
$\quad q \vee \Box \neg p$
$= \quad \langle (61)\ \text{Dual of } \Diamond \rangle$
$\quad q \vee \neg \Diamond p$
$= \quad \langle (3.59)\ \text{Implication},\ p \Rightarrow q \equiv \neg p \vee q \text{ with } p := \Diamond p \rangle$
$\quad \Diamond p \Rightarrow q \quad \blacksquare$

(S91) $q \mathcal{W} \Box \neg q \Rightarrow (\Box (\circ q \Rightarrow q) \Rightarrow (\Diamond q \Rightarrow q))$

Proof:

$\quad q \mathcal{W} \Box \neg q \Rightarrow (\Box (\circ q \Rightarrow q) \Rightarrow (\Diamond q \Rightarrow q))$
$= \quad \langle (3.65)\ \text{Shunting},\ p \wedge q \Rightarrow r \equiv p \Rightarrow (q \Rightarrow r) \rangle$
$\quad q \mathcal{W} \Box \neg q \wedge \Box (\circ q \Rightarrow q) \Rightarrow (\Diamond q \Rightarrow q)$

And now,

$\quad q \mathcal{W} \Box \neg q \wedge \Box (\circ q \Rightarrow q)$
$\Rightarrow \quad \langle (3.76b)\ \text{Strengthening},\ p \wedge q \Rightarrow p \rangle$
$\quad q \mathcal{W} \Box \neg q$
$\Rightarrow \quad \langle (S90)\ \text{with } p := q \rangle$
$\quad \Diamond q \Rightarrow q \quad \blacksquare$

The next five theorems, (S92), (S93), (S94), (S95), (S96) come from Spot documentation [4]. Theorem (S92) could also be placed in section 2.4 with the Absorption theorems.

(S92) $p \mathcal{W} \Box p \equiv \Box p$

Proof:

$\quad p \mathcal{W} \Box p$
$= \quad \langle (169)\ \text{Definition of } \mathcal{W} \rangle$
$\quad \Box p \vee p \mathcal{U} \Box p$
$= \quad \langle (141)\ \text{Absorption of } \mathcal{U} \text{ into } \Box \rangle$
$\quad \Box p \vee \Box p$
$= \quad \langle (3.26)\ \text{Idempotency of } \vee,\ p \vee p \equiv p \rangle$
$\quad \Box p \quad \blacksquare$

(S93) $\square p \mathcal{W} q \equiv \square p \vee q$

Proof:

$\quad\quad \square p \mathcal{W} q$
$= \quad \langle (169) \text{ Definition of } \mathcal{W} \rangle$
$\quad\quad \square \square p \vee \square p \mathcal{U} q$
$= \quad \langle (72) \text{ Absorption of } \square \rangle$
$\quad\quad \square p \vee \square p \mathcal{U} q$
$= \quad \langle (31) \text{ Absorption} \rangle$
$\quad\quad \square p \vee q \quad \blacksquare$

(S94) $\square(p \mathcal{W} q) \equiv \square(p \vee q)$

Proof: The proof is by (4.7) Mutual implication.
The proof in the first direction follows.

$\quad\quad true$
$= \quad \langle (206) \ p \mathcal{W} q \Rightarrow p \vee q \rangle$
$\quad\quad p \mathcal{W} q \Rightarrow p \vee q$
$= \quad \langle (139) \text{ Metatheorem } \square \text{ with the above theorem} \rangle$
$\quad\quad \square(p \mathcal{W} q) \Rightarrow \square(p \vee q)$

The proof in the second direction follows.

$\quad\quad true$
$= \quad \langle (207) \ \square(p \vee q) \Rightarrow p \mathcal{W} q \rangle$
$\quad\quad \square(p \vee q) \Rightarrow p \mathcal{W} q$
$= \quad \langle (139) \text{ Metatheorem } \square \text{ with the above theorem} \rangle$
$\quad\quad \square \square(p \vee q) \Rightarrow \square(p \mathcal{W} q)$
$= \quad \langle (72) \text{ Absorption of } \square \rangle$
$\quad\quad \square(p \vee q) \Rightarrow \square(p \mathcal{W} q) \quad \blacksquare$

(S95) $p \mathcal{W} q \equiv p \mathcal{W} (q \vee \square p)$

Proof:

$\quad\quad p \mathcal{W} (q \vee \square p)$
$= \quad \langle (184) \text{ Left distributivity of } \mathcal{W} \text{ over } \vee \rangle$
$\quad\quad p \mathcal{W} q \vee p \mathcal{W} \square p$
$= \quad \langle (S92) \ p \mathcal{W} \square p \equiv \square p \rangle$
$\quad\quad p \mathcal{W} q \vee \square p$
$= \quad \langle (233) \text{ Absorption of } \square \text{ into } \mathcal{W} \rangle$
$\quad\quad p \mathcal{W} q \quad \blacksquare$

Pepperdine Papers on LTL

(S96) $\quad p\,\mathcal{W}\,(q\vee\Diamond r)\equiv p\,\mathcal{W}\,q\vee\Diamond r$

Proof:

$\quad\quad p\,\mathcal{W}\,(q\vee\Diamond r)$
$=\quad \langle(184)\text{ Left distributivity of }\mathcal{W}\text{ over }\vee\rangle$
$\quad\quad p\,\mathcal{W}\,q\vee p\,\mathcal{W}\,\Diamond r$
$=\quad \langle(176)\;\mathcal{W}\Diamond\text{ equivalence}\rangle$
$\quad\quad p\,\mathcal{W}\,q\vee\Box p\vee\Diamond r$
$=\quad \langle(233)\text{ Absorption of }\Box\text{ into }\mathcal{W}\,\rangle$
$\quad\quad p\,\mathcal{W}\,q\vee\Diamond r\quad\blacksquare$

Theorems (S97) (S98) come from Spot documentation [4].

(S97) $\quad p\,\mathcal{U}\,r\wedge q\,\mathcal{W}\,r\equiv (p\wedge q)\,\mathcal{U}\,r$

Proof:

$\quad\quad (p\wedge q)\,\mathcal{U}\,r$
$=\quad \langle(15)\text{ Right distributivity of }\mathcal{U}\text{ over }\wedge\rangle$
$\quad\quad p\,\mathcal{U}\,r\wedge q\,\mathcal{U}\,r$
$=\quad \langle(171)\;\mathcal{U}\text{ in terms of }\mathcal{W}\,\rangle$
$\quad\quad p\,\mathcal{U}\,r\wedge q\,\mathcal{W}\,r\wedge\Diamond r$
$=\quad \langle(39)\text{Absorption of }\Diamond\text{ into }\mathcal{U}\,\rangle$
$\quad\quad p\,\mathcal{U}\,r\wedge q\,\mathcal{W}\,r\quad\blacksquare$

(S98) $\quad p\,\mathcal{U}\,q\vee p\,\mathcal{W}\,r\equiv p\,\mathcal{W}\,(q\vee r)$

Proof:

$\quad\quad p\,\mathcal{W}\,(q\vee r)$
$=\quad \langle(184)\text{ Left distributivity of }\mathcal{W}\text{ over }\vee\rangle$
$\quad\quad p\,\mathcal{W}\,q\vee p\,\mathcal{W}\,r$
$=\quad \langle(169)\text{ Definition of }\mathcal{W}\,\rangle$
$\quad\quad \Box p\vee p\,\mathcal{U}\,q\vee p\,\mathcal{W}\,r$
$=\quad \langle(233)\text{ Absorption of }\Box\text{ into }\mathcal{W}\,\rangle$
$\quad\quad p\,\mathcal{U}\,q\vee p\,\mathcal{W}\,r\quad\blacksquare$

(S99) $\quad p\,\mathcal{W}\,q\vee\Diamond q\equiv \Box p\vee\Diamond q$

Proof:

$$p \mathcal{W} q \vee \Diamond q$$
$= \quad \langle (169) \text{ Definition of } \mathcal{W} \rangle$
$$\Box p \vee p \,\mathcal{U}\, q \vee \Diamond q$$
$= \quad \langle (171) \,\mathcal{U}\, \text{ in terms of } \mathcal{W} \rangle$
$$\Box p \vee (p \mathcal{W} q \wedge \Diamond q) \vee \Diamond q$$
$= \quad \langle (3.43\text{b}) \text{ Absorption}, p \vee (p \wedge q) \equiv p \text{ with } p, q := \Diamond q, p \mathcal{W} q \rangle$
$$\Box p \vee \Diamond q \quad \blacksquare$$

The next theorem is a corollary of (S99).

(S100) $\quad p \mathcal{W} \Diamond q \equiv p \mathcal{W} q \vee \Diamond q$

Proof:

$$p \mathcal{W} q \vee \Diamond q$$
$= \quad \langle (\text{S99}) \quad p \mathcal{W} q \vee \Diamond q \equiv \Box p \vee \Diamond q \rangle$
$$\Box p \vee \Diamond q$$
$= \quad \langle (176) \, \mathcal{W} \Diamond \text{ equivalence}, p \mathcal{W} \Diamond q \equiv \Box p \vee \Diamond q \rangle$
$$p \mathcal{W} \Diamond q \quad \blacksquare$$

The following theorems (S101), (S102) and (S103) all rewrite temporal $\mathcal{W}$ formulas into non-temporal propositional formulas.

(S101) $\quad p \mathcal{W} q \vee q \mathcal{W} p \equiv p \vee q$

Proof: The proof is by (4.7) Mutual implication.
The proof in the first direction follows.

$$p \mathcal{W} q \vee q \mathcal{W} p$$
$\Rightarrow \quad \langle (206) \, p \mathcal{W} q \Rightarrow p \vee q \text{ and } (4.2) \text{ Monotonicity of } \vee, \text{ twice} \rangle$
$$(p \vee q) \vee (q \vee p)$$
$= \quad \langle (3.25) \text{ Associativity of } \vee, (p \vee q) \vee r \equiv p \vee (q \vee r)$
$\quad \quad \text{and } (3.26) \text{ Idempotency of } \vee, p \vee p \equiv p \rangle$
$$p \vee q$$

The proof in the second direction follows.

$$p \vee q$$
$\Rightarrow \quad \langle (209) \, \mathcal{W} \text{ insertion and } (4.2) \text{ Monotonicity of } \vee, \text{ twice} \rangle$
$$q \mathcal{W} p \vee p \mathcal{W} q$$
$= \quad \langle (3.24) \text{ Symmetry of } \vee, p \vee q \equiv q \vee p \rangle$
$$p \mathcal{W} q \vee q \mathcal{W} p \quad \blacksquare$$

(S102) $\neg p \mathcal{W} q \vee q \mathcal{W} \neg p \equiv p \Rightarrow q$

Proof:

$\qquad \neg p \mathcal{W} q \vee q \mathcal{W} \neg p$
$= \quad \langle (S101) \text{ with } p := \neg p \rangle$
$\qquad \neg p \vee q$
$= \quad \langle (3.59) \text{ Implication}, p \Rightarrow q \equiv \neg p \vee q \rangle$
$\qquad p \Rightarrow q \quad \blacksquare$

(S103) $(\neg p \mathcal{W} q \vee q \mathcal{W} \neg p) \wedge (\neg q \mathcal{W} p \vee p \mathcal{W} \neg q) \equiv (p \equiv q)$

Proof:

$\qquad (\neg p \mathcal{W} q \vee q \mathcal{W} \neg p) \wedge (\neg q \mathcal{W} p \vee p \mathcal{W} \neg q)$
$= \quad \langle (S102), \text{twice} \rangle$
$\qquad (p \Rightarrow q) \wedge (q \Rightarrow p)$
$= \quad \langle (3.80) \text{ Mutual implication}, (p \Rightarrow q) \wedge (q \Rightarrow p) \equiv (p \equiv q) \rangle$
$\qquad p \equiv q \quad \blacksquare$

To add a theorem to the collection, (S101), (S102) and (S103), which rewrites a similar $\mathcal{W}$ formula to $p \wedge q$, (2.3b) Metatheorem Duality can be used. Starting with (S101) $p \mathcal{W} q \vee q \mathcal{W} p \equiv p \vee q$, by (2.3b) Metatheorem Duality and theorem (S41) Definition of strong release $\mathcal{M}$, the following is a theorem.

(S104) $q \mathcal{U} (p \wedge q) \wedge p \mathcal{U} (p \wedge q) \equiv p \wedge q$

Theorems (S104) and (171) $\mathcal{U}$ in terms of $\mathcal{W}$ lead to the following theorem (S105) which shows an equivalent theorem to (S104) in terms of $\mathcal{W}$

(S105) $q \mathcal{W} (p \wedge q) \wedge p \mathcal{W} (p \wedge q) \wedge \Diamond (p \wedge q) \equiv p \wedge q$

Finally, we can show an even simpler theorem (S106) for rewriting to $p \wedge q$ from a $\mathcal{W}$ formula as follows.

(S106) $p \mathcal{W} (p \wedge q) \wedge q \mathcal{W} (p \wedge q) \equiv p \wedge q$

Proof:

$\qquad p \mathcal{W} (p \wedge q) \wedge q \mathcal{W} (p \wedge q)$
$= \quad \langle (187) \text{ Right distributivity of } \mathcal{W} \text{ over } \wedge \rangle$
$\qquad (p \wedge q) \mathcal{W} (p \wedge q)$
$= \quad \langle (203) \text{ Idempotency of } \mathcal{W} \rangle$
$\qquad p \wedge q \quad \blacksquare$

2.13 Variations of the Dummett Formula

There are five versions of the Dummett formula presented in [7]. These same variations are found also in [10]. The variation of the Dummett formula most often cited in the literature is as follows.

$$\Box(\Box(p \Rightarrow \Box p) \Rightarrow \Box p) \Rightarrow (\Diamond \Box p \Rightarrow \Box p)$$

This formula has a long history of study in the modal logic literature and is considered quite unique as the following two quotes indicate. (*Dum* is Dummett and *Grz* is Grzegorczyk)

> "The formulae known as *Dum* and *Grz* are two of the most bizarre formulae that occur in the literature."[7]

> "It is the only scheme of linear-time discrete temporal logic in which modalities occur nested three deep."[10]

As discussed in section 4.1 (Modal logic systems) in [16], the Lemmon formula (255) imposes linearity of the time line. The Dummett formula imposes discreteness of the time line [10] for a temporal system without the *next* operator $\bigcirc$. In CDS4LTL discreteness of the time line is implicit in the definition of the *next* operator $\bigcirc$. There is no state of p defined between p and $\bigcirc p$ or in general between $\underbrace{\bigcirc \bigcirc \ldots \bigcirc}_{n \text{ times}} p$ and $\underbrace{\bigcirc \bigcirc \ldots \bigcirc}_{n+1 \text{ times}} p$.

The Lemmon formula (255) is a theorem in CDS4LTL. That the Dummett formula is valid in LTL is supported by our own work in ACL2 and Spot. In a model of LTL built in ACL2 (this work is described in the last chapter of this collection) the automated theorem prover of ACL2 has constructed a proof confirming the Dummett formula is a theorem. Prior to that, ACL2 was used in an extensive search for a counterexample to Dummett. No counterexample was found. We have also used the Spot on-line translator to confirm the validity of the Dummett formula. Using basic rewriting, and reductions based on syntactic implication, pure eventualities, pure universalities, and automata-based containments Spot confirms that the formula is a tautology and (syntactically) stutter-invariant. Spot rewrites the Dummett formula

$$\Box(\Box(p \Rightarrow \Box p) \Rightarrow \Box p) \Rightarrow (\Diamond \Box p \Rightarrow \Box p)$$

into what they call *negative normal form*.

$$\Box p \vee \neg \Diamond \Box p \vee \neg \Box (\Box p \vee \neg \Box (\neg p \vee \Box p))$$

This is as far as the Spot translator gets with showing the results of rewriting and simplification before it goes on to apply reductions based on automata-based containments.

Pepperdine Papers on LTL

Other work that confirms the Dummett formula is a tautology can be found in Goré [7] and Schmitt [13]. Like our work with ACL2 and Spot, Goré and Schmitt used automated theorem provers in their work on the Dummett formula. Goré and his colleagues from the University of Bern used an automated theorem prover called Logics Workbench which is capable of proving theorems in numerous non-classical logics. These logics do not have an explicit *next* operator $\circ$ defined. These systems are axomitized using only the $\square$ and $\diamond$ operators. In the appendix of the paper they show LWB output confirming *Dum* is valid.

In the paper by Schmitt, a Tableaux system for LTL with only $\square$ and $\diamond$ as temporal operators (there is no *next* operator $\circ$) is used to prove the Dummett formula. The system is implemented in the HimML byte-coded interpreter, an implementation of standard ML. In Section 7 (pages 141-142), Figure 10 provides a detailed example of a Tableaux proof using the Dummett formula. Figure 11 (page 143), summarizes the experimental results of the work showing which formulas analyzed were valid, and which were not. All this activity leads to the hypothesis that Dummett formula is a theorem in CDS4LTL. And, it further suggests that in CDS4LTL there may be both an explicit and implicit version of the theorem depending on use of the *next* operator $\circ$.

Before presenting the proof of the Dummett formula in CDS4LTL, we show some analysis that leads to the proof of theorems (S107) and (S108). This is based on a four page internet document by Peter Hancock [10].

Write the Dummett formula as follows.

$$\diamond \square p \wedge \square (\square (p \Rightarrow \square p) \Rightarrow \square p) \Rightarrow \square p$$

and let

$$Precessive(p) \equiv \square (\square (p \Rightarrow \square p) \Rightarrow \square p)$$

Then the Dummett formula can be written as follows.

$$\diamond \square p \wedge Precessive(p) \Rightarrow \square p$$

Which is described in English as follows from [10].

> "In words, if eventually p holds forever, and p is *Precessive* in a certain sense, then p must hold already."

The obivous LTL formula to choose for *Precessive(p)* to make the previous formula a tautology would be $\square p$. First show that if we assume $Precessive(p) \equiv \square p$, this results in a theorem.

$$\Diamond \Box p \land \mathit{Precessive}(p) \Rightarrow \Box p$$

Proof:

$$\Diamond \Box p \land \mathit{Precessive}(p) \Rightarrow \Box p$$
$$= \quad \langle \text{Assume } \mathit{Precessive}(p) \equiv \Box p \rangle$$
$$\Diamond \Box p \land \Box p \Rightarrow \Box p) \quad -(3.76b) \text{ Strengthening, } p \land q \Rightarrow p \quad \blacksquare$$

Next, show that if we assume $\mathit{Precessive}(p) \Rightarrow \Box p$, this also results in a theorem.

$$\Diamond \Box p \land \mathit{Precessive}(p) \Rightarrow \Box p$$

Proof:

$$\Diamond \Box p \land \mathit{Precessive}(p)$$
$$\Rightarrow \quad \langle (3.76b) \text{ Strengthening, } p \land q \Rightarrow p \rangle$$
$$\mathit{Precessive}(p)$$
$$\Rightarrow \quad \langle \text{Assume } \mathit{Precessive}(p) \Rightarrow \Box p \rangle$$
$$\Box p \quad \blacksquare$$

The proof of the first Dummett variant will be implicit. That is, there will be no use of the *next* operator $\bigcirc$. Following the process outlined above, define $\mathit{Precessive}(p)$ based on theorem (S76) as follows.

$$\mathit{Precessive}(p) \equiv \Box((p \Rightarrow \Box p) \Rightarrow \Box p)$$

An implicit variant of the Dummett formula (S107) is now stated and proven in CDS4LTL as follows.

(S107) Dummett variant: $\quad \Box((p \Rightarrow \Box p) \Rightarrow \Box p) \Rightarrow (\Diamond \Box p \Rightarrow \Box p)$

Proof:

$$\Box((p \Rightarrow \Box p) \Rightarrow \Box p) \Rightarrow (\Diamond \Box p \Rightarrow \Box p)$$
$$= \quad \langle (3.65) \text{ Shunting, } p \land q \Rightarrow r \equiv p \Rightarrow (q \Rightarrow r) \rangle$$
$$\Diamond \Box p \land \Box((p \Rightarrow \Box p) \Rightarrow \Box p) \Rightarrow \Box p$$

And now,

$$\Diamond \Box p \land \Box((p \Rightarrow \Box p) \Rightarrow \Box p)$$
$$\Rightarrow \quad \langle (3.76b) \text{ Strengthening, } p \land q \Rightarrow p \text{ with } p, q := \Box((p \Rightarrow \Box p) \Rightarrow \Box p), \Diamond \Box p \rangle$$
$$\Box((p \Rightarrow \Box p) \Rightarrow \Box p)$$
$$= \quad \langle (S76) \; \Box((p \Rightarrow \Box p) \Rightarrow \Box p) \equiv \Box p \rangle$$
$$\Box p \quad \blacksquare$$

Pepperdine Papers on LTL

The next theorem (S108) is a CDS4LTL explicit variant of the Dummett formula. This form of the theorem shows how the *next* operator, $\circ$, is used with induction to setup a *Precessive(p)* that gives a theorem with the Dummett formula consequent.

$$Precessive(p) \equiv (\Box(p \Rightarrow \circ p) \land \Box(\Box(p \Rightarrow \Box p) \Rightarrow \Box p))$$

The explicit version of the Dummett formula (S108) is now stated and proven in CDS4LTL as follows.

$$\Diamond \Box p \land Precessive(p) \Rightarrow \Box p$$

The definition of *Precessive* given above is one "sense"in which CDS4LTL is precessive. Note how the following proof continues to follow the model of Peter Hancock described above.

(S108) **Dummett variant:** $\quad \Box((p \Rightarrow \circ p) \land (\Box(p \Rightarrow \Box p) \Rightarrow \Box p)) \Rightarrow (\Diamond \Box p \Rightarrow \Box p)$

Proof:

$\qquad \Box((p \Rightarrow \circ p) \land (\Box(p \Rightarrow \Box p) \Rightarrow \Box p)) \Rightarrow (\Diamond \Box p \Rightarrow \Box p)$

$= \quad \langle (99) \text{ Distributivity of } \Box \text{ over } \land \rangle$

$\qquad \Box(p \Rightarrow \circ p) \land \Box(\Box(p \Rightarrow \Box p) \Rightarrow \Box p) \Rightarrow (\Diamond \Box p \Rightarrow \Box p)$

$= \quad \langle (3.65) \text{ Shunting, } p \land q \Rightarrow r \equiv p \Rightarrow (q \Rightarrow r) \rangle$

$\qquad \Box(p \Rightarrow \circ p) \land \Box(\Box(p \Rightarrow \Box p) \Rightarrow \Box p) \land \Diamond \Box p \Rightarrow \Box p$

And now,

$\qquad \Box(p \Rightarrow \circ p) \land \Box(\Box(p \Rightarrow \Box p) \Rightarrow \Box p) \land \Diamond \Box p$

$\Rightarrow \quad \langle (S25) \ \Box(p \Rightarrow \circ p) \Rightarrow \Box(p \Rightarrow \Box p) \text{ and (4.3) Monotonicity of } \land \rangle$

$\qquad \Box(p \Rightarrow \Box p) \land \Box(\Box(p \Rightarrow \Box p) \Rightarrow \Box p) \land \Diamond \Box p$

$\Rightarrow \quad \langle (76) \text{ Strengthening of } \Box \text{ and (4.3) Monotonicity of } \land \rangle$

$\qquad \Box(p \Rightarrow \Box p) \land (\Box(p \Rightarrow \Box p) \Rightarrow \Box p) \land \Diamond \Box p$

$\Rightarrow \quad \langle (3.77) \text{ Modus ponens, } p \land (p \Rightarrow q) \Rightarrow q \text{ with } p, q := \Box(p \Rightarrow \Box p), \Box p$
$\qquad \text{ and (4.3) Monotonicity of } \land \rangle$

$\qquad \Box p \land \Diamond \Box p$

$= \quad \langle (49) \text{ Absorption of } \Diamond \text{ into } \land \text{ with } p := \Box p \rangle$

$\qquad \Box p \quad \blacksquare$

The next two theorems (S109) (S110) were proved during the course of working on the Dummett formula proof. Theorem (S109) provides another candidate for the *Precessive(p)* club. Theorem (S110) has the same antecedent as the Dummett formula while the consequent includes the antecedent of (57) $\Box$ induction. They are added here in the spirit of the paper.

(S109) **Dummett variant:** $\Box(\Diamond(p \Rightarrow \Box p) \Rightarrow \Box p) \Rightarrow (\Diamond \Box p \Rightarrow \Box p)$

Proof:

$\quad \Box(\Diamond(p \Rightarrow \Box p) \Rightarrow \Box p) \Rightarrow (\Diamond \Box p \Rightarrow \Box p)$
$= \quad \langle(3.65) \text{ Shunting, } p \land q \Rightarrow r \equiv p \Rightarrow (q \Rightarrow r)\rangle$
$\quad \Diamond \Box p \land \Box(\Diamond(p \Rightarrow \Box p) \Rightarrow \Box p) \Rightarrow \Box p$

And now,

$\quad \Diamond \Box p \land \Box(\Diamond(p \Rightarrow \Box p) \Rightarrow \Box p)$
$\Rightarrow \quad \langle(S27) \Diamond \Box p \Rightarrow \Diamond(p \Rightarrow \Box p) \text{ and (4.3) Monotonicity of } \land\rangle$
$\quad \Diamond(p \Rightarrow \Box p) \land \Box(\Diamond(p \Rightarrow \Box p) \Rightarrow \Box p)$
$\Rightarrow \quad \langle(76) \text{ Strengthening of } \Box \text{ and (4.3) Monotonicity of } \land\rangle$
$\quad \Diamond(p \Rightarrow \Box p) \land (\Diamond(p \Rightarrow \Box p) \Rightarrow \Box p)$
$\Rightarrow \quad \langle(3.77) \text{ Modus ponens, } p \land (p \Rightarrow q) \Rightarrow q \text{ with } p, q := \Diamond(p \Rightarrow \Box p), \Box p\rangle$
$\quad \Box p \quad \blacksquare$

(S110) **Dummett variant:** $\Box(\Box(p \Rightarrow \Box p) \Rightarrow \Box p) \Rightarrow (\Box(p \Rightarrow \circ p) \Rightarrow \Box p)$

Proof:

$\quad \Box(\Box(p \Rightarrow \Box p) \Rightarrow \Box p) \Rightarrow (\Box(p \Rightarrow \circ p) \Rightarrow \Box p)$
$= \quad \langle(3.65) \text{ Shunting, } p \land q \Rightarrow r \equiv p \Rightarrow (q \Rightarrow r)\rangle$
$\quad \Box(p \Rightarrow \circ p) \land \Box(\Box(p \Rightarrow \Box p) \Rightarrow \Box p) \Rightarrow \Box p$

And now,

$\quad \Box(p \Rightarrow \circ p) \land \Box(\Box(p \Rightarrow \Box p) \Rightarrow \Box p)$
$\Rightarrow \quad \langle(S25) \Box(p \Rightarrow \circ p) \Rightarrow \Box(p \Rightarrow \Box p) \text{ and (4.3) Monotonicity of } \land\rangle$
$\quad \Box(p \Rightarrow \Box p) \land \Box(\Box(p \Rightarrow \Box p) \Rightarrow \Box p)$
$\Rightarrow \quad \langle(76) \text{ Strengthening of } \Box \text{ and (4.3) Monotonicity of } \land\rangle$
$\quad \Box(p \Rightarrow \Box p) \land (\Box(p \Rightarrow \Box p) \Rightarrow \Box p)$
$\Rightarrow \quad \langle(3.77) \text{ Modus ponens, } p \land (p \Rightarrow q) \Rightarrow q \text{ with } p, q := \Box(p \Rightarrow \Box p), \Box p\rangle$
$\quad \Box p \quad \blacksquare$

2.14 Proof of the Dummett Formula

And now, for the last proof of a theorem in this paper: the proof of the Dummett Formula folllows.

(S111) **Dummett formula:** $\Box(\Box(p \Rightarrow \Box p) \Rightarrow \Box p) \Rightarrow (\Diamond \Box p \Rightarrow \Box p)$

Proof: The proof is by (82) Temporal deduction and (4.9) Proof by contradiction, where R is defined as the expression

Pepperdine Papers on LTL

$$R : \Box(p \Rightarrow \Box p) \Rightarrow \Box p$$

With this definition, the theorem to be proved is

$$\Box R \Rightarrow (\Diamond \Box p \Rightarrow \Box p)$$

Now rewrite this into an expression so that (82) Temporal deduction can be used.

$$\begin{aligned}
& \Box R \Rightarrow (\Diamond \Box p \Rightarrow \Box p) \\
=\ & \langle (3.65)\ \text{Shunting},\ p \wedge q \Rightarrow r \equiv p \Rightarrow (q \Rightarrow r) \rangle \\
& \Box R \wedge \Diamond \Box p \Rightarrow \Box p \\
=\ & \langle (152)\ \text{Absorption of } \Box \text{ into } \Diamond \Box \rangle \\
& \Box R \wedge \Box(\Diamond \Box p) \Rightarrow \Box p
\end{aligned}$$

And now, assume R and $\Diamond \Box p$ and prove $\Box p$ using (4.9) Proof by contradiction. By way of contradiction assume $\Box p \equiv false$ and show this results in a contradiction.

$$\begin{aligned}
& true \\
=\ & \langle \text{Assume } \Diamond \Box p \rangle \\
& \Diamond \Box p \\
=\ & \langle \text{Assume } \Box p \equiv false \rangle \\
& \Diamond\, false \\
=\ & \langle (44)\ \text{Falsehood of } \Diamond \rangle \\
& false
\end{aligned}$$

which is a contradiction, therefore $\Box p$ must be *true*. ■

This completes the proof of the Dummett formula.

3 Theorem Counterexamples

We discovered two expressions in the literature claimed to be theorems but for which we found counterexamples. The first is from the book by Schneider, *On Concurrent Programming*, labeled (3.38) $\mathcal{W}$ Introduction-L Rule on page 66. [14]

$$\Box(q \vee (p \wedge \circ r) \Rightarrow r) \Rightarrow (\Box p \vee p\, \mathcal{U}\, q \Rightarrow r)$$

Counterexample:

$$\begin{aligned}
& \Box(q \vee (p \wedge \circ r) \Rightarrow r) \Rightarrow (\Box p \vee p\, \mathcal{U}\, q \Rightarrow r) \\
=\ & \langle \text{Counterexample state } p, q, r := true, false, false \rangle \\
& \Box(false \vee (true \wedge \circ\, false) \Rightarrow false) \Rightarrow (\Box\, true \vee true\, \mathcal{U}\, false \Rightarrow false)
\end{aligned}$$

$=$ ⟨(8) Falsehood of ○⟩

$\Box(false \lor (true \land false) \Rightarrow false) \Rightarrow (\Box true \lor true \,\mathcal{U}\, false \Rightarrow false)$

$=$ ⟨(3.40) Zero of $\land$, $p \land false \equiv false$⟩

$\Box(false \lor false \Rightarrow false) \Rightarrow (\Box true \lor true \,\mathcal{U}\, false \Rightarrow false)$

$=$ ⟨(3.30) Identity of $\lor$, $p \lor false \equiv p$⟩

$\Box(false \Rightarrow false) \Rightarrow (\Box true \lor true \,\mathcal{U}\, false \Rightarrow false)$

$=$ ⟨(3.74) $p \Rightarrow false \equiv \neg p$, twice⟩

$\Box(\neg false) \Rightarrow \neg(\Box true \lor true \,\mathcal{U}\, false)$

$=$ ⟨(3.13) Negation of $false$, $\neg false \equiv true$⟩

$\Box true \Rightarrow \neg(\Box true \lor true \,\mathcal{U}\, false)$

$=$ ⟨(11) Right zero of $\mathcal{U}$⟩

$\Box true \Rightarrow \neg(\Box true \lor false)$

$=$ ⟨(3.30) Identity of $\lor$, $p \lor false \equiv p$⟩

$\Box true \Rightarrow \neg(\Box true)$

$=$ ⟨(64) Truth of $\Box$, twice⟩

$true \Rightarrow \neg true$

$=$ ⟨(3.8) Definition of $false$, $false \equiv \neg true$⟩

$true \Rightarrow false$

$=$ ⟨(3.73) Left identity of $\Rightarrow$, $true \Rightarrow p \equiv p$⟩

$false$ ∎

The second expression is from the book by BenAri, *Principles of Concurrent and Distributed Programming*, named Progress Proof Rule on page 81. [1]

$$\Diamond \Box p \land (\Box p \Rightarrow \Diamond q) \Rightarrow \Diamond q$$

Exercise 4 on page 90 asks the reader to prove the validity of the implication.
Although the theorem seems to be valid intuitively, we tried to prove the theorem formally for quite some time and were never successful. Finally, one of our students began to suspect the validity of the proposed theorem and devised the following counterexample.
Counterexample: (Alessandro Monteros)

	σ	s_0	s_1	s_2	s_3	s_4	s_5	s_6	s_7	...
1.	x	0	1	2	3	4	5	6	7	...
2.	$p : x = 3 \lor x > 5$	F	F	F	T	F	F	T	T	...
3.	$q : x = 1$	F	T	F	F	F	F	F	F	...
4.	$\Box p$	F	F	F	F	F	F	T	T	...
5.	$\Diamond \Box p$	T	T	T	T	T	T	T	T	...
6.	$\Diamond q$	T	T	F	F	F	F	F	F	...
7.	$\Box p \Rightarrow \Diamond q$	T	T	T	T	T	T	F	F	...
8.	$\Diamond \Box p \land (\Box p \Rightarrow \Diamond q)$	T	T	T	T	T	T	F	F	...
9.	$\Diamond \Box p \land (\Box p \Rightarrow \Diamond q) \Rightarrow \Diamond q$	T	T	F	F	F	F	T	T	...

Pepperdine Papers on LTL

x is an integer variable that increments by one from one state to the next. Boolean expression p is true in state s_3 and from states s_6 and onward. Boolean expression q is true only in state s_1. The LTL expression $\Box p$ is true from state s_6 and onward, so the expression $\Box \Diamond p$ is true in every state. Because q is true only in state s_1, $\Diamond q$ is true only in states s_0 and s_1. Now, $\Box p \Rightarrow \Diamond q$ is true up until s_5 because the antecedent of the implication is false for those states. However, it is false from states s_6 and onward, because for those states $\Box p$ is true but $\Diamond q$ is false. Row 8 comes from the state-wise conjunction of rows 5 and 7. Row 9 comes from the state-wise implication of rows 8 and 6.

The counterexample follows from the F entries in states $s_3..s_5$ in row 9. ∎

Our version of the Progress proof rule is

$$(168) \quad \Diamond \Box p \land \Box (\Box p \Rightarrow \Diamond q) \Rightarrow \Diamond q$$

4 Left-over Lemmas

In this section lemmas that were proved during the work, but not used in any subsequent theorems, are collected here. These might be of use in future proofs, and seemed interesting enough to document them as part of this work. The first two lemmas are not temporal formulas, but propositional.

Lemma: $(p \Rightarrow q) \Rightarrow p \equiv p$
Proof:

$\quad (p \Rightarrow q) \Rightarrow p$
$= \quad \langle (3.59) \text{ Implication}, p \Rightarrow q \equiv \neg p \lor q, \text{twice} \rangle$
$\quad \neg(\neg p \lor q) \lor q$
$= \quad \langle (3.47b) \text{ De Morgan}, \neg(p \lor q) \equiv \neg p \land \neg q \rangle$
$\quad (p \land \neg q) \lor p$
$= \quad \langle (3.43b) \text{ Absorption}, p \lor (p \land q) \equiv p \rangle$
$\quad p$ ∎

Lemma: $(p \Rightarrow q) \Rightarrow q \equiv p \lor q$
Proof:

$\quad (p \Rightarrow q) \Rightarrow q$
$= \quad \langle (3.59) \text{ Implication}, p \Rightarrow q \equiv \neg p \lor q, \text{twice} \rangle$
$\quad \neg(\neg p \lor q) \lor q$
$= \quad \langle (3.47b) \text{ De Morgan}, \neg(p \lor q) \equiv \neg p \land \neg q \rangle$
$\quad (p \land \neg q) \lor q$
$= \quad \langle (3.44b) \text{ Absorption}, p \lor (\neg p \land q) \equiv p \lor q \rangle$
$\quad p \lor q$ ∎

Temporal versions of the last two lemmas are also possible as shown in the next two lemmas.

Lemma: $(\Box(p \Rightarrow q) \Rightarrow p) \Rightarrow \Diamond p$
Proof:

$\quad\quad \Box(p \Rightarrow q) \Rightarrow p$
$= \quad \langle(3.59)\text{ Implication}, p \Rightarrow q \equiv \neg p \vee q, \text{twice}\rangle$
$\quad\quad \neg\Box(\neg p \vee q) \vee p$
$= \quad \langle(60)\text{ Dual of }\Box\rangle$
$\quad\quad \Diamond\neg(\neg p \vee q) \vee p$
$= \quad \langle(3.47\text{b})\text{ De Morgan}, \neg(p \vee q) \equiv \neg p \wedge \neg q\rangle$
$\quad\quad \Diamond(p \wedge \neg q) \vee p$
$\Rightarrow \quad \langle(46)\text{ Weakening of }\Diamond\text{ and }(4.2)\text{ Monotonicity of }\vee\rangle$
$\quad\quad \Diamond(p \wedge \neg q) \vee \Diamond p$
$= \quad \langle(52)\text{ Distributivity of }\Diamond\text{ over }\vee\rangle$
$\quad\quad \Diamond((p \wedge \neg q) \vee p)$
$= \quad \langle(3.43\text{b})\text{ Absorption}, p \vee (p \wedge q) \equiv p\rangle$
$\quad\quad \Diamond p \quad \blacksquare$

Lemma: $(\Box(p \Rightarrow q) \Rightarrow q) \Rightarrow \Diamond(p \vee q)$
Proof:

$\quad\quad \Box(p \Rightarrow q) \Rightarrow q$
$= \quad \langle(3.59)\text{ Implication}, p \Rightarrow q \equiv \neg p \vee q, \text{twice}\rangle$
$\quad\quad \neg\Box(\neg p \vee q) \vee q$
$= \quad \langle(60)\text{ Dual of }\Box\rangle$
$\quad\quad \Diamond\neg(\neg p \vee q) \vee q$
$= \quad \langle(3.47\text{b})\text{ De Morgan}, \neg(p \vee q) \equiv \neg p \wedge \neg q\rangle$
$\quad\quad \Diamond(p \wedge \neg q) \vee q$
$\Rightarrow \quad \langle(46)\text{ Weakening of }\Diamond\text{ and }(4.2)\text{ Monotonicity of }\vee\rangle$
$\quad\quad \Diamond(p \wedge \neg q) \vee \Diamond q$
$= \quad \langle(52)\text{ Distributivity of }\Diamond\text{ over }\vee\rangle$
$\quad\quad \Diamond((p \wedge \neg q) \vee q)$
$= \quad \langle(3.44\text{b})\text{ Absorption}, p \vee (\neg p \wedge q) \equiv p \vee q\rangle$
$\quad\quad \Diamond(p \vee q) \quad \blacksquare$

The next two lemmas come from work on the Dummett formula.

Pepperdine Papers on LTL 197

Lemma: $\Box(p \Rightarrow \bigcirc p) \land \Box(\bigcirc p \Rightarrow \Box p) \Rightarrow \Box(p \Rightarrow \Box p)$

Proof:

$\quad\quad \Box(p \Rightarrow \bigcirc p) \land \Box(\bigcirc p \Rightarrow \Box p)$
$= \quad \langle$(99) Distributivity of $\Box$ over $\land\rangle$
$\quad\quad \Box((p \Rightarrow \bigcirc p) \land (\bigcirc p \Rightarrow \Box p))$
$\Rightarrow \quad \langle$(3.82a) Transitivity, $(p \Rightarrow q) \land (q \Rightarrow r) \Rightarrow (p \Rightarrow r)$ with $p,q,r := p, \bigcirc p, \Box p$ and
$\quad\quad$ (120) Monotonicity of $\Box$ with $p,q := (p \Rightarrow \bigcirc p) \land (\bigcirc p \Rightarrow \Box p), (p \Rightarrow \Box p)\rangle$
$\quad\quad \Box(p \Rightarrow \Box p)$ ∎

Lemma: $\Box(p \Rightarrow \bigcirc p) \land \Box(\bigcirc p \Rightarrow \Box(p \Rightarrow \Box p) \Rightarrow \Box p) \Rightarrow (\Box(p \Rightarrow \Box p) \Rightarrow \Box p)$

Proof:

$\quad\quad \Box(p \Rightarrow \bigcirc p) \land \Box(\bigcirc p \Rightarrow \Box(p \Rightarrow \Box p) \Rightarrow \Box p)$
$= \quad \langle$(99) Distributivity of $\Box$ over $\land\rangle$
$\quad\quad \Box((p \Rightarrow \bigcirc p) \land (\bigcirc p \Rightarrow \Box(p \Rightarrow \Box p) \Rightarrow \Box p))$
$\Rightarrow \quad \langle$(3.82a) Transitivity, $(p \Rightarrow q) \land (q \Rightarrow r) \Rightarrow (p \Rightarrow r)$ with
$\quad\quad p,q,r := p, \bigcirc p, \Box(p \Rightarrow \Box p) \Rightarrow \Box p$ and (120) Monotonicity of $\Box$
$\quad\quad$ with $p,q := (p \Rightarrow \bigcirc p) \land (\bigcirc p \Rightarrow \Box p), (p \Rightarrow \Box p)\rangle$
$\quad\quad \Box(p \Rightarrow \Box p) \Rightarrow \Box p$ ∎

Lemma: $\Box(p \land q) \Rightarrow \Box(p \equiv q)$

Proof: The proof is by (4.7.1) Truth implication.

$\quad\quad true$
$= \quad \langle$(99) Distributivity of $\Box$ over $\land\rangle$
$\quad\quad \Box(p \land q) \equiv \Box p \land \Box q$
$= \quad \langle$(3.39) Identity of $\land$, $p \land true \equiv p\rangle$
$\quad\quad \Box(p \land q) \equiv (\Box p \land \Box q) \land true$
$= \quad \langle$(117) $\equiv$ frame law of $\Box$ and
$\quad\quad$ (3.65) Shunting, $p \land q \Rightarrow r \equiv p \Rightarrow (q \Rightarrow r)$ with the above theorem$\rangle$
$\quad\quad (\Box(p \land q) \equiv \Box p \land \Box q) \land (\Box p \land \Box q \Rightarrow \Box(p \equiv q))$
$= \quad \langle$(3.82b) Transitivity, $(p \equiv q) \land (q \Rightarrow r) \Rightarrow (p \Rightarrow r)$ with p,q,r :=
$\quad\quad \Box(p \land q), \Box p \land \Box q, \Box(p \equiv q)\rangle$
$\quad\quad \Box(p \land q) \Rightarrow \Box(p \equiv q)$ ∎

5 Summary

The purpose of this work is to collect together theorems of CDS4LTL that were not included in the CDS4LTL paper [16] for various reasons, or theorems that were proven subsequent to submitting the CDS4LTL paper for peer review. An additional 111 theorems and their proofs are included in this paper, which will be available as a Technical Report from the Computer Science Department, Pepperdine University.

6 Acknowledgements

This work has been a follow-on to the work that resulted in the CDS4LTL paper. I would like to thank Professor Warford of Pepperdine University, first for teaching me formal methods through his iTunes University courses, second for continued encouragement in working on aspects of LTL, and third for serving as an editor and reviewer for this paper. I also thank Julee Staley for LATEX typesetting a large number of proofs in this paper.

7 References

[1] Mordechai Ben-Ari. *Principles of Concurrent and Distributed Programming*. 2nd. Harlow, England: Addison-Wesley Pearson, 2006.

[2] StackExchange Community. *Various*. https://stackexchange.com. 2019. (Accessed: 2019).

[3] Fabio Corpina. "The Ancient Master Argument and Some Examples of Tense Logic". In: *ARGUMENTA* 1.2 (2016), pp. 245–258.

[4] Alexandre Duret-Lutz. *Spot's Temporal Logic Formulas*. Nov. 2019.

[5] Alexandre Duret-Lutz et al. "Spot 2.0 — a Framework for LTL and ω-automata Manipulation". In: *Proceedings of the 14th International Symposium on Automated Technology for Verification and Analysis (ATVA'16)*. Vol. 9938. Lecture Notes in Computer Science. Springer, Oct. 2016, pp. 122–129.

[6] Thomson Gale. *Modal Logic*. Encyclopedia.com. 2006. URL: `%7Bhttps://www.encyclopedia.com/humanities/encyclopedias-almanacs-transcripts-and-maps/modal-logic%7D` (visited on 08/11/2019). (Accessed: 2019).

[7] Rajeev Goré, Wolfgang Heinle, and Alain Heuerding. *Relations Between Propositional Normal Modal Logics: An Overview*. Tech. rep. TR-ARP-16-95. Australian National University and University of Bern, Sept. 1995.

[8] Rajeev Goré, Wolfgang Heinle, and Alain Heuerding. "Relations Between Propositional Normal Modal Logics: An Overview". In: *Journal of Logic Computation* 7.5 (1997), pp. 649–658.

[9] David Gries and Fred B. Schneider. *A Logical Approach to Discrete Math*. New York: Springer-Verlag, 1994.

[10] Peter G. Hancock. *The Discreteness of Time*. June 2003. URL: http://www.dcs.ed.ac.uk/home/pgh/dummet.html (visited on 06/19/2003).

[11] Fred Kröger and Stephan Merz. *Temporal Logic and State Systems*. First. Berlin: Springer-Verlag, 2008.

[12] Zohar Manna and Amir Pnueli. *Verification of Concurrent Programs: A Temporal Proof System*. Tech. rep. STAN-CS-83-967. Dept of Computer Science, Stanford CA: Stanford University, June 1983. URL: http://apps.dtic.mil.

[13] Peter H. Schmitt and Jean Goubault-Larrecq. "A Tableau System for Linear-Time Temporal Logic". In: *Tools and Algorithms for the Construction and Analysis of Systems*. Ed. by Ed Brinksma. LNCS 1217 Third International Workshop, TACAS'97. University of Twente. New York: Springer-Verlag, Apr. 1997, pp. 130–144.

[14] Fred B. Schneider. *On Concurrent Programming*. First. New York: Springer-Verlag, 1997.

[15] *Spot: a Platform for LTL and ω-automata Manipulation*. Nov. 2019. URL: https://spot.lrde.epita.fr.

[16] J. Stanley Warford, David Vega, and Scott M. Staley. "A Calculational Deductive System for Linear Temporal Logic". In: *ACM Comput. Surv.* 53.3 (June 2020). ISSN: 0360-0300. DOI: 10.1145/3387109. URL: https://doi.org/10.1145/3387109.

[17] J. Stanley Warford, David Vega, and Scott M. Staley. *Theorems from CDS4LTL (Expanded)*. Pepperdine University Research Report, Natural Science Division, 2019. (Version: 2019).

Theorems from CDS4LTL (Expanded)

J. Stanley Warford
Computer Science Department
Pepperdine University
Malibu, CA 90263

David Vega *
The Aerospace Corporation
El Segundo, CA 90245

Scott M. Staley
Ford Motor Company Research Labs (retired)
Dearborn, MI 48124

Abstract

The first section of this document is a collection of the axioms and theorems of the propositional calculus in Gries and Schneider's book *A Logical Approach to Discrete Math*, Springer-Verlag, 1993 (LADM). The numbering is consistent with that text with the chapter number followed by the equation number separated by a period. Additional theorems, either not included in LADM or included but not numbered, are indicated by a three-part number with two period separators. The second section is a collection of the axioms and theorems of linear temporal logic in Warford, Vega, and Staley's paper *A Calculational Deductive System for Linear Temporal Logic* (CDS4LTL), Pepperdine University Research Report, Natural Science Division, 2019. And, the third section is a collection of the axioms and theorems of linear temporal logic in Staley's paper *A Calculational Deductive System for Linear Temporal Logic: Additional Theorems* (CDS4LTL), Pepperdine University Research Report, Natural Science Division, 2020.

---

*Research supported by Tooma Undergraduate Research Fellowship Program, Pepperdine University, Summer 2009 and academic year 2009-10.

Table of Precedences

$[x := e]$ (textual substitution)	Highest precedence
$\neg\ \ \circ\ \ \Diamond\ \ \Box$	
$\mathcal{U}\ \ \mathcal{W}$	
$=$ (conjunctional)	
$\vee\ \ \wedge$	
$\Rightarrow\ \ \Leftarrow$	
$\equiv$ (associative)	Lowest precedence

Theorems of the Propositional Calculus

Equivalence and *true*

(3.1) **Axiom, Associativity of $\equiv$:** $((p \equiv q) \equiv r) \equiv (p \equiv (q \equiv r))$
(3.2) **Axiom, Symmetry of $\equiv$:** $p \equiv q \equiv q \equiv p$
(3.3) **Axiom, Identity of $\equiv$:** $true \equiv q \equiv q$
(3.4) $true$
(3.5) **Reflexivity of $\equiv$:** $p \equiv p$

Negation, inequivalence, and *false*

(3.8) **Definition of $false$:** $false \equiv \neg true$
(3.9) **Axiom, Distributivity of $\neg$ over $\equiv$:** $\neg(p \equiv q) \equiv \neg p \equiv q$
(3.10) **Definition of $\not\equiv$:** $(p \not\equiv q) \equiv \neg(p \equiv q)$
(3.11) $\neg p \equiv q \equiv p \equiv \neg q$
(3.12) **Double negation:** $\neg\neg p \equiv p$
(3.13) **Negation of $false$:** $\neg false \equiv true$
(3.14) $(p \not\equiv q) \equiv \neg p \equiv q$
(3.15) $\neg p \equiv p \equiv false$
(3.16) **Symmetry of $\not\equiv$:** $(p \not\equiv q) \equiv (q \not\equiv p)$
(3.17) **Associativity of $\not\equiv$:** $((p \not\equiv q) \not\equiv r) \equiv (p \not\equiv (q \not\equiv r))$
(3.18) **Mutual associativity:** $((p \not\equiv q) \equiv r) \equiv (p \not\equiv (q \equiv r))$
(3.19) **Mutual interchangeability:** $p \not\equiv q \equiv r \ \equiv\ p \equiv q \not\equiv r$
(3.19.1) $p \not\equiv p \not\equiv q \equiv q$

Disjunction

(3.24) **Axiom, Symmetry of** $\vee$: $\quad p \vee q \equiv q \vee p$
(3.25) **Axiom, Associativity of** $\vee$: $\quad (p \vee q) \vee r \equiv p \vee (q \vee r)$
(3.26) **Axiom, Idempotency of** $\vee$: $\quad p \vee p \equiv p$
(3.27) **Axiom, Distributivity of** $\vee$ **over** $\equiv$: $\quad p \vee (q \equiv r) \equiv p \vee q \equiv p \vee r$
(3.28) **Axiom, Excluded middle:** $\quad p \vee \neg p$
(3.29) **Zero of** $\vee$: $\quad p \vee true \equiv true$
(3.30) **Identity of** $\vee$: $\quad p \vee false \equiv p$
(3.31) **Distributivity of** $\vee$ **over** $\vee$: $\quad p \vee (q \vee r) \equiv (p \vee q) \vee (p \vee r)$
(3.32) $\quad p \vee q \equiv p \vee \neg q \equiv p$

Conjunction

(3.35) **Axiom, Golden rule:** $\quad p \wedge q \equiv p \equiv q \equiv p \vee q$
(3.36) **Symmetry of** $\wedge$: $\quad p \wedge q \equiv q \wedge p$
(3.37) **Associativity of** $\wedge$: $\quad (p \wedge q) \wedge r \equiv p \wedge (q \wedge r)$
(3.38) **Idempotency of** $\wedge$: $\quad p \wedge p \equiv p$
(3.39) **Identity of** $\wedge$: $\quad p \wedge true \equiv p$
(3.40) **Zero of** $\wedge$: $\quad p \wedge false \equiv false$
(3.41) **Distributivity of** $\wedge$ **over** $\wedge$: $\quad p \wedge (q \wedge r) \equiv (p \wedge q) \wedge (p \wedge r)$
(3.42) **Contradiction:** $\quad p \wedge \neg p \equiv false$
(3.43) **Absorption:**
 (a) $p \wedge (p \vee q) \equiv p$
 (b) $p \vee (p \wedge q) \equiv p$
(3.44) **Absorption:**
 (a) $p \wedge (\neg p \vee q) \equiv p \wedge q$
 (b) $p \vee (\neg p \wedge q) \equiv p \vee q$
(3.45) **Distributivity of** $\vee$ **over** $\wedge$: $\quad p \vee (q \wedge r) \equiv (p \vee q) \wedge (p \vee r)$
(3.46) **Distributivity of** $\wedge$ **over** $\vee$: $\quad p \wedge (q \vee r) \equiv (p \wedge q) \vee (p \wedge r)$
(3.47) **De Morgan:**
 (a) $\neg(p \wedge q) \equiv \neg p \vee \neg q$
 (b) $\neg(p \vee q) \equiv \neg p \wedge \neg q$
(3.48) $\quad p \wedge q \equiv p \wedge \neg q \equiv \neg p$
(3.49) $\quad p \wedge (q \equiv r) \equiv p \wedge q \equiv p \wedge r \equiv p$
(3.50) $\quad p \wedge (q \equiv p) \equiv p \wedge q$
(3.51) **Replacement:** $\quad (p \equiv q) \wedge (r \equiv p) \equiv (p \equiv q) \wedge (r \equiv q)$

(3.52) **Equivalence:** $p \equiv q \equiv (p \wedge q) \vee (\neg p \wedge \neg q)$
(3.53) **Exclusive or:** $p \not\equiv q \equiv (\neg p \wedge q) \vee (p \wedge \neg q)$
(3.55) $(p \wedge q) \wedge r \equiv p \equiv q \equiv r \equiv p \vee q \equiv q \vee r \equiv r \vee p \equiv p \vee q \vee r$

Implication

(3.57) **Definition of Implication:** $p \Rightarrow q \equiv p \vee q \equiv q$
(3.58) **Axiom, Consequence:** $p \Leftarrow q \equiv q \Rightarrow p$
(3.59) **Implication:** $p \Rightarrow q \equiv \neg p \vee q$
(3.60) **Implication:** $p \Rightarrow q \equiv p \wedge q \equiv p$
(3.61) **Contrapositive:** $p \Rightarrow q \equiv \neg q \Rightarrow \neg p$
(3.62) $p \Rightarrow (q \equiv r) \equiv p \wedge q \equiv p \wedge r$
(3.63) **Distributivity of $\Rightarrow$ over $\equiv$:** $p \Rightarrow (q \equiv r) \equiv (p \Rightarrow q) \equiv (p \Rightarrow r)$
(3.63.1) **Distributivity of $\Rightarrow$ over $\wedge$:** $p \Rightarrow q \wedge r \equiv (p \Rightarrow q) \wedge (p \Rightarrow r)$
(3.63.2) **Distributivity of $\Rightarrow$ over $\vee$:** $p \Rightarrow q \vee r \equiv (p \Rightarrow q) \vee (p \Rightarrow r)$
(3.64) $p \Rightarrow (q \Rightarrow r) \equiv (p \Rightarrow q) \Rightarrow (p \Rightarrow r)$
(3.65) **Shunting:** $p \wedge q \Rightarrow r \equiv p \Rightarrow (q \Rightarrow r)$
(3.66) $p \wedge (p \Rightarrow q) \equiv p \wedge q$
(3.67) $p \wedge (q \Rightarrow p) \equiv p$
(3.68) $p \vee (p \Rightarrow q) \equiv true$
(3.69) $p \vee (q \Rightarrow p) \equiv q \Rightarrow p$
(3.70) $p \vee q \Rightarrow p \wedge q \equiv p \equiv q$
(3.71) **Reflexivity of $\Rightarrow$:** $p \Rightarrow p$
(3.72) **Right zero of $\Rightarrow$:** $p \Rightarrow true \equiv true$
(3.73) **Left identity of $\Rightarrow$:** $true \Rightarrow p \equiv p$
(3.74) $p \Rightarrow false \equiv \neg p$
(3.74.1) $\neg p \Rightarrow false \equiv p$
(3.75) $false \Rightarrow p \equiv true$
(3.76) **Weakening/strengthening:**
  (a) $p \Rightarrow p \vee q$ (Weakening the consequent)
  (b) $p \wedge q \Rightarrow p$ (Strengthening the antecedent)
  (c) $p \wedge q \Rightarrow p \vee q$ (Weakening/strengthening)
  (d) $p \vee (q \wedge r) \Rightarrow p \vee q$
  (e) $p \wedge q \Rightarrow p \wedge (q \vee r)$
(3.76.1) $p \wedge q \Rightarrow p \vee r$ (Weakening/strengthening)
(3.76.2) $(p \Rightarrow q) \Rightarrow ((q \Rightarrow r) \Rightarrow (p \Rightarrow r))$
(3.76.3) $(p \vee q) \wedge (q \Rightarrow r) \Rightarrow p \vee r$

(3.77)   **Modus ponens:** $p \wedge (p \Rightarrow q) \Rightarrow q$
(3.77.1) **Modus tollens:** $(p \Rightarrow q) \wedge \neg q \Rightarrow \neg p$
(3.77.2) $((p \Rightarrow q) \Rightarrow (r \Rightarrow s)) \wedge (s \Rightarrow t) \Rightarrow ((p \Rightarrow q) \Rightarrow (r \Rightarrow t))$
(3.77.3) $((p \Rightarrow (q \Rightarrow r)) \wedge (r \Rightarrow s)) \Rightarrow (p \Rightarrow (q \Rightarrow s))$
(3.78) $(p \Rightarrow r) \wedge (q \Rightarrow r) \equiv p \vee q \Rightarrow r$
(3.78.1) $(p \Rightarrow r) \vee (q \Rightarrow r) \equiv p \wedge q \Rightarrow r$
(3.79) $(p \Rightarrow r) \wedge (\neg p \Rightarrow r) \equiv r$
(3.80)   **Mutual implication:** $(p \Rightarrow q) \wedge (q \Rightarrow p) \equiv (p \equiv q)$
(3.81)   **Antisymmetry:** $(p \Rightarrow q) \wedge (q \Rightarrow p) \Rightarrow (p \equiv q)$
(3.82)   **Transitivity:**
    (a) $(p \Rightarrow q) \wedge (q \Rightarrow r) \Rightarrow (p \Rightarrow r)$
    (b) $(p \equiv q) \wedge (q \Rightarrow r) \Rightarrow (p \Rightarrow r)$
    (c) $(p \Rightarrow q) \wedge (q \equiv r) \Rightarrow (p \Rightarrow r)$
(3.82.1) **Transitivity of $\equiv$:** $(p \equiv q) \wedge (q \equiv r) \Rightarrow (p \equiv r)$
(3.82.2) $(p \equiv q) \Rightarrow (p \Rightarrow q)$

Leibniz as an axiom

This section uses the following notation: E_X^z means $E[z := X]$.

(3.83)   **Axiom, Leibniz:** $e = f \Rightarrow E_e^z = E_f^z$
(3.84)   **Substitution:**
    (a) $(e = f) \wedge E_e^z \equiv (e = f) \wedge E_f^z$
    (b) $(e = f) \Rightarrow E_e^z \equiv (e = f) \Rightarrow E_f^z$
    (c) $q \wedge (e = f) \Rightarrow E_e^z \equiv q \wedge (e = f) \Rightarrow E_f^z$
(3.85)   **Replace by** *true*:
    (a) $p \Rightarrow E_p^z \equiv p \Rightarrow E_{true}^z$
    (b) $q \wedge p \Rightarrow E_p^z \equiv q \wedge p \Rightarrow E_{true}^z$
(3.86)   **Replace by** *false*:
    (a) $E_p^z \Rightarrow p \equiv E_{false}^z \Rightarrow p$
    (b) $E_p^z \Rightarrow p \vee q \equiv E_{false}^z \Rightarrow p \vee q$
(3.87)   **Replace by** *true*: $p \wedge E_p^z \equiv p \wedge E_{true}^z$
(3.88)   **Replace by** *false*: $p \vee E_p^z \equiv p \vee E_{false}^z$
(3.89)   **Shannon:** $E_p^z \equiv (p \wedge E_{true}^z) \vee (\neg p \wedge E_{false}^z)$
(3.89.1) $E_{true}^z \wedge E_{false}^z \Rightarrow E_p^z$

Additional theorems concerning implication

(4.1) $\quad p \Rightarrow (q \Rightarrow p)$

(4.2) $\quad$ **Monotonicity of $\vee$:** $\quad (p \Rightarrow q) \Rightarrow (p \vee r \Rightarrow q \vee r)$

(4.3) $\quad$ **Monotonicity of $\wedge$:** $\quad (p \Rightarrow q) \Rightarrow (p \wedge r \Rightarrow q \wedge r)$

Proof technique metatheorems

(4.4) $\quad$ **Deduction (assume conjuncts of antecedent):**
$\quad\quad$ To prove $P_1 \wedge P_2 \Rightarrow Q$, assume P_1 and P_2, and prove Q
$\quad\quad$ You cannot use textual substitution in P_1 or P_2.

(4.7) $\quad$ **Mutual implication:** $\quad$ To prove $P \equiv Q$, prove $P \Rightarrow Q$ and $Q \Rightarrow P$

(4.7.1) $\quad$ **Truth implication:** $\quad$ To prove P, prove $true \Rightarrow P$

(4.9) $\quad$ **Proof by contradiction:** $\quad$ To prove P, prove $\neg P \Rightarrow false$

(4.12) $\quad$ **Proof by contrapositive:** $\quad$ To prove $P \Rightarrow Q$, prove $\neg Q \Rightarrow \neg P$

Theorems of Linear Temporal Logic

Next $\quad \circ$

(1) $\quad$ **Axiom, Self-dual:** $\quad \circ \neg p \equiv \neg \circ p$

(2) $\quad$ **Axiom, Distributivity of $\circ$ over $\Rightarrow$:** $\quad \circ(p \Rightarrow q) \equiv \circ p \Rightarrow \circ q$

(3) $\quad$ **Linearity:** $\quad \circ p \equiv \neg \circ \neg p$

(4) $\quad$ **Distributivity of $\circ$ over $\vee$:** $\quad \circ(p \vee q) \equiv \circ p \vee \circ q$

(5) $\quad$ **Distributivity of $\circ$ over $\wedge$:** $\quad \circ(p \wedge q) \equiv \circ p \wedge \circ q$

(6) $\quad$ **Distributivity of $\circ$ over $\equiv$:** $\quad \circ(p \equiv q) \equiv \circ p \equiv \circ q$

(7) $\quad$ **Truth of $\circ$:** $\quad \circ\, true \equiv true$

(8) $\quad$ **Falsehood of $\circ$:** $\quad \circ\, false \equiv false$

Until $\mathcal{U}$

(9) **Axiom, Distributivity of $\circ$ over $\mathcal{U}$:** $\circ(p\,\mathcal{U}\,q) \equiv \circ p\,\mathcal{U}\,\circ q$

(10) **Axiom, Expansion of $\mathcal{U}$:** $p\,\mathcal{U}\,q \equiv q \vee (p \wedge \circ(p\,\mathcal{U}\,q))$

(11) **Axiom, Right zero of $\mathcal{U}$:** $p\,\mathcal{U}\,false \equiv false$

(12) **Axiom, Left distributivity of $\mathcal{U}$ over $\vee$:** $p\,\mathcal{U}\,(q \vee r) \equiv p\,\mathcal{U}\,q \vee p\,\mathcal{U}\,r$

(13) **Axiom, Right distributivity of $\mathcal{U}$ over $\vee$:** $p\,\mathcal{U}\,r \vee q\,\mathcal{U}\,r \Rightarrow (p \vee q)\,\mathcal{U}\,r$

(14) **Axiom, Left distributivity of $\mathcal{U}$ over $\wedge$:** $p\,\mathcal{U}\,(q \wedge r) \Rightarrow p\,\mathcal{U}\,q \wedge p\,\mathcal{U}\,r$

(15) **Axiom, Right distributivity of $\mathcal{U}$ over $\wedge$:** $(p \wedge q)\,\mathcal{U}\,r \equiv p\,\mathcal{U}\,r \wedge q\,\mathcal{U}\,r$

(16) **Axiom, $\mathcal{U}$ implication ordering:** $p\,\mathcal{U}\,q \wedge \neg q\,\mathcal{U}\,r \Rightarrow p\,\mathcal{U}\,r$

(17) **Axiom, Right $\mathcal{U} \vee$ ordering:** $p\,\mathcal{U}\,(q\,\mathcal{U}\,r) \Rightarrow (p \vee q)\,\mathcal{U}\,r$

(18) **Axiom, Right $\wedge\,\mathcal{U}$ ordering:** $p\,\mathcal{U}\,(q \wedge r) \Rightarrow (p\,\mathcal{U}\,q)\,\mathcal{U}\,r$

(19) **Right distributivity of $\mathcal{U}$ over $\Rightarrow$:** $(p \Rightarrow q)\,\mathcal{U}\,r \Rightarrow (p\,\mathcal{U}\,r \Rightarrow q\,\mathcal{U}\,r)$

(20) **Right zero of $\mathcal{U}$:** $p\,\mathcal{U}\,true \equiv true$

(21) **Left identity of $\mathcal{U}$:** $false\,\mathcal{U}\,q \equiv q$

(22) **Idempotency of $\mathcal{U}$:** $p\,\mathcal{U}\,p \equiv p$

(23) **$\mathcal{U}$ excluded middle:** $p\,\mathcal{U}\,q \vee p\,\mathcal{U}\,\neg q$

(24) $\neg p\,\mathcal{U}\,(q\,\mathcal{U}\,r) \wedge p\,\mathcal{U}\,r \Rightarrow q\,\mathcal{U}\,r$

(25) $p\,\mathcal{U}\,(\neg q\,\mathcal{U}\,r) \wedge q\,\mathcal{U}\,r \Rightarrow p\,\mathcal{U}\,r$

(26) $p\,\mathcal{U}\,q \wedge \neg q\,\mathcal{U}\,p \Rightarrow p$

(27) $p \wedge \neg p\,\mathcal{U}\,q \Rightarrow q$

(28) $p\,\mathcal{U}\,q \Rightarrow p \vee q$

(29) **$\mathcal{U}$ insertion:** $q \Rightarrow p\,\mathcal{U}\,q$

(30) $p \wedge q \Rightarrow p\,\mathcal{U}\,q$

(31) **Absorption:** $p \vee p\,\mathcal{U}\,q \equiv p \vee q$

(32) **Absorption:** $p\,\mathcal{U}\,q \vee q \equiv p\,\mathcal{U}\,q$

(33) **Absorption:** $p\,\mathcal{U}\,q \wedge q \equiv q$

(34) **Absorption:** $p\,\mathcal{U}\,q \vee (p \wedge q) \equiv p\,\mathcal{U}\,q$

(35) **Absorption:** $p\,\mathcal{U}\,q \wedge (p \vee q) \equiv p\,\mathcal{U}\,q$

(36) **Left absorption of $\mathcal{U}$:** $p\,\mathcal{U}\,(p\,\mathcal{U}\,q) \equiv p\,\mathcal{U}\,q$

(37) **Right absorption of $\mathcal{U}$:** $(p\,\mathcal{U}\,q)\,\mathcal{U}\,q \equiv p\,\mathcal{U}\,q$

Eventually $\Diamond$

(38) **Definition of** $\Diamond$: $\quad \Diamond q \equiv true\ \mathcal{U}\ q$

(39) **Absorption of** $\Diamond$ **into** $\mathcal{U}$: $\quad p\ \mathcal{U}\ q \wedge \Diamond q \equiv p\ \mathcal{U}\ q$

(40) **Absorption of** $\mathcal{U}$ **into** $\Diamond$: $\quad p\ \mathcal{U}\ q \vee \Diamond q \equiv \Diamond q$

(41) **Absorption of** $\mathcal{U}$ **into** $\Diamond$: $\quad p\ \mathcal{U}\ \Diamond q \equiv \Diamond q$

(42) **Eventuality:** $\quad p\ \mathcal{U}\ q \Rightarrow \Diamond q$

(43) **Truth of** $\Diamond$: $\quad \Diamond\, true \equiv true$

(44) **Falsehood of** $\Diamond$: $\quad \Diamond\, false \equiv false$

(45) **Expansion of** $\Diamond$: $\quad \Diamond p \equiv p \vee \circ \Diamond p$

(46) **Weakening of** $\Diamond$: $\quad p \Rightarrow \Diamond p$

(47) **Weakening of** $\Diamond$: $\quad \circ p \Rightarrow \Diamond p$

(48) **Absorption of** $\vee$ **into** $\Diamond$: $\quad p \vee \Diamond p \equiv \Diamond p$

(49) **Absorption of** $\Diamond$ **into** $\wedge$: $\quad \Diamond p \wedge p \equiv p$

(50) **Absorption of** $\Diamond$: $\quad \Diamond \Diamond p \equiv \Diamond p$

(51) **Exchange of** $\circ$ **and** $\Diamond$: $\quad \circ \Diamond p \equiv \Diamond \circ p$

(52) **Distributivity of** $\Diamond$ **over** $\vee$: $\quad \Diamond(p \vee q) \equiv \Diamond p \vee \Diamond q$

(53) **Distributivity of** $\Diamond$ **over** $\wedge$: $\quad \Diamond(p \wedge q) \Rightarrow \Diamond p \wedge \Diamond q$

Always □

(54) **Definition of □:** $\Box p \equiv \neg \Diamond \neg p$

(55) **Axiom, $\mathcal{U}$ induction:** $\Box(p \Rightarrow (\bigcirc p \land q) \lor r) \Rightarrow (p \Rightarrow \Box q \lor q \, \mathcal{U} \, r)$

(56) **Axiom, $\mathcal{U}$ induction:** $\Box(p \Rightarrow \bigcirc(p \lor q)) \Rightarrow (p \Rightarrow \Box p \lor p \, \mathcal{U} \, q)$

(57) **□ induction:** $\Box(p \Rightarrow \bigcirc p) \Rightarrow (p \Rightarrow \Box p)$

(58) **◇ induction:** $\Box(\bigcirc p \Rightarrow p) \Rightarrow (\Diamond p \Rightarrow p)$

(59) $\Diamond p \equiv \neg \Box \neg p$

(60) **Dual of □:** $\neg \Box p \equiv \Diamond \neg p$

(61) **Dual of ◇:** $\neg \Diamond p \equiv \Box \neg p$

(62) **Dual of ◇□:** $\neg \Diamond \Box p \equiv \Box \Diamond \neg p$

(63) **Dual of □◇:** $\neg \Box \Diamond p \equiv \Diamond \Box \neg p$

(64) **Truth of □:** $\Box \textit{true} \equiv \textit{true}$

(65) **Falsehood of □:** $\Box \textit{false} \equiv \textit{false}$

(66) **Expansion of □:** $\Box p \equiv p \land \bigcirc \Box p$

(67) **Expansion of □:** $\Box p \equiv p \land \bigcirc p \land \bigcirc \Box p$

(68) **Absorption of $\land$ into □:** $p \land \Box p \equiv \Box p$

(69) **Absorption of □ into $\lor$:** $\Box p \lor p \equiv p$

(70) **Absorption of ◇ into □:** $\Diamond p \land \Box p \equiv \Box p$

(71) **Absorption of □ into ◇:** $\Box p \lor \Diamond p \equiv \Diamond p$

(72) **Absorption of □:** $\Box \Box p \equiv \Box p$

(73) **Exchange of ○ and □:** $\bigcirc \Box p \equiv \Box \bigcirc p$

(74) $p \Rightarrow \Box p \equiv p \Rightarrow \bigcirc \Box p$

(75) $p \land \Diamond \neg p \Rightarrow \Diamond(p \land \bigcirc \neg p)$

(76) **Strengthening of □:** $\Box p \Rightarrow p$

(77) **Strengthening of □:** $\Box p \Rightarrow \Diamond p$

(78) **Strengthening of □:** $\Box p \Rightarrow \bigcirc p$

(79) **Strengthening of □:** $\Box p \Rightarrow \bigcirc \Box p$

(80) **○ generalization:** $\Box p \Rightarrow \Box \bigcirc p$

(81) $\Box p \Rightarrow \neg(q \, \mathcal{U} \, \neg p)$

Temporal Deduction

(82) **Temporal deduction:**

To prove $\Box P_1 \land \Box P_2 \Rightarrow Q$, assume P_1 and P_2, and prove Q.
You cannot use textual substitution in P_1 or P_2.

Always, continued

(83) **Distributivity of $\land$ over $\mathcal{U}$:** $\quad \Box p \land q \, \mathcal{U} \, r \Rightarrow (p \land q) \, \mathcal{U} \, (p \land r)$

(84) **$\mathcal{U}$ implication:** $\quad \Box p \land \Diamond q \Rightarrow p \, \mathcal{U} \, q$

(85) **Right monotonicity of $\mathcal{U}$:** $\quad \Box(p \Rightarrow q) \Rightarrow (r \, \mathcal{U} \, p \Rightarrow r \, \mathcal{U} \, q)$

(86) **Left monotonicity of $\mathcal{U}$:** $\quad \Box(p \Rightarrow q) \Rightarrow (p \, \mathcal{U} \, r \Rightarrow q \, \mathcal{U} \, r)$

(87) **Distributivity of $\neg$ over $\Box$:** $\quad \Box \neg p \Rightarrow \neg \Box p$

(88) **Distributivity of $\Diamond$ over $\land$:** $\quad \Box p \land \Diamond q \Rightarrow \Diamond(p \land q)$

(89) **$\Diamond$ excluded middle:** $\quad \Diamond p \lor \Box \neg p$

(90) **$\Box$ excluded middle:** $\quad \Box p \lor \Diamond \neg p$

(91) **Temporal excluded middle:** $\quad \Diamond p \lor \Diamond \neg p$

(92) **$\Diamond$ contradiction:** $\quad \Diamond p \land \Box \neg p \equiv false$

(93) **$\Box$ contradiction:** $\quad \Box p \land \Diamond \neg p \equiv false$

(94) **Temporal contradiction:** $\quad \Box p \land \Box \neg p \equiv false$

(95) **$\Box \Diamond$ excluded middle:** $\quad \Box \Diamond p \lor \Diamond \Box \neg p$

(96) **$\Diamond \Box$ excluded middle:** $\quad \Diamond \Box p \lor \Box \Diamond \neg p$

(97) **□◇ contradiction:** $\Box\Diamond p \land \Diamond\Box \neg p \equiv false$
(98) **◇□ contradiction:** $\Diamond\Box p \land \Box\Diamond \neg p \equiv false$
(99) **Distributivity of □ over ∧:** $\Box(p \land q) \equiv \Box p \land \Box q$
(100) **Distributivity of □ over ∨:** $\Box p \lor \Box q \Rightarrow \Box(p \lor q)$
(101) **Logical equivalence law of ○:** $\Box(p \equiv q) \Rightarrow (\bigcirc p \equiv \bigcirc q)$
(102) **Logical equivalence law of ◇:** $\Box(p \equiv q) \Rightarrow (\Diamond p \equiv \Diamond q)$
(103) **Logical equivalence law of □:** $\Box(p \equiv q) \Rightarrow (\Box p \equiv \Box q)$
(104) **Distributivity of ◇ over ⇒:** $\Diamond(p \Rightarrow q) \equiv (\Box p \Rightarrow \Diamond q)$
(105) **Distributivity of ◇ over ⇒:** $(\Diamond p \Rightarrow \Diamond q) \Rightarrow \Diamond(p \Rightarrow q)$
(106) **∧ frame law of ○:** $\Box p \Rightarrow (\bigcirc q \Rightarrow \bigcirc(p \land q))$
(107) **∧ frame law of ◇:** $\Box p \Rightarrow (\Diamond q \Rightarrow \Diamond(p \land q))$
(108) **∧ frame law of □:** $\Box p \Rightarrow (\Box q \Rightarrow \Box(p \land q))$
(109) **∨ frame law of ○:** $\Box p \Rightarrow (\bigcirc q \Rightarrow \bigcirc(p \lor q))$
(110) **∨ frame law of ◇:** $\Box p \Rightarrow (\Diamond q \Rightarrow \Diamond(p \lor q))$
(111) **∨ frame law of □:** $\Box p \Rightarrow (\Box q \Rightarrow \Box(p \lor q))$
(112) **⇒ frame law of ○:** $\Box p \Rightarrow (\bigcirc q \Rightarrow \bigcirc(p \Rightarrow q))$
(113) **⇒ frame law of ◇:** $\Box p \Rightarrow (\Diamond q \Rightarrow \Diamond(p \Rightarrow q))$
(114) **⇒ frame law of □:** $\Box p \Rightarrow (\Box q \Rightarrow \Box(p \Rightarrow q))$
(115) **≡ frame law of ○:** $\Box p \Rightarrow (\bigcirc q \Rightarrow \bigcirc(p \equiv q))$
(116) **≡ frame law of ◇:** $\Box p \Rightarrow (\Diamond q \Rightarrow \Diamond(p \equiv q))$
(117) **≡ frame law of □:** $\Box p \Rightarrow (\Box q \Rightarrow \Box(p \equiv q))$
(118) **Monotonicity of ○:** $\Box(p \Rightarrow q) \Rightarrow (\bigcirc p \Rightarrow \bigcirc q)$
(119) **Monotonicity of ◇:** $\Box(p \Rightarrow q) \Rightarrow (\Diamond p \Rightarrow \Diamond q)$
(120) **Monotonicity of □:** $\Box(p \Rightarrow q) \Rightarrow (\Box p \Rightarrow \Box q)$
(121) **Consequence rule of ○:** $\Box((p \Rightarrow q) \land (q \Rightarrow \bigcirc r) \land (r \Rightarrow s)) \Rightarrow (p \Rightarrow \bigcirc s)$
(122) **Consequence rule of ◇:** $\Box((p \Rightarrow q) \land (q \Rightarrow \Diamond r) \land (r \Rightarrow s)) \Rightarrow (p \Rightarrow \Diamond s)$
(123) **Consequence rule of □:** $\Box((p \Rightarrow q) \land (q \Rightarrow \Box r) \land (r \Rightarrow s)) \Rightarrow (p \Rightarrow \Box s)$
(124) **Catenation rule of ◇:** $\Box((p \Rightarrow \Diamond q) \land (q \Rightarrow \Diamond r)) \Rightarrow (p \Rightarrow \Diamond r)$
(125) **Catenation rule of □:** $\Box((p \Rightarrow \Box q) \land (q \Rightarrow \Box r)) \Rightarrow (p \Rightarrow \Box r)$
(126) **Catenation rule of 𝒰:** $\Box((p \Rightarrow q\,\mathcal{U}\,r) \land (r \Rightarrow q\,\mathcal{U}\,s)) \Rightarrow (p \Rightarrow q\,\mathcal{U}\,s)$
(127) **𝒰 strengthening rule:** $\Box((p \Rightarrow r) \land (q \Rightarrow s)) \Rightarrow (p\,\mathcal{U}\,q \Rightarrow r\,\mathcal{U}\,s)$
(128) **Induction rule ◇:** $\Box(p \lor \bigcirc q \Rightarrow q) \Rightarrow (\Diamond p \Rightarrow q)$
(129) **Induction rule □:** $\Box(p \Rightarrow q \land \bigcirc p) \Rightarrow (p \Rightarrow \Box q)$
(130) **Induction rule 𝒰:** $\Box(p \Rightarrow \neg q \land \bigcirc p) \Rightarrow (p \Rightarrow \neg(r\,\mathcal{U}\,q))$
(131) **◇ confluence:** $\Box((p \Rightarrow \Diamond(q \lor r)) \land (q \Rightarrow \Diamond t) \land (r \Rightarrow \Diamond t)) \Rightarrow (p \Rightarrow \Diamond t)$
(132) **Temporal generalization law:** $\Box(\Box p \Rightarrow q) \Rightarrow (\Box p \Rightarrow \Box q)$
(133) **Temporal particularization law:** $\Box(p \Rightarrow \Diamond q) \Rightarrow (\Diamond p \Rightarrow \Diamond q)$
(134) $\Box(p \Rightarrow \bigcirc q) \Rightarrow (p \Rightarrow \Diamond q)$
(135) $\Box(p \Rightarrow \bigcirc \neg p) \Rightarrow (p \Rightarrow \neg \Box p)$

Proof Metatheorems

(136) **Metatheorem:** P is a theorem iff $\Box P$ is a theorem.
(137) **Metatheorem** $\circ$: If $P \Rightarrow Q$ is a theorem then $\circ P \Rightarrow \circ Q$ is a theorem.
(138) **Metatheorem** $\Diamond$: If $P \Rightarrow Q$ is a theorem then $\Diamond P \Rightarrow \Diamond Q$ is a theorem.
(139) **Metatheorem** $\Box$: If $P \Rightarrow Q$ is a theorem then $\Box P \Rightarrow \Box Q$ is a theorem.

Always, continued

(140) $\mathcal{U}\,\Box$ **implication:** $p\,\mathcal{U}\,\Box q \Rightarrow \Box(p\,\mathcal{U}\,q)$
(141) **Absorption of** $\mathcal{U}$ **into** $\Box$: $p\,\mathcal{U}\,\Box p \equiv \Box p$
(142) **Right** $\wedge\,\mathcal{U}$ **strengthening:** $p\,\mathcal{U}\,(q \wedge r) \Rightarrow p\,\mathcal{U}\,(q\,\mathcal{U}\,r)$
(143) **Left** $\wedge\,\mathcal{U}$ **strengthening:** $(p \wedge q)\,\mathcal{U}\,r \Rightarrow (p\,\mathcal{U}\,q)\,\mathcal{U}\,r$
(144) **Left** $\wedge\,\mathcal{U}$ **ordering:** $(p \wedge q)\,\mathcal{U}\,r \Rightarrow p\,\mathcal{U}\,(q\,\mathcal{U}\,r)$
(145) $\Diamond\Box$ **implication:** $\Diamond\Box p \Rightarrow \Box\Diamond p$
(146) $\Box\Diamond$ **excluded middle:** $\Box\Diamond p \vee \Box\Diamond\neg p$
(147) $\Diamond\Box$ **contradiction:** $\Diamond\Box p \wedge \Diamond\Box\neg p \equiv false$
(148) $\mathcal{U}$ **frame law of** $\circ$: $\Box p \Rightarrow (\circ q \Rightarrow \circ(p\,\mathcal{U}\,q))$
(149) $\mathcal{U}$ **frame law of** $\Diamond$: $\Box p \Rightarrow (\Diamond q \Rightarrow \Diamond(p\,\mathcal{U}\,q))$
(150) $\mathcal{U}$ **frame law of** $\Box$: $\Box p \Rightarrow (\Box q \Rightarrow \Box(p\,\mathcal{U}\,q))$
(151) **Absorption of** $\Diamond$ **into** $\Box\Diamond$: $\Diamond\Box\Diamond p \equiv \Box\Diamond p$
(152) **Absorption of** $\Box$ **into** $\Diamond\Box$: $\Box\Diamond\Box p \equiv \Diamond\Box p$
(153) **Absorption of** $\Box\Diamond$: $\Box\Diamond\Box\Diamond p \equiv \Box\Diamond p$
(154) **Absorption of** $\Diamond\Box$: $\Diamond\Box\Diamond\Box p \equiv \Diamond\Box p$
(155) **Absorption of** $\circ$ **into** $\Box\Diamond$: $\circ\Box\Diamond p \equiv \Box\Diamond p$
(156) **Absorption of** $\circ$ **into** $\Diamond\Box$: $\circ\Diamond\Box p \equiv \Diamond\Box p$
(157) **Monotonicity of** $\Box\Diamond$: $\Box(p \Rightarrow q) \Rightarrow (\Box\Diamond p \Rightarrow \Box\Diamond q)$
(158) **Monotonicity of** $\Diamond\Box$: $\Box(p \Rightarrow q) \Rightarrow (\Diamond\Box p \Rightarrow \Diamond\Box q)$
(159) **Distributivity of** $\Box\Diamond$ **over** $\wedge$: $\Box\Diamond(p \wedge q) \Rightarrow \Box\Diamond p \wedge \Box\Diamond q$
(160) **Distributivity of** $\Diamond\Box$ **over** $\vee$: $\Diamond\Box p \vee \Diamond\Box q \Rightarrow \Diamond\Box(p \vee q)$
(161) **Distributivity of** $\Box\Diamond$ **over** $\vee$: $\Box\Diamond(p \vee q) \equiv \Box\Diamond p \vee \Box\Diamond q$
(162) **Distributivity of** $\Diamond\Box$ **over** $\wedge$: $\Diamond\Box(p \wedge q) \equiv \Diamond\Box p \wedge \Diamond\Box q$
(163) **Eventual latching:** $\Diamond\Box(p \Rightarrow \Box q) \equiv \Diamond\Box\neg p \vee \Diamond\Box q$
(164) $\Box(\Box\Diamond p \Rightarrow \Diamond q) \equiv \Diamond\Box\neg p \vee \Box\Diamond q$
(165) $\Box((p \vee \Box q) \wedge (\Box p \vee q)) \equiv \Box p \vee \Box q$
(166) $\Diamond\Box p \wedge \Box\Diamond q \Rightarrow \Box\Diamond(p \wedge q)$
(167) $\Box((\Box p \Rightarrow \Diamond q) \wedge (q \Rightarrow \circ r)) \Rightarrow (\Box p \Rightarrow \circ\Box\Diamond r)$
(168) **Progress proof rule:** $\Diamond\Box p \wedge \Box(\Box p \Rightarrow \Diamond q) \Rightarrow \Diamond q$

Wait $\mathcal{W}$

(169) **Definition of** $\mathcal{W}$: $p \mathcal{W} q \equiv \Box p \lor p \mathcal{U} q$

(170) **Axiom, Distributivity of** $\neg$ **over** $\mathcal{W}$: $\neg(p \mathcal{W} q) \equiv \neg q \mathcal{U} (\neg p \land \neg q)$

(171) $\mathcal{U}$ **in terms of** $\mathcal{W}$: $p \mathcal{U} q \equiv p \mathcal{W} q \land \Diamond q$

(172) $p \mathcal{W} q \equiv \Box(p \land \neg q) \lor p \mathcal{U} q$

(173) **Distributivity of** $\neg$ **over** $\mathcal{U}$: $\neg(p \mathcal{U} q) \equiv \neg q \mathcal{W} (\neg p \land \neg q)$

(174) $\mathcal{U}$ **implication:** $p \mathcal{U} q \Rightarrow p \mathcal{W} q$

(175) **Distributivity of** $\land$ **over** $\mathcal{W}$: $\Box p \land q \mathcal{W} r \Rightarrow (p \land q) \mathcal{W} (p \land r)$

(176) $\mathcal{W} \Diamond$ **equivalence:** $p \mathcal{W} \Diamond q \equiv \Box p \lor \Diamond q$

(177) $\mathcal{W} \Box$ **implication:** $p \mathcal{W} \Box q \Rightarrow \Box(p \mathcal{W} q)$

(178) **Absorption of** $\mathcal{W}$ **into** $\Box$: $p \mathcal{W} \Box p \equiv \Box p$

(179) **Perpetuity:** $\Box p \Rightarrow p \mathcal{W} q$

(180) **Distributivity of** $\circ$ **over** $\mathcal{W}$: $\circ (p \mathcal{W} q) \equiv \circ p \mathcal{W} \circ q$

(181) **Expansion of** $\mathcal{W}$: $p \mathcal{W} q \equiv q \lor (p \land \circ (p \mathcal{W} q))$

(182) $\mathcal{W}$ **excluded middle:** $p \mathcal{W} q \lor p \mathcal{W} \neg q$

(183) **Left zero of** $\mathcal{W}$: $true \mathcal{W} q \equiv true$

(184) **Left distributivity of** $\mathcal{W}$ **over** $\lor$: $p \mathcal{W} (q \lor r) \equiv p \mathcal{W} q \lor p \mathcal{W} r$

(185) **Right distributivity of** $\mathcal{W}$ **over** $\lor$: $p \mathcal{W} r \lor q \mathcal{W} r \Rightarrow (p \lor q) \mathcal{W} r$

(186) **Left distributivity of** $\mathcal{W}$ **over** $\land$: $p \mathcal{W} (q \land r) \Rightarrow p \mathcal{W} q \land p \mathcal{W} r$

(187) **Right distributivity of $\mathcal{W}$ over $\wedge$:** $(p \wedge q) \mathcal{W} r \equiv p \mathcal{W} r \wedge q \mathcal{W} r$

(188) **Right distributivity of $\mathcal{W}$ over $\Rightarrow$:** $(p \Rightarrow q) \mathcal{W} r \Rightarrow (p \mathcal{W} r \Rightarrow q \mathcal{W} r)$

(189) **Disjunction rule of $\mathcal{W}$:** $p \mathcal{W} q \equiv (p \vee q) \mathcal{W} q$

(190) **Disjunction rule of $\mathcal{U}$:** $p \mathcal{U} q \equiv (p \vee q) \mathcal{U} q$

(191) **Rule of $\mathcal{W}$:** $\neg q \mathcal{W} q$

(192) **Rule of $\mathcal{U}$:** $\neg q \mathcal{U} q \equiv \Diamond q$

(193) $(p \Rightarrow q) \mathcal{W} p$

(194) $\Diamond p \Rightarrow (p \Rightarrow q) \mathcal{U} p$

(195) **Conjunction rule of $\mathcal{W}$:** $p \mathcal{W} q \equiv (p \wedge \neg q) \mathcal{W} q$

(196) **Conjunction rule of $\mathcal{U}$:** $p \mathcal{U} q \equiv (p \wedge \neg q) \mathcal{U} q$

(197) **Distributivity of $\neg$ over $\mathcal{W}$:** $\neg(p \mathcal{W} q) \equiv (p \wedge \neg q) \mathcal{U} (\neg p \wedge \neg q)$

(198) **Distributivity of $\neg$ over $\mathcal{U}$:** $\neg(p \mathcal{U} q) \equiv (p \wedge \neg q) \mathcal{W} (\neg p \wedge \neg q)$

(199) **Dual of $\mathcal{U}$:** $\neg(\neg p \mathcal{U} \neg q) \equiv q \mathcal{W} (p \wedge q)$

(200) **Dual of $\mathcal{U}$:** $\neg(\neg p \mathcal{U} \neg q) \equiv (\neg p \wedge q) \mathcal{W} (p \wedge q)$

(201) **Dual of $\mathcal{W}$:** $\neg(\neg p \mathcal{W} \neg q) \equiv q \mathcal{U} (p \wedge q)$

(202) **Dual of $\mathcal{W}$:** $\neg(\neg p \mathcal{W} \neg q) \equiv (\neg p \wedge q) \mathcal{U} (p \wedge q)$

(203) **Idempotency of $\mathcal{W}$:** $p \mathcal{W} p \equiv p$

(204) **Right zero of $\mathcal{W}$:** $p \mathcal{W} \mathit{true} \equiv \mathit{true}$

Pepperdine Papers on LTL 215

(205) **Left identity of** $\mathcal{W}$: $\mathit{false}\,\mathcal{W}\,q \equiv q$

(206) $p\,\mathcal{W}\,q \Rightarrow p \vee q$

(207) $\Box(p \vee q) \Rightarrow p\,\mathcal{W}\,q$

(208) $\Box(\neg q \Rightarrow p) \Rightarrow p\,\mathcal{W}\,q$

(209) $\mathcal{W}$ **insertion:** $q \Rightarrow p\,\mathcal{W}\,q$

(210) $\mathcal{W}$ **frame law of** $\circ$: $\Box p \Rightarrow (\circ q \Rightarrow \circ(p\,\mathcal{W}\,q))$

(211) $\mathcal{W}$ **frame law of** $\Diamond$: $\Box p \Rightarrow (\Diamond q \Rightarrow \Diamond(p\,\mathcal{W}\,q))$

(212) $\mathcal{W}$ **frame law of** $\Box$: $\Box p \Rightarrow (\Box q \Rightarrow \Box(p\,\mathcal{W}\,q))$

(213) $\mathcal{W}$ **induction:** $\Box(p \Rightarrow (\circ p \wedge q) \vee r) \Rightarrow (p \Rightarrow q\,\mathcal{W}\,r)$

(214) $\mathcal{W}$ **induction:** $\Box(p \Rightarrow \circ(p \vee q)) \Rightarrow (p \Rightarrow p\,\mathcal{W}\,q)$

(215) $\mathcal{W}$ **induction:** $\Box(p \Rightarrow \circ p) \Rightarrow (p \Rightarrow p\,\mathcal{W}\,q)$

(216) $\mathcal{W}$ **induction:** $\Box(p \Rightarrow q \wedge \circ p) \Rightarrow (p \Rightarrow p\,\mathcal{W}\,q)$

(217) **Absorption:** $p \vee p\,\mathcal{W}\,q \equiv p \vee q$

(218) **Absorption:** $p\,\mathcal{W}\,q \vee q \equiv p\,\mathcal{W}\,q$

(219) **Absorption:** $p\,\mathcal{W}\,q \wedge q \equiv q$

(220) **Absorption:** $p\,\mathcal{W}\,q \wedge (p \vee q) \equiv p\,\mathcal{W}\,q$

(221) **Absorption:** $p\,\mathcal{W}\,q \vee (p \wedge q) \equiv p\,\mathcal{W}\,q$

(222) **Left absorption of** $\mathcal{W}$: $p\,\mathcal{W}\,(p\,\mathcal{W}\,q) \equiv p\,\mathcal{W}\,q$

(223) **Right absorption of** $\mathcal{W}$: $(p\,\mathcal{W}\,q)\,\mathcal{W}\,q \equiv p\,\mathcal{W}\,q$

(224) $\Box$ **to** $\mathcal{W}$ **law:** $\Box p \equiv p\,\mathcal{W}\,\mathit{false}$

(225) $\Diamond$ **to** $\mathcal{W}$ **law:** $\Diamond p \equiv \neg(\neg p\,\mathcal{W}\,\mathit{false})$

(226) $\mathcal{W}$ **implication:** $p\,\mathcal{W}\,q \Rightarrow \Box p \vee \Diamond q$

(227) **Absorption:** $p\,\mathcal{W}\,(p\,\mathcal{U}\,q) \equiv p\,\mathcal{W}\,q$

(228) **Absorption:** $(p\,\mathcal{U}\,q)\,\mathcal{W}\,q \equiv p\,\mathcal{U}\,q$

(229) **Absorption:** $p\,\mathcal{U}\,(p\,\mathcal{W}\,q) \equiv p\,\mathcal{W}\,q$

(230) **Absorption:** $(p\,\mathcal{W}\,q)\,\mathcal{U}\,q \equiv p\,\mathcal{U}\,q$

(231) **Absorption of** $\mathcal{W}$ **into** $\Diamond$: $\Diamond q\,\mathcal{W}\,q \equiv \Diamond q$

(232) **Absorption of** $\mathcal{W}$ **into** $\Box$: $\Box p \wedge p\,\mathcal{W}\,q \equiv \Box p$

(233) **Absorption of** $\Box$ **into** $\mathcal{W}$: $\Box p \vee p\,\mathcal{W}\,q \equiv p\,\mathcal{W}\,q$

(234) $p\,\mathcal{W}\,q \wedge \Box \neg q \Rightarrow \Box p$

(235) $\Box p \Rightarrow p\,\mathcal{U}\,q \vee \Box \neg q$

(236) $\neg \Box p \wedge p\,\mathcal{W}\,q \Rightarrow \Diamond q$

(237) $\Diamond q \Rightarrow \neg \Box p \vee p\,\mathcal{U}\,q$

(238) $\Box \neg p \wedge p\,\mathcal{U}\,q \Rightarrow q$

(239) **Left monotonicity of** $\mathcal{W}$: $\Box(p \Rightarrow q) \Rightarrow (p\,\mathcal{W}\,r \Rightarrow q\,\mathcal{W}\,r)$

(240) **Right monotonicity of** $\mathcal{W}$: $\Box(p \Rightarrow q) \Rightarrow (r\,\mathcal{W}\,p \Rightarrow r\,\mathcal{W}\,q)$

(241) $\mathcal{W}$ **strengthening rule:** $\Box((p \Rightarrow r) \wedge (q \Rightarrow s)) \Rightarrow (p\,\mathcal{W}\,q \Rightarrow r\,\mathcal{W}\,s)$

(242) $\mathcal{W}$ **catenation rule:** $\Box((p \Rightarrow q\,\mathcal{W}\,r) \wedge (r \Rightarrow q\,\mathcal{W}\,s)) \Rightarrow (p \Rightarrow q\,\mathcal{W}\,s)$

(243) **Left** $\mathcal{U}\,\mathcal{W}$ **implication:** $(p\,\mathcal{U}\,q)\,\mathcal{W}\,r \Rightarrow (p\,\mathcal{W}\,q)\,\mathcal{W}\,r$

(244) **Right** $\mathcal{W}\,\mathcal{U}$ **implication:** $p\,\mathcal{W}\,(q\,\mathcal{U}\,r) \Rightarrow p\,\mathcal{W}\,(q\,\mathcal{W}\,r)$

(245) **Right** $\mathcal{U}\,\mathcal{U}$ **implication:** $p\,\mathcal{U}\,(q\,\mathcal{U}\,r) \Rightarrow p\,\mathcal{U}\,(q\,\mathcal{W}\,r)$

(246) **Left** $\mathcal{U}\,\mathcal{U}$ **implication:** $(p\,\mathcal{U}\,q)\,\mathcal{U}\,r \Rightarrow (p\,\mathcal{W}\,q)\,\mathcal{U}\,r$

(247) **Left** $\mathcal{U} \vee$ **strengthening:** $(p\,\mathcal{U}\,q)\,\mathcal{U}\,r \Rightarrow (p \vee q)\,\mathcal{U}\,r$

(248) **Left** $\mathcal{W} \vee$ **strengthening:** $(p\,\mathcal{W}\,q)\,\mathcal{W}\,r \Rightarrow (p \vee q)\,\mathcal{W}\,r$

(249) **Right** $\mathcal{W} \vee$ **strengthening:** $p\,\mathcal{W}\,(q\,\mathcal{W}\,r) \Rightarrow p\,\mathcal{W}\,(q \vee r)$

(250) **Right** $\mathcal{W} \vee$ **ordering:** $p\,\mathcal{W}\,(q\,\mathcal{W}\,r) \Rightarrow (p \vee q)\,\mathcal{W}\,r$

(251) **Right** $\wedge\,\mathcal{W}$ **ordering:** $p\,\mathcal{W}\,(q \wedge r) \Rightarrow (p\,\mathcal{W}\,q)\,\mathcal{W}\,r$

(252) $\mathcal{U}$ **ordering:** $\neg p\,\mathcal{U}\,q \vee \neg q\,\mathcal{U}\,p \equiv \Diamond(p \vee q)$

(253) $\mathcal{W}$ **ordering:** $\neg p\,\mathcal{W}\,q \vee \neg q\,\mathcal{W}\,p$

(254) $\mathcal{W}$ **implication ordering:** $p\,\mathcal{W}\,q \wedge \neg q\,\mathcal{W}\,r \Rightarrow p\,\mathcal{W}\,r$

(255) **Lemmon formula:** $\Box(\Box p \Rightarrow q) \vee \Box(\Box q \Rightarrow p)$

Additional Theorems of Linear Temporal Logic

Temporal Modus Ponens and Modus Tollens

(S1) **Negation of $\Rightarrow$:** $\quad \neg(p \Rightarrow q) \equiv p \wedge \neg q$

(S2) **Temporal modus ponens of $\bigcirc$:** $\quad \Box(p \Rightarrow q) \wedge \bigcirc p \Rightarrow \bigcirc q$

(S3) **Temporal modus ponens of $\Diamond$:** $\quad \Box(p \Rightarrow q) \wedge \Diamond p \Rightarrow \Diamond q$

(S4) **Temporal modus ponens of $\Box$:** $\quad \Box(p \Rightarrow q) \wedge \Box p \Rightarrow \Box q$

(S5) **Temporal modus ponens of $\Box\Diamond$:** $\quad \Box(p \Rightarrow q) \wedge \Box\Diamond p \Rightarrow \Box\Diamond q$

(S6) **Temporal modus ponens of $\Diamond\Box$:** $\quad \Box(p \Rightarrow q) \wedge \Diamond\Box p \Rightarrow \Diamond\Box q$

(S7) **Prior formula:** $\quad \neg\Diamond q \Rightarrow (\Box(p \Rightarrow q) \Rightarrow \neg\Diamond p)$

(S8) **Temporal modus tollens of $\Box$:** $\quad \Box(p \Rightarrow q) \wedge \neg\Box q \Rightarrow \neg\Box p$

(S9) **Temporal modus tollens of $\Box$:** $\quad \Box(p \Rightarrow q) \wedge \Box\neg q \Rightarrow \Box\neg p$

(S10) **Temporal modus tollens of $\Diamond$:** $\quad \Box(p \Rightarrow q) \wedge \neg\Diamond q \Rightarrow \neg\Diamond p$

(S11) **Temporal modus tollens of $\Diamond$:** $\quad \Box(p \Rightarrow q) \wedge \Diamond\neg q \Rightarrow \Diamond\neg p$

(S12) **Temporal modus tollens of $\bigcirc$:** $\quad \Box(p \Rightarrow q) \wedge \neg\bigcirc q \Rightarrow \neg\bigcirc p$

(S13) **Temporal modus tollens of $\bigcirc$:** $\quad \Box(p \Rightarrow q) \wedge \bigcirc\neg q \Rightarrow \bigcirc\neg p$

Spot

Theorems from Spot are distributed throughout the paper.

StackExchange.com

(S14) $\quad \Box p \Rightarrow \Diamond q \equiv p \, \mathcal{U} \, (q \vee \neg p)$

(S15) $\quad \Diamond (p \Rightarrow q) \equiv p \, \mathcal{U} \, (q \vee \neg p)$

(S16) $\quad p \, \mathcal{U} \, \neg p \equiv \neg \Box p$

(S17) $\quad p \, \mathcal{U} \, (q \vee \neg p) \equiv \Box p \Rightarrow p \, \mathcal{U} \, q$

(S18) $\quad \Box p \Rightarrow (p \, \mathcal{U} \, q) \, \mathcal{U} \, p$

(S19) $\quad p \, \mathcal{U} \, (q \, \mathcal{U} \, p) \equiv q \, \mathcal{U} \, p$

(S20) $\quad (p \, \mathcal{U} \, q) \, \mathcal{U} \, p \equiv q \, \mathcal{U} \, p$

Induction

(S21) **(165) Lemma F:** $\quad \Box p \vee \Box q \Rightarrow \circ (\Box p \vee \Box q)$

(S22) $\quad \Box p \wedge \Box q \Rightarrow \circ (\Box p \wedge \Box q)$

(S23) $\quad \circ (\Diamond p \vee \Diamond q) \Rightarrow (\Diamond p \vee \Diamond q)$

(S24) $\quad \circ (\Diamond p \wedge \Diamond q) \Rightarrow (\Diamond p \wedge \Diamond q)$

(S25) $\quad \Box (p \Rightarrow \circ p) \Rightarrow \Box (p \Rightarrow \Box p)$

(S26) $\quad \Box p \Rightarrow (p \Rightarrow \Box p)$

(S27) $\quad \Diamond \Box p \Rightarrow \Diamond (p \Rightarrow \Box p)$

(S28) $\quad \Box p \Rightarrow \Diamond (p \Rightarrow \Box p)$

(S29) $\quad \Box \Diamond$ **induction:** $\quad \Box ((p \Rightarrow q \wedge \circ p) \wedge (q \Rightarrow \Diamond p)) \Rightarrow (p \Rightarrow \Box \Diamond p)$

(S30) $\quad \Box \Diamond$ **induction:** $\quad \Box ((p \Rightarrow \circ p) \wedge (p \Rightarrow \Diamond p)) \Rightarrow (p \Rightarrow \Box \Diamond p)$

(S31) **Obligation induction:** $\quad \Box (p \Rightarrow \circ p) \vee \Box (p \Rightarrow \circ q) \Rightarrow (p \Rightarrow p \, \mathcal{W} \, \Diamond q)$

(S32) **Induction rule** $\mathcal{U}$: $\quad \Box (p \Rightarrow \circ (p \wedge q)) \Rightarrow (p \Rightarrow p \, \mathcal{U} \, q)$

(S33) **Dummett induction:** $\quad \Box (p \equiv \circ p) \Rightarrow (\Diamond \Box p \Rightarrow \Box p)$

Absorption

(S34) **Absorption of $\Diamond$ into $\neg\Box$:** $\quad \Diamond \neg \Box p \equiv \neg \Box p$

(S35) **Absorption of $\Box$ into $\neg\Diamond$:** $\quad \Box \neg \Diamond p \equiv \neg \Diamond p$

(S36) $p \mathcal{W} q \vee \neg q \mathcal{W} p \equiv q \vee \neg q \mathcal{W} p$

Frame Rules

(S37) **Manna-Pnueli frame rule $\Diamond$:** $\quad \Box(p \Rightarrow \Diamond q) \Rightarrow (\Box w \wedge p \Rightarrow \Diamond(\Box w \wedge q))$

(S38) **Manna-Pnueli frame rule $\Box$:** $\quad \Box(p \Rightarrow \Box q) \Rightarrow (\Box w \wedge p \Rightarrow \Box(\Box w \wedge q))$

(S39) **Manna-Pnueli frame rule $\circ$:** $\quad \Box(p \Rightarrow \circ q) \Rightarrow (\Box w \wedge p \Rightarrow \circ(\Box w \wedge q))$

Duality

(S40) **Definition of weak release $\mathcal{R}$:** $\quad p \mathcal{R} q \equiv q \mathcal{W} (p \wedge q)$

(S41) **Definition of strong release $\mathcal{M}$:** $\quad p \mathcal{M} q \equiv q \mathcal{U} (p \wedge q)$

(S42) $\neg(q \mathcal{U} (\neg p \wedge q)) \vee \neg(p \mathcal{U} (\neg q \wedge p))$

(S43) $\Diamond((p \wedge \Diamond q) \vee (\Diamond p \wedge q)) \equiv \Diamond p \wedge \Diamond q$

(S44) $\Diamond p \wedge \Diamond q \Rightarrow (\Diamond(p \wedge q) \vee \Diamond(p \wedge \Diamond q) \vee \Diamond(\Diamond p \wedge q))$

(S45) $p \not\Leftarrow q \equiv \neg(p \Leftarrow q)$

(S46) $p \Leftarrow q \equiv \neg q \vee p$

(S47) $p \not\Leftarrow q \equiv \neg(q \Rightarrow p)$

Next $\circ$

(S48) $\Box(p \Rightarrow \circ p) \Rightarrow \Box(p \Rightarrow \circ(p \vee q))$

(S49) $\Box(p \Rightarrow \circ(p \Rightarrow q) \wedge \circ p) \equiv \Box(p \Rightarrow \circ(p \wedge q))$

(S50) $p \wedge \circ p \equiv \neg(\circ p \Rightarrow \neg p)$

Until $\mathcal{U}$

(S51) $\square(p \wedge q) \Rightarrow p \, \mathcal{U} \, q$

(S52) $p \, \mathcal{U} \, (q \vee \Diamond r) \equiv p \, \mathcal{U} \, q \vee \Diamond r$

(S53) $\square p \, \mathcal{U} \, q \Rightarrow p \, \mathcal{U} \, q \vee \square \neg q$

(S54) $\neg(p \, \mathcal{U} \, q) \Rightarrow p \, \mathcal{U} \, \neg q$

(S55) **Distributivity of $\neg$ over $\mathcal{W}$:** $\neg(p \, \mathcal{W} \, q) \equiv \neg q \, \mathcal{W} \, (\neg p \wedge \neg q) \wedge \neg \square p$

(S56) **Distributivity of $\neg$ over $\mathcal{U}$:** $\neg(p \, \mathcal{U} \, q) \equiv \neg q \, \mathcal{U} \, (\neg p \wedge \neg q) \vee \square \neg q$

(S57) **Weak symmetry of $\mathcal{U}$:** $(p \vee q) \, \mathcal{U} \, q \Rightarrow q \, \mathcal{U} \, (p \vee q)$

(S58) **Generalized $\mathcal{U}$ excluded middle:** $p \, \mathcal{U} \, q \vee r \, \mathcal{U} \, \neg q$

(S59) $((\square p \vee \Diamond q) \equiv p \, \mathcal{W} \, \Diamond q) \equiv \square p \vee \Diamond q \vee \square \neg q \, \mathcal{U} \, (\neg p \wedge \square \neg q)$

In-state Expansion of $\mathcal{U}$

(S60) **In-state next-state equivalence:** $q \vee (p \wedge p \, \mathcal{U} \, q) \equiv q \vee (p \wedge \bigcirc (p \, \mathcal{U} \, q))$

(S61) **In-state expansion of $\mathcal{U}$:** $p \, \mathcal{U} \, q \equiv q \vee (p \wedge p \, \mathcal{U} \, q)$

Nested Insertion

(S62) **Nested insertion:** $r \Rightarrow p \, \mathcal{U} \, (q \, \mathcal{U} \, r)$

(S63) **Nested insertion:** $r \Rightarrow (p \, \mathcal{U} \, q) \, \mathcal{U} \, r$

(S64) **Indefinite nested insertion:** $x_n \Rightarrow x_1 \, \mathcal{U} \, (x_2 \, \mathcal{U} \, (\ldots \, \mathcal{U} \, (x_{n-1} \, \mathcal{U} \, x_n \underbrace{)\ldots))}_{n-2 \text{ times}}$ for $n \geq 3$

(S65) **Indefinite nested insertion:** $x_n \Rightarrow \underbrace{(\ldots(}_{n-2 \text{ times}} x_1 \, \mathcal{U} \, x_2) \, \mathcal{U} \, x_3) \ldots \, \mathcal{U} \, x_{n-1}) \, \mathcal{U} \, x_n$

Eventually $\diamond$

(S66) $\diamond p \wedge \diamond q \Rightarrow \diamond (p \wedge \diamond q) \vee \diamond (\diamond p \wedge q)$

(S67) $\square p \wedge \diamond q \Rightarrow \diamond (\square p \wedge q)$

(S68) $\square (p \Rightarrow \circ (p \Rightarrow q)) \Rightarrow \diamond (p \Rightarrow q)$

(S69) $\diamond (p \, \mathcal{U} \, q) \equiv \diamond q$

(S70) $p \, \mathcal{U} \diamond q \equiv \diamond (p \, \mathcal{U} \, q)$

(S71) $\diamond q \Rightarrow (p \, \mathcal{W} \, q \equiv p \, \mathcal{U} \, q)$

Always $\square$

(S72) $\square (p \vee q) \wedge \square (\square p \vee q) \wedge \square (p \vee \square q) \Rightarrow \square p \vee \square q$

(S73) $\square (\square p \vee q) \wedge \square (p \vee \square q) \Rightarrow \square p \vee \square q$

(S74) $\square (p \wedge \square p \Rightarrow q) \vee \square (q \wedge \square q \Rightarrow p)$

(S75) $\square (\square p \Rightarrow \square q) \vee \square (\square q \Rightarrow \square p)$

(S76) $\square ((p \Rightarrow \square p) \Rightarrow \square p) \equiv \square p$

Always Eventually $\square \diamond$ and its Dual $\diamond \square$

(S77) $\square (p \Rightarrow \circ (p \Rightarrow q)) \Rightarrow \square \diamond (p \Rightarrow q)$

(S78) $\square \diamond (p \vee \circ q) \equiv \square \diamond (p \vee q)$

(S79) $\diamond \square (p \wedge \circ q) \equiv \diamond \square (p \wedge q)$

(S80) $\square \diamond (p \vee \diamond q) \equiv \square \diamond (p \vee q)$

(S81) $\diamond \square (p \wedge \square q) \equiv \diamond \square (p \wedge q)$

(S82) $\circ p \vee \square \diamond q \equiv \circ (p \vee \square \diamond q)$

(S83) $\circ p \wedge \diamond \square q \equiv \circ (p \wedge \diamond \square q)$

Wait $\mathcal{W}$

(S84) **In-state expansion of** $\mathcal{W}$: $\quad p \mathcal{W} q \equiv q \vee (p \wedge p \mathcal{W} q)$

(S85) $\neg(p \mathcal{W} q) \Rightarrow p \mathcal{W} \neg q$

(S86) $p \mathcal{W} \square q \wedge \Diamond \square q \Rightarrow \square (p \mathcal{W} q)$

(S87) **Generalized** $\mathcal{W}$ **excluded middle:** $\quad p \mathcal{W} q \vee r \mathcal{W} \neg q$

(S88) $p \mathcal{W} q \equiv \Diamond \neg p \Rightarrow p \mathcal{U} q$

(S89) $p \mathcal{W} q \equiv p \mathcal{U} (q \vee \square p)$

(S90) $q \mathcal{W} \square \neg p \Rightarrow (\Diamond p \Rightarrow q)$

(S91) $q \mathcal{W} \square \neg q \Rightarrow (\square (\circ q \Rightarrow q) \Rightarrow (\Diamond q \Rightarrow q))$

(S92) $p \mathcal{W} \square p \equiv \square p$

(S93) $\square p \mathcal{W} q \equiv \square p \vee q$

(S94) $\square (p \mathcal{W} q) \equiv \square (p \vee q)$

(S95) $p \mathcal{W} q \equiv p \mathcal{W} (q \vee \square p)$

(S96) $p \mathcal{W} (q \vee \Diamond r) \equiv p \mathcal{W} q \vee \Diamond r$

(S97) $p \mathcal{U} r \wedge q \mathcal{W} r \equiv (p \wedge q) \mathcal{U} r$

(S98) $p \mathcal{U} q \vee p \mathcal{W} r \equiv p \mathcal{W} (q \vee r)$

(S99) $p \mathcal{W} q \vee \Diamond q \equiv \square p \vee \Diamond q$

(S100) $p \mathcal{W} \Diamond q \equiv p \mathcal{W} q \vee \Diamond q$

(S101) $p \mathcal{W} q \vee q \mathcal{W} p \equiv p \vee q$

(S102) $\neg p \mathcal{W} q \vee q \mathcal{W} \neg p \equiv p \Rightarrow q$

(S103) $(\neg p \mathcal{W} q \vee q \mathcal{W} \neg p) \wedge (\neg q \mathcal{W} p \vee p \mathcal{W} \neg q) \equiv (p \equiv q)$

(S104) $q \mathcal{U} (p \wedge q) \wedge p \mathcal{U} (p \wedge q) \equiv p \wedge q$

(S105) $q \mathcal{W} (p \wedge q) \wedge p \mathcal{W} (p \wedge q) \wedge \Diamond (p \wedge q) \equiv p \wedge q$

(S106) $p \mathcal{W} (p \wedge q) \wedge q \mathcal{W} (p \wedge q) \equiv p \wedge q$

Proof and Variations of the Dummett Formula

(S107) **Dummett variant:** $\quad \square ((p \Rightarrow \square p) \Rightarrow \square p) \Rightarrow (\Diamond \square p \Rightarrow \square p)$

(S108) **Dummett variant:** $\quad \square ((p \Rightarrow \circ p) \wedge (\square (p \Rightarrow \square p) \Rightarrow \square p)) \Rightarrow (\Diamond \square p \Rightarrow \square p)$

(S109) **Dummett variant:** $\quad \square (\Diamond (p \Rightarrow \square p) \Rightarrow \square p) \Rightarrow (\Diamond \square p \Rightarrow \square p)$

(S110) **Dummett variant:** $\quad \square (\square (p \Rightarrow \square p) \Rightarrow \square p) \Rightarrow (\square (p \Rightarrow \circ p) \Rightarrow \square p)$

(S111) **Dummett formula:** $\quad \square (\square (p \Rightarrow \square p) \Rightarrow \square p) \Rightarrow (\Diamond \square p \Rightarrow \square p)$

Using ACL2 to Confirm Theorems in CDS4LTL

Scott M. Staley
Ford Motor Company Research Labs (retired)
Seattle, WA 98198

Abstract

This paper reports on work done to use ACL2 to prove all the axioms and theorems stated in the paper A Calculational Deductive System for Linear Temporal Logic (Complete) by Warford, Vega and Staley [3]. The theorems of CDS4LTL are collected together in a supporting document [4]. ACL2 (A Computational Logic for Applicative Common Lisp) is a software system freely available from the University of Texas at Austin which allows modeling in an extensible first-order theory and includes an automated theorem prover for reasoning about the models [1]. In this work, ACL2 is used to extend the operators of the propositional calculus (IFF, NOT, OR, AND, IMPLIES) to take arbitrary (but finite) length lists and return lists which are the result of applying the operators to the elements of the list pairwise. Next, using these operators, ACL2 is used to prove the axioms and theorems of the propositional calculus [(3.1) - (3.82.2) and (4.1) - (4.3) and (4.6) from the equations sheet that accompanies the paper]. Finally, using the semantics for the Linear Temporal Logic (LTL) operators presented in the paper, ACL2 functions are defined for each LTL operator and the theorems in the CDS4LTL paper are proved using ACL2.

1 Preface to this Paper (June 2020)

This paper was written at the end of 2015. At that time I thought we were nearly done with our work on CDS4LTL which included 154 axioms, definitions and theorems. However the work continued on for another 4 years and added 100 theorems to the survey. I never returned to the work of confirming this additional 100 theorems by producing an ACL2 proof for each. Perhaps some motivated student studying CDS4LTL or ACL2 can take this unfinished project on as a learning exercise. That is what proving the first 154 theorems was for me at the time. The theorems in this paper hold for traces modeled as arbitrarily long (but finite) boolean lists. My plan is to revisit this work using a model of infinite length traces proposed by Sandip Ray in his book *Scalable Techniques for Formal Verification* [2].

2 Preliminaries

This paper is not intended to be a tutorial for either LTL or ACL2. There is plenty of tutorial information readily available about LTL and ACL2 both in books and from searching the internet. What is presented is simply how ACL2 was used to extend the propositional calculus operators to handle state sequences (also called traces), and how LTL operators were modeled in ACL2 and the axioms and theorems of CDS4LTL proved using the automated theorem prover in ACL2. Normally, axioms and definitions are taken as given and are not proved. This is the case in the hand proofs of LTL theorems in the CDS4LTL paper. However, in ACL2 the computational functions used to model the operators of propositional calculus and LTL are admitted to the ACL2 logic as axioms allowing the axioms of the paper to be checked for validity (proved) based on the functions modeling the operator semantics defined in the CDS4LTL paper.

3 True and False

In the ACL2 logic false is denoted by nil or (). True is anything that is non-nil. In LTL true is defined as the infinite sequence consisting of $(t\ t\ t\ t \ldots)$, and false is the infinite sequence $(nil\ nil\ nil\ nil \ldots)$. So we need to model LTL true and false in ACL2. This is done using arbitrarily long (but finite) lists. The following four functions construct true and false objects and define recognizers for these objects. First for true.

```
;;; Function Definitions for True and Truep
;;;
;;; True - returns a boolean-listp of length n,
;;;        where 0 < n, with all elements set to t.
;;;
(defun true (n)
 (declare (xargs :measure (acl2-count n)
          :guard (natp n)
          :verify-guards t))
 (if (zp n)
   nil
   (cons t (true (1- n)))))
;;;
;;; Truep - returns t if x is a boolean-listp
;;;         with all elements t. Atoms are truep,
;;;         for example, (truep '()) is t.
;;;
(defun truep (x)
 (declare (xargs :measure (acl2-count x)
          :guard (boolean-listp x)
          :verify-guards t))
 (if (endp x)
   t
   (acl2::and (car x) (truep (cdr x)))))
```

With these definitions admitted into the ACL2 logic it is now possible to prove properties of these two functions using the theorem prover. For example, it is a theorem that for any natural number n, the (cdr (true n)) is truep. The cdr is what remains in a list when the first element is removed. For example the (cdr '(a b c d)) is '(b c d). Here is the command in ACL2 to attempt to prove this theorem.

```
(acl2::implies (natp n)
        (truep (cdr (true n))))
:hints (("Goal" :expand ((true (+ -1 n))))))
```

ACL2 successfully proved this theorem and stored results in its database of theorems. Note that this defthm event is not a function definition and does not change the logic. It proves a property within the logic. The next two functions, false and falsep do extend the logic by introducing functions that create false objects and recognize false objects.

```
;;;
;;; Function Definitions for False and Falsep
;;;
;;; False - returns a boolean-listp of length n,
;;;         where 0 < n, with all elements set to nil.
;;;
(defun false (n)
 (declare (xargs :guard (natp n)
         :verify-guards t))
 (not (true n)))

;;;
;;; Falsep - returns t if x is a boolean-listp,
;;;         of any length, with all elements nil.
;;;         Atoms are falsep, so for example, (falsep '()) is t.
;;;
(defun falsep (x)
 (declare (xargs :measure (acl2-count x)
         :guard (boolean-listp x)
         :verify-guards t))
 (if (endp x)
   t
  (acl2::and (acl2::not (car x))
       (falsep (cdr x)))))
```

```
;;;
;;; Theorem (3.13) Negation of False
;;;
;;; ACL2 proved this.
;;;
(defthm negation-of-false
 (equal (not (false n)) (true n))
 :rule-classes ((:rewrite))
```

By the function definition of false it is a theorem now in ACL2 that (false n) is equivalent to (not (true n)). Using the theorem prover, (not (false n)) is equivalent to (true n) is a theorem.

4 Propositional Operators

The propositional operators are defined in ACL2 to operate on the state sequences of the CDS4LTL paper. The state sequences are simply modeled in ACL2 using the internal data type boolean-list (which also has the built-in recognizer boolean-listp). The variables in propositional calculus and LTL formulas are sequences of the ACL2 atoms t and nil. That is, the truth value of a variable changes over time. A variable, say p, with a value of '(t t nil nil) is defined over 4 time steps. p is true at time 0 and 1 and false at time 2 and 3.

The operators are defined in the order they are introduced in the equation sheet that accompanies the CDS4LTL paper. All of these operators exist in ACL2, but do not take state sequences as input. So these functions, defined in the LTL package use the ACL2 versions and map them over the state sequences. The first function is equivales (also called iff).

```
;;;
;;; Equivales (iff) operator
;;;
;;; This works on state sequences (lists).
;;; Basically iffs two lists together by acl2::iff-ing
;;; pairwise the elements in each at location i.
;;; If the input lists are of different sizes, the output
;;; length is the length of the shorter list.
;;;
(defun iff (p q)
 (declare (xargs :measure (acl2-count p)
         :guard (acl2::and (boolean-listp p)
                   (boolean-listp q))
         :verify-guards t))
 (cond ((acl2::or (endp p) (endp q)) nil)
    (t (cons (acl2::iff (car p) (car q))
         (iff (cdr p) (cdr q))))))
```

The next function implements the negation (not) operator.

```
;;;
;;; Not Operator
;;;
;;; The not operator for state sequences uses
;;; acl2::not on each element.
;;; This not operator operates on the entire
;;; sequence of states of p. For example if
;;; p is (t nil nil t) then (not p) is (nil t t nil)
;;;
(defun not (p)
 (declare (xargs :measure (acl2-count p)
```

```
              :guard (boolean-listp p)
              :verify-guards t))
  (cond ((endp p) nil)
        (t (cons (acl2::not (car p))
                 (not (cdr p))))))
```

While not strictly needed, defining a niff (not (iff p q)) function, makes some proofs easier.

```
;;;
;;; Niff Operator
;;;
;;; Definition of niff as a recursive function like iff.
;;; This form is better for proving some theorems than
;;; defining it as (not (iff p q)).
;;;
(defun niff (p q)
  (declare (xargs :measure (acl2-count p)
              :guard (acl2::and (boolean-listp p)
                                (boolean-listp q))
              :verify-guards t))
  (cond ((acl2::or (endp p) (endp q)) nil)
        (t (cons (acl2::not (acl2::iff (car p) (car q)))
                 (niff (cdr p) (cdr q))))))
```

Next we complete the propositional operators by defining disjunction, conjunction and implication.

```
;;;
;;; Or operator
;;;
;;; This is a binary operator, unlike the acl2::or
;;; which takes any number of arguments.
;;; This ltl::or works on lists which are used
;;; to model state sequences. This function
;;; or-s two lists together comparing pairwise
;;; the elements in location i. The ith location in
;;; the result contains (or (nth i p) (nth i q))
;;;
(defun or (p q)
  (declare (xargs :measure (acl2-count p)
              :guard (acl2::and (boolean-listp p)
                                (boolean-listp q))
              :verify-guards t))
  (cond ((acl2::or (endp p) (endp q)) nil)
        (t (cons (acl2::or (car p) (car q))
                 (or (cdr p) (cdr q))))))

;;;
;;; And operator
;;;
;;; This is a binary operator, unlike the acl2::and
;;; which takes any number of arguments. This ltl::and
```

```
;;; works on lists which are used to model state sequences.
;;; This function and-s two lists together
;;; comparing pairwise the elements in location i.
;;; The ith location in the result contains
;;; (and (nth i p) (nth i q))
;;;
(defun and (p q)
 (declare (xargs :measure (acl2-count p)
          :guard (acl2::and (boolean-listp p)
                      (boolean-listp q))
          :verify-guards t))
 (cond ((acl2::or (endp p) (endp q)) nil)
    (t (cons (acl2::and (car p) (car q))
         (and (cdr p) (cdr q))))))

;;;
;;; Implies operator
;;;
;;; This works on state sequences (lists).
;;; Basically implies two lists together comparing
;;; pairwise the elements in each at location i.
;;;
(defun implies (p q)
 (declare (xargs :measure (acl2-count p)
          :guard (acl2::and (boolean-listp p)
                      (boolean-listp q))
          :verify-guards t))
 (cond ((acl2::or (endp p) (endp q)) nil)
    (t (cons (acl2::implies (car p) (car q))
         (implies (cdr p) (cdr q))))))
```

With these operators the propositional calculus in ACL2 is extended to work on the state sequences of the CDS4LTL paper.

5 Propositional Calculus Proofs

All the theorems on pages 2 - 4 of the equation sheet have been proven for the above state sequence operator definitions using ACL2. In addition ACL2 proved (4.1) - (4.3) and (4.6) on page 5 for a total of 94 theorems. The procedure for reproducing these results involves the following steps.

1. Install a Common Lisp implementation on your computer. Clozure Common Lisp is recommended, and is freely available. Learn more at https://ccl.clozure.com.

2. Install the latest version of ACL2 on your computer. It is freely available. Learn more at https://www.cs.utexas.edu/users/moore/acl2/.

Pepperdine Papers on LTL 229

3. (Recommended) Install an emacs editor on your computer and run ACL2 in a shell buffer in emacs. emacs is freely available. Learn more at https://www.gnu.org/software/emacs/.

4. With ACL2 running, either as in Step 3, or otherwise, load the following files into ACL2. These files contain all the function definitions above, as well as instructions to ACL2 to prove all the theorems of the propositional calculus. The directory structure of the filenames shown will have to be modified to reflect where the files really are on a particular machine.

```
;;;
;;; Run ACL2 at the shell prompt.
;;;

bash-3.2$ ~/acl2/v6-4/acl2-sources/saved_acl2

;;;
;;; This results in the following system information
;;; being printed out.
;;;

Welcome to Clozure Common Lisp Version 1.9-r15759 (DarwinX8664)!

ACL2 Version 6.4 built January 15, 2014 15:01:45.
Copyright (C) 2014, Regents of the University of Texas
ACL2 comes with ABSOLUTELY NO WARRANTY. This is free software
and you are welcome to redistribute it under certain conditions.
For details, see the LICENSE file distributed with ACL2.

Initialized with
(INITIALIZE-ACL2 'INCLUDE-BOOK *ACL2-PASS-2-FILES*).
See the documentation topic note-6-4 for recent changes.
Note: We have modified the prompt in some underlying Lisps
to further distinguish it from the ACL2 prompt.

ACL2 Version 6.4. Level 1. Cbd "/".
System books directory "~/acl2/v6-4/acl2-sources/books/".
Type :help for help.
Type (good-bye) to quit completely out of ACL2.

ACL2 !>

;;;
;;; That is the ACL2 prompt above.
;;;
;;; Load Package Definition for LTL at the ACL2 prompt.
;;;
ACL2 !>(ld "~/Warford Copy/ltl-defpkg.lisp")

;;;
;;; Now move to LTL package.
;;;
```

```
(in-package "LTL")

;;;
;;; This changes the ACL2 prompt to
;;;
LTL !>

;;;
;;; Now, at this prompt load all the following files.
;;;
;;; Equivalence and True
;;;
(ld "~/Warford Copy/ltl-functions/equiv-and-true.lisp")

;;;
;;; Negation, Inequivalence and False
;;;
(ld "~/Warford Copy/ltl-functions/negation-niff-and-false.lisp")

;;;
;;; Disjunction
;;;
(ld "~/Warford Copy/ltl-functions/disjunction.lisp")

;;;
;;; Conjunction
;;;
(ld "~/Warford Copy/ltl-functions/conjunction.lisp")

;;;
;;; Implication
;;;
(ld "~/Warford Copy/ltl-functions/implication.lisp")
```

ACL2 should run, defining all the operators and proving all the theorems. All these files can be obtained from Professor Warford of Pepperdine University.

6 Linear Temporal Logic Operators

In Section 2.2 of the CDS4LTL paper the following LTL operators are defined: Next $\bigcirc$, Until $\mathcal{U}$, Eventually $\diamond$, Always $\square$, and Wait $\mathcal{W}$. Based on the semantics of these operators given in the CDS4LTL paper, here are the ACL2 function definitions.

```
;;;
;;; Next operator
;;;
;;; (next p) holds in state j iff p holds in state j+1
;;; (next p) has the effect of shifting the states of p ahead
;;; one time increment. For example if p is (t nil nil t) then
;;; (next p) is (nil nil t). Notice that the length of the
```

```
;;; sequence is smaller by one. This means that we are now only
;;; able to reason over a smaller set of states. (Unless a way
;;; is devised to update the sequence of values as we move
;;; ahead in time.)
;;; As we look at next states (next (next (next p)))
;;; or we increment the clock the future unknown states move into
;;; the system description. For example, if p is (nil t t t nil),
;;; then (next (next (next (next p)))) is (nil). Conveniently,
;;; this turns out to be the Lisp function called ''cdr.''
;;;
(defun next (p)
 (declare (xargs :guard (acl2::and (consp p)
                    (boolean-listp p))))
 (cdr p))


;;;
;;; Until operator (binary operator)
;;;
;;; (until p q) holds in state j iff there exists a
;;; future state k (k >= j) that satisfies q,
;;; and for all states i between j and k, state i satisfies p.
;;;
;;; For example if p is
;;; (nil t t t t nil nil t nil t t t t t t)
;;; and, q is
;;; (t nil nil nil nil t nil nil t nil nil t nil nil nil nil)
;;; then (until p q) is
;;; (t t t t t t nil t t t t t nil nil nil nil)
;;;
(defun until (p q)
 (declare (xargs :measure (acl2-count p)
          :guard (acl2::and (boolean-listp p)
                    (boolean-listp q))
          :verify-guards t))
 (cond ((acl2::or (endp p) (endp q)) nil)
    (t (let* ((next-until (until (cdr p) (cdr q))))
       (cond ((eq (car q) t) (cons t next-until))
          ((eq (car p) nil) (cons nil next-until))
          (t (cons (car next-until) next-until)))))))


;;;
;;; Eventually operator
;;;
;;; (eventually q) holds in state j iff q holds in state j
;;; or in any other state k, s.t. k >= j.
;;; If q is (nil nil t t t nil nil) then
;;; (eventually q) is (t t t t t nil nil).
;;; (eventually nil) = nil.
;;;
(defun eventually (q)
 (declare (xargs :measure (acl2-count q)
         :guard (boolean-listp q)
         :verify-guards t))
 (if (consp q)
   (if (falsep q) (false (len q))
```

```
      (cons t (eventually (cdr q))))
  nil))

;;;
;;; Always operator
;;;
;;; (always p) holds in state j iff p holds in state j
;;; and in all future states k, k > j.
;;; If p is (t t t nil nil t t t t) then
;;; (always p) is (nil nil nil nil nil t t t t).
;;; (always nil) = nil
;;;
(defun always (p)
 (declare (xargs :measure (acl2-count p)
         :guard (boolean-listp p)
         :verify-guards t))
 (if (consp p)
   (if (truep p) (true (len p))
     (cons nil (always (cdr p))))
  nil))

;;;
;;; Wait operator (binary operator)
;;;
;;; (wait p q) holds in state j iff (or (until p q) (always p))
;;; holds in state j
;;;
;;; The wait operator is weaker than the until operator,
;;; since (wait p q) does not require q to ever be true
;;; while (until p q) does.
;;;
(defun wait (p q)
 (declare (xargs :measure (acl2-count p)
         :guard (acl2::and (boolean-listp p)
                 (boolean-listp q))
         :verify-guards t))
 (cond ((acl2::or (endp p) (endp q)) nil)
    (t (let* ((next-wait (wait (cdr p) (cdr q)))
          (p-until-q (until p q)))
      (cond ((truep p) (cons t next-wait))
         (t (cons (car p-until-q) next-wait)))))))
```

Using these function definitions, and the ACL2 theorem prover, proofs have been constructed for all the axioms, definitions and theorems in the CDS4LTL paper. Recall that the function definitions implement the semantics of the operators as defined in the CDS4LTL paper, and the fact that the axioms and definitions of CDS4LTL are theorems in ACL2 confirms the correctness of the function definitions. All axioms of CDS4LTL have been confirmed with ACL2 proofs.

Next, one ACL2 proof from each operator section of the CDS4LTL paper equation sheet will be shown to provide some flavor for how theorems are stated in ACL2 and how the theorem prover output looks.

6.1 Next

For (2) Axiom, Distributivity of Next over Implies here is the statement of the axiom in ACL2.

```
;;;
;;; (2) Axiom, Distributivity of Next over Implies
;;;
;;; ACL2 proved this.
;;;
(defthm ltl-2-axiom-distributivity-of-next-over-implies
 (equal (next (implies p q))
    (implies (next p) (next q))))
```

And here is the proof summary output from ACL2 indicating the proof attempt was successful.

```
Q.E.D.

Summary
Form: ( DEFTHM LTL-2-AXIOM-DISTRIBUTIVITY-OF-NEXT-OVER-IMPLIES)
Rules: ((:DEFINITION IMPLIES)
    (:DEFINITION NEXT)
    (:EXECUTABLE-COUNTERPART BOOLEAN-LISTP)
    (:EXECUTABLE-COUNTERPART EQUAL)
    (:EXECUTABLE-COUNTERPART FALSEP)
    (:EXECUTABLE-COUNTERPART LEN)
    (:EXECUTABLE-COUNTERPART NOT)
    (:EXECUTABLE-COUNTERPART TRUEP)
    (:FAKE-RUNE-FOR-LINEAR NIL)
    (:REWRITE CDR-CONS)
    (:REWRITE DEFAULT-CDR)
    (:REWRITE IMPLIES-3-75-LEMMA)
    (:REWRITE RIGHT-ZERO-OF-IMPLIES-LEMMA)
    (:TYPE-PRESCRIPTION LEN))
Splitter rules (see :DOC splitter):
 if-intro: ((:DEFINITION IMPLIES))
Warnings: Subsume and Non-rec
Time: 0.01 seconds (prove: 0.01, print: 0.00, other: 0.00)
Prover steps counted: 1598
LTL-2-AXIOM-DISTRIBUTIVITY-OF-NEXT-OVER-IMPLIES
LTL !>
```

6.2 Until

For (10) Axiom, Expansion of Until here is the statement of the axiom in ACL2.

```
;;;
;;; (10) Axiom, Expansion of Until
;;;
;;; ACL2 proved this.
;;;
(defthm ltl-10-axiom-expansion-of-until
 (acl2::implies (boolean-listp q)
        (truep (iff (until p q)
              (or q (and p (next (until p q))))))))
```

And here is the proof summary output from ACL2 indicating the proof attempt was successful.

```
Q.E.D.

The storage of LTL-10-AXIOM-EXPANSION-OF-UNTIL depends upon
the :type-prescription rule TRUEP.

Summary
Form:  ( DEFTHM LTL-10-AXIOM-EXPANSION-OF-UNTIL ...)
Rules: ((:DEFINITION AND)
    (:DEFINITION BOOLEAN-LISTP)
    (:DEFINITION ENDP)
    (:DEFINITION FALSEP)
    (:DEFINITION ACL2::IFF)
    (:DEFINITION IFF)
    (:DEFINITION LEN)
    (:DEFINITION NEXT)
    (:DEFINITION COMMON-LISP::NOT)
    (:DEFINITION OR)
    (:DEFINITION SYNP)
    (:DEFINITION TRUEP)
    (:DEFINITION UNTIL)
    (:ELIM CAR-CDR-ELIM)
    (:EQUIVALENCE IFF-IS-AN-EQUIVALENCE)
    (:EXECUTABLE-COUNTERPART <)
    (:EXECUTABLE-COUNTERPART BINARY-+)
    (:EXECUTABLE-COUNTERPART BOOLEAN-LISTP)
    (:EXECUTABLE-COUNTERPART CAR)
    (:EXECUTABLE-COUNTERPART CDR)
    (:EXECUTABLE-COUNTERPART CONS)
    (:EXECUTABLE-COUNTERPART CONSP)
    (:EXECUTABLE-COUNTERPART ENDP)
    (:EXECUTABLE-COUNTERPART EQUAL)
    (:EXECUTABLE-COUNTERPART FALSEP)
    (:EXECUTABLE-COUNTERPART ACL2::IFF)
    (:EXECUTABLE-COUNTERPART IFF)
    (:EXECUTABLE-COUNTERPART LEN)
    (:EXECUTABLE-COUNTERPART COMMON-LISP::NOT)
```

```
    (:EXECUTABLE-COUNTERPART OR)
    (:EXECUTABLE-COUNTERPART TRUEP)
    (:FAKE-RUNE-FOR-LINEAR NIL)
    (:FAKE-RUNE-FOR-TYPE-SET NIL)
    (:FORWARD-CHAINING BOOLEAN-LISTP-FORWARD-TO-SYMBOL-LISTP)
    (:FORWARD-CHAINING SYMBOL-LISTP-FORWARD-TO-TRUE-LISTP)
    (:INDUCTION AND)
    (:INDUCTION BOOLEAN-LISTP)
    (:INDUCTION OR)
    (:INDUCTION UNTIL)
    (:REWRITE ACL2::|(+ c (+ d x))|)
    (:REWRITE AND-IS-CONSP)
    (:REWRITE CAR-CONS)
    (:REWRITE CDR-CONS)
    (:REWRITE DEFAULT-CDR)
    (:REWRITE IDENTITY-OF-OR-LEMMA-COMMUTED)
    (:REWRITE LTL-9-AXIOM-DISTRIBUTIVITY-OF-NEXT-OVER-UNTIL)
    (:REWRITE OR-IS-CONSP)
    (:REWRITE SYMMETRY-OF-IFF)
    (:REWRITE ZERO-OF-AND-LEMMA)
    (:REWRITE ZERO-OF-OR-LEMMA)
    (:TYPE-PRESCRIPTION AND)
    (:TYPE-PRESCRIPTION BOOLEAN-LISTP)
    (:TYPE-PRESCRIPTION IFF)
    (:TYPE-PRESCRIPTION LEN)
    (:TYPE-PRESCRIPTION OR)
    (:TYPE-PRESCRIPTION SYMBOL-LISTP)
    (:TYPE-PRESCRIPTION TRUEP)
    (:TYPE-PRESCRIPTION UNTIL))
Splitter rules (see :DOC splitter):
 if-intro: (((:DEFINITION AND)
       (:DEFINITION BOOLEAN-LISTP)
       (:DEFINITION ENDP)
       (:DEFINITION ACL2::IFF)
       (:DEFINITION IFF)
       (:DEFINITION TRUEP)
       (:DEFINITION UNTIL))
Warnings: Non-rec
Time: 0.69 seconds (prove: 0.68, print: 0.01, other: 0.00)
Prover steps counted: 186040
LTL-10-AXIOM-EXPANSION-OF-UNTIL
LTL !>
```

6.3 Eventually

For (41) Distributivity of Eventually over Or here is the statement of the theorem in ACL2.

```
;;;
;;; (41) Distributivity of Eventually over Or
;;;
;;; ACL2 proved this.
;;;
```

```
(defthm ltl-41-distributivity-of-eventually-over-or
 (acl2::implies (acl2::and (boolean-listp p)
                           (boolean-listp q)
                           (= (len p) (len q)))
          (equal (eventually (or p q))
                 (or (eventually p) (eventually q))))
 :hints (("Goal" :in-theory (enable ltl-29-defn-of-eventually))))
```

And here is the proof summary output from ACL2 indicating the proof attempt was successful.

```
Q.E.D.

Summary
Form: ( DEFTHM LTL-41-DISTRIBUTIVITY-OF-EVENTUALLY-OVER-OR ...)
Rules: ((:DEFINITION =)
    (:DEFINITION BOOLEAN-LISTP)
    (:DEFINITION LEN)
    (:DEFINITION MIN)
    (:DEFINITION SYNP)
    (:EXECUTABLE-COUNTERPART <)
    (:EXECUTABLE-COUNTERPART CONSP)
    (:EXECUTABLE-COUNTERPART EQUAL)
    (:EXECUTABLE-COUNTERPART EVENTUALLY)
    (:EXECUTABLE-COUNTERPART FALSEP)
    (:EXECUTABLE-COUNTERPART LEN)
    (:EXECUTABLE-COUNTERPART MIN)
    (:EXECUTABLE-COUNTERPART OR)
    (:EXECUTABLE-COUNTERPART TRUE)
    (:EXECUTABLE-COUNTERPART UNARY-/)
    (:FAKE-RUNE-FOR-TYPE-SET NIL)
    (:FORWARD-CHAINING BOOLEAN-LISTP-FORWARD-TO-SYMBOL-LISTP)
    (:FORWARD-CHAINING SYMBOL-LISTP-FORWARD-TO-TRUE-LISTP)
    (:REWRITE ACL2::|(* 0 x)|)
    (:REWRITE ACL2::|(* a (/ a))|)
    (:REWRITE ACL2::|(* y x)|)
    (:REWRITE ACL2::|(/ (/ x))|)
    (:REWRITE ACL2::|(< (if a b c) x)|)
    (:REWRITE ACL2::|(< x (if a b c))|)
    (:REWRITE ACL2::|(equal (/ x) c)|)
    (:REWRITE IDENTITY-OF-OR-LEMMA)
    (:REWRITE IDENTITY-OF-OR-LEMMA-COMMUTED)
    (:REWRITE LEN-OF-CDR)
    (:REWRITE LEN-OF-OR)
    (:REWRITE LTL-12-AXIOM-LEFT-DISTRIBUTIVITY-
    OF-UNTIL-OVER-OR)
    (:REWRITE LTL-21-LEFT-IDENTITY-OF-UNTIL-LEMMA)
    (:REWRITE LTL-29-DEFN-OF-EVENTUALLY)
    (:REWRITE OR-IS-BOOLEAN-LISTP)
    (:REWRITE ACL2::SIMPLIFY-PRODUCTS-GATHER-EXPONENTS-<)
    (:TYPE-PRESCRIPTION BOOLEAN-LISTP)
    (:TYPE-PRESCRIPTION LEN)
    (:TYPE-PRESCRIPTION SYMBOL-LISTP))
Splitter rules (see :DOC splitter):
```

```
    case-split: ((:REWRITE ACL2::SIMPLIFY-PRODUCTS-
            GATHER-EXPONENTS-<))
Time: 0.12 seconds (prove: 0.11, print: 0.00, other: 0.00)
Prover steps counted: 16934
LTL-41-DISTRIBUTIVITY-OF-EVENTUALLY-OVER-OR
LTL !>
```

6.4 Always

For (63) (un-named) also Schneider's (3.22e) here is the statement of the theorem in ACL2.

```
;;;
;;; (63)  LTL-63 (un-named) also Schneider's (3.22e)
;;;
;;; ACL2 proved this.
;;;
;- - - - - - - Lemmas needed for (LTL-63) - - - - - - - - - - - -
;;;
;;; First, some pre-existing helper lemmas were pulled-ahead
;;; from later theorems as they are needed here, earlier in
;;; the proof sequence. This occurs often when theorems are
;;; moved earlier in the system sequence.
;;;

;;;
;;; This lemma is needed: truep-or
;;;
(defthm truep-or
 (acl2::implies (acl2::or (truep p) (truep q))
         (truep (or p q))))

;;;
;;; This is not needed here, but added for later use.
;;;
;;; From (3.32) on the equation sheet
(defthmd truep-or-2
 (acl2::implies (equal p (or p (not q)))
         (truep (or p q))))

;;;
;;; Theorem on Length of And.
;;;
(defthm len-of-and
 (equal (len (and p q))
     (min (len p) (len q))))

;;;
;;; Two theorems on Implies.
;;;
(defthm truep-implies
 (acl2::implies (acl2::or (falsep p) (truep q))
         (truep (implies p q))))
```

```
;;;
;;;   From (3.59) on the equation sheet
(defthm truep-implies-2
 (acl2::implies (truep (or (not p) q))
        (truep (implies p q))))


;;;
;;; Next are two lemmas that support proof of (63).
;;;
;;; Proof Strategy:
;;;   Sometimes ACL2 will not directly prove the formula
;;;   that we want to show is a theorem. So we get ACL2 to
;;;   prove some equivalent formula is a theorem and also
;;;   prove both formulas are equivalent. Then we prove the
;;;   formula we wanted to show initially. We use this strategy
;;;   to prove (63) is a theorem.
;;

;;;
;;; First, prove Lemma 1.
;;; Lemma 1 shows that an equivalent form is a theorem.
;;;
(defthm ltl-63-lemma-1
 (acl2::implies (boolean-listp p)
        (truep (implies (and p (not (always p)))
          (eventually (and p (not (next p)))))))
 :hints (("Goal" :in-theory (e/d ()
          (LTL-50-EVENTUALLY-IN-TERMS-OF-ALWAYS)))))


;;;
;;; Next, prove Lemma 2.
;;; Lemma 2 is Proof of Equivalence of the Lemma 1
;;; form to the form in the CDS4LTL paper.
;;;
(defthm ltl-63-lemma-1-equiv-ltl-63
 (acl2::implies (boolean-listp p)
   (equal (implies (and p (eventually (not p)))
          (eventually (and p (next (not p)))))
   (implies (and p (not (always p)))
      (eventually (and p (not (next p)))))
   ))
 :hints (("Goal" :in-theory (union-theories '(ltl-63-lemma-1
    LTL-51-DUAL-OF-ALWAYS LTL-1-AXIOM-SELF-DUAL-2)
          (theory 'ground-zero))
     :do-not-induct t)))

;- - - - - - - End of Lemmas - - - - - - - - - - - - - - - - - - -
;;;
;;; Main Result - LTL-63
;;; Now, proof of the original form.
;;;
(defthm ltl-63
 (acl2::implies (boolean-listp p)
      (truep (implies (and p (eventually (not p)))
          (eventually (and p (next (not p))))))))
```

Pepperdine Papers on LTL

```
:hints (("Goal" :in-theory (union-theories '(ltl-63-lemma-1
            ltl-63-lemma-1-equiv-ltl-63)
         (theory 'ground-zero))
  :do-not-induct t)))
```

And here is the proof summary output from ACL2 indicating the proof attempt was successful.

Note: ACL2 output associated with processing the helper lemmas has been deleted to shorten up this presentation (which is still long). Note that in the output for LTL-63-LEMMA-1 contrapositive is used as it was in the hand proof provided in the CDS4LTL paper by David Vega.

Output from proof of LTL-63-LEMMA-1

```
Q.E.D.

The storage of LTL-63-LEMMA-1 depends upon the
:type-prescription rule TRUEP.

Summary
Form:  ( DEFTHM LTL-63-LEMMA-1 ...)
Rules: ((:COMPOUND-RECOGNIZER ACL2::NATP-COMPOUND-RECOGNIZER)
        (:COMPOUND-RECOGNIZER ACL2::ZP-COMPOUND-RECOGNIZER)
        (:DEFINITION ALWAYS)
        (:DEFINITION AND)
        (:DEFINITION ATOM)
        (:DEFINITION BOOLEAN-LISTP)
        (:DEFINITION ENDP)
        (:DEFINITION EQ)
        (:DEFINITION EVENTUALLY)
        (:DEFINITION FALSEP)
        (:DEFINITION IMPLIES)
        (:DEFINITION LEN)
        (:DEFINITION MIN)
        (:DEFINITION NEXT)
        (:DEFINITION COMMON-LISP::NOT)
        (:DEFINITION NOT)
        (:DEFINITION OR)
        (:DEFINITION SYNP)
        (:DEFINITION TRUE)
        (:DEFINITION TRUE-LISTP)
        (:DEFINITION TRUEP)
        (:ELIM CAR-CDR-ELIM)
        (:EXECUTABLE-COUNTERPART <)
        (:EXECUTABLE-COUNTERPART ALWAYS)
        (:EXECUTABLE-COUNTERPART AND)
        (:EXECUTABLE-COUNTERPART BINARY-+)
        (:EXECUTABLE-COUNTERPART BOOLEAN-LISTP)
        (:EXECUTABLE-COUNTERPART BOOLEANP)
        (:EXECUTABLE-COUNTERPART CAR)
        (:EXECUTABLE-COUNTERPART CDR)
        (:EXECUTABLE-COUNTERPART CONS)
```

```
(:EXECUTABLE-COUNTERPART CONSP)
(:EXECUTABLE-COUNTERPART ENDP)
(:EXECUTABLE-COUNTERPART EQUAL)
(:EXECUTABLE-COUNTERPART EVENTUALLY)
(:EXECUTABLE-COUNTERPART FALSEP)
(:EXECUTABLE-COUNTERPART IF)
(:EXECUTABLE-COUNTERPART IMPLIES)
(:EXECUTABLE-COUNTERPART LEN)
(:EXECUTABLE-COUNTERPART MIN)
(:EXECUTABLE-COUNTERPART COMMON-LISP::NOT)
(:EXECUTABLE-COUNTERPART NOT)
(:EXECUTABLE-COUNTERPART OR)
(:EXECUTABLE-COUNTERPART TAU-SYSTEM)
(:EXECUTABLE-COUNTERPART TRUE)
(:EXECUTABLE-COUNTERPART TRUEP)
(:EXECUTABLE-COUNTERPART UNARY-/)
(:FAKE-RUNE-FOR-LINEAR NIL)
(:FAKE-RUNE-FOR-TYPE-SET NIL)
(:FORWARD-CHAINING BOOLEAN-LISTP-FORWARD-TO-SYMBOL-LISTP)
(:FORWARD-CHAINING SYMBOL-LISTP-FORWARD-TO-TRUE-LISTP)
(:INDUCTION ALWAYS)
(:INDUCTION AND)
(:INDUCTION BOOLEAN-LISTP)
(:INDUCTION IMPLIES)
(:INDUCTION LEN)
(:INDUCTION NOT)
(:INDUCTION TRUE-LISTP)
(:INDUCTION TRUEP)
(:REWRITE ACL2::|(* 0 x)|)
(:REWRITE ACL2::|(* a (/ a))|)
(:REWRITE ACL2::|(* y x)|)
(:REWRITE ACL2::|(+ 0 x)|)
(:REWRITE ACL2::|(+ c (+ d x))|)
(:REWRITE ACL2::|(+ x (- x))|)
(:REWRITE ACL2::|(+ x (if a b c))|)
(:REWRITE ACL2::|(+ y (+ x z))|)
(:REWRITE ACL2::|(+ y x)|)
(:REWRITE ACL2::|(/ (/ x))|)
(:REWRITE ACL2::|(< 0 (/ x))|)
(:REWRITE ACL2::|(< x (if a b c))|)
(:REWRITE ACL2::|(equal (/ x) c)|)
(:REWRITE ALWAYS-IS-BOOLEAN-LISTP)
(:REWRITE AND-IS-CONSP)
(:REWRITE BOOLEAN-LISTP-CONS)
(:REWRITE ACL2::BUBBLE-DOWN-+-MATCH-1)
(:REWRITE CAR-CONS)
(:REWRITE CAR-EVENTUALLY-1)
(:REWRITE CDR-CONS)
(:REWRITE CDR-OF-AND)
(:REWRITE CONTRAPOSITIVE)        <<<< Use of Contrapositive
(:REWRITE DEFN-OF-FALSE)
(:REWRITE DEMORGAN-3-47-A)
(:REWRITE EQUAL-ALWAYS-FALSE)
(:REWRITE EQUAL-ALWAYS-TRUE)
(:REWRITE EVENTUALLY-IS-CONSP)
```

```
          (:REWRITE IDENTITY-OF-AND-LEMMA)
          (:REWRITE IMPLIES-3-74-LEMMA)
          (:REWRITE IMPLIES-3-75-LEMMA)
          (:REWRITE LEN-OF-ALWAYS)
          (:REWRITE LEN-OF-AND)
          (:REWRITE LEN-OF-CDR)
          (:REWRITE LEN-OF-NOT)
          (:REWRITE LEN-OF-TRUE)
          (:REWRITE LTL-53-TRUTH-OF-ALWAYS)
          (:REWRITE LTL-NOT-DOUBLE-NEGATION)
          (:REWRITE ACL2::NORMALIZE-ADDENDS)
          (:REWRITE NOT-IS-BOOLEAN-LISTP)
          (:REWRITE NOT-IS-CONSP)
          (:REWRITE ACL2::REDUCE-ADDITIVE-CONSTANT-<)
          (:REWRITE REFLEXIVITY-OF-IMPLIES-2)
          (:REWRITE ACL2::REMOVE-STRICT-INEQUALITIES)
          (:REWRITE ACL2::REMOVE-WEAK-INEQUALITIES)
          (:REWRITE RIGHT-ZERO-OF-IMPLIES-LEMMA)
          (:REWRITE ACL2::SIMPLIFY-PRODUCTS-GATHER-EXPONENTS-<)
          (:REWRITE ACL2::SIMPLIFY-SUMS-<)
          (:REWRITE SYMMETRY-OF-AND)
          (:REWRITE SYMMETRY-OF-OR)
          (:REWRITE TRUE-IS-BOOLEAN-LISTP)
          (:REWRITE TRUE-IS-CONSP)
          (:REWRITE TRUE-IS-TRUEP)
          (:REWRITE TRUEP-IMPLIES-2)
          (:REWRITE TRUEP-OR)
          (:REWRITE ZERO-OF-AND-LEMMA)
          (:REWRITE ACL2::ZP-OPEN)
          (:TYPE-PRESCRIPTION ALWAYS)
          (:TYPE-PRESCRIPTION AND)
          (:TYPE-PRESCRIPTION BOOLEAN-LISTP)
          (:TYPE-PRESCRIPTION EVENTUALLY)
          (:TYPE-PRESCRIPTION FALSEP)
          (:TYPE-PRESCRIPTION IMPLIES)
          (:TYPE-PRESCRIPTION LEN)
          (:TYPE-PRESCRIPTION MIN)
          (:TYPE-PRESCRIPTION NOT)
          (:TYPE-PRESCRIPTION OR)
          (:TYPE-PRESCRIPTION SYMBOL-LISTP)
          (:TYPE-PRESCRIPTION TRUE)
          (:TYPE-PRESCRIPTION TRUEP))
Splitter rules (see :DOC splitter):
  case-split: ((:REWRITE ACL2::SIMPLIFY-PRODUCTS-
              GATHER-EXPONENTS-<))
  if-intro: ((:DEFINITION ALWAYS)
       (:DEFINITION AND)
       (:DEFINITION BOOLEAN-LISTP)
       (:DEFINITION ENDP)
       (:DEFINITION EVENTUALLY)
       (:DEFINITION IMPLIES)
       (:DEFINITION LEN)
       (:DEFINITION MIN)
       (:DEFINITION COMMON-LISP::NOT)
       (:DEFINITION NOT)
```

```
                    (:DEFINITION OR)
                    (:DEFINITION TRUE)
                    (:DEFINITION TRUE-LISTP)
                    (:DEFINITION TRUEP)
                    (:REWRITE ACL2::|(+ x (if a b c))|)
                    (:REWRITE ACL2::|(< x (if a b c))|)
                    (:REWRITE ACL2::ZP-OPEN))
Warnings: Non-rec
Time: 3.21 seconds (prove: 3.17, print: 0.04, other: 0.01)
Prover steps counted: 641610
 LTL-63-LEMMA-1
LTL !>

Output from proof of LTL-63-LEMMA-1-EQUIV-LTL-63

ACL2 Warning [Non-rec] in (DEFTHM LTL-63-LEMMA-1-EQUIV-LTL-63):
A :REWRITE rule generated from LTL-63-LEMMA-1-EQUIV-LTL-63
will be triggered only by terms containing the
non-recursive function symbol NEXT.
Unless this function is disabled, this rule is unlikely
ever to be used.

Goal'

Q.E.D.

Summary
Form:  ( DEFTHM LTL-63-LEMMA-1-EQUIV-LTL-63 ...)
Rules: ((:FAKE-RUNE-FOR-TYPE-SET NIL)
        (:REWRITE LTL-1-AXIOM-SELF-DUAL-2)
        (:REWRITE LTL-51-DUAL-OF-ALWAYS)
        (:TYPE-PRESCRIPTION BOOLEAN-LISTP))
Warnings: Non-rec
Time: 0.01 seconds (prove: 0.00, print: 0.00, other: 0.00)
Prover steps counted: 123
 LTL-63-LEMMA-1-EQUIV-LTL-63
LTL !>

Output from proof of LTL-63

ACL2 Warning [Non-rec] in ( DEFTHM LTL-63 ...): A :REWRITE rule
generated from LTL-63 will be triggered only by terms containing
the non-recursive function symbol NEXT. Unless this function
is disabled, this rule is unlikely ever to be used.

Goal'

Q.E.D.

The storage of LTL-63 depends upon the :type-prescription rule
TRUEP.

Summary
Form:  ( DEFTHM LTL-63 ...)
Rules: ((:REWRITE LTL-63-LEMMA-1)
```

Pepperdine Papers on LTL

```
      (:REWRITE LTL-63-LEMMA-1-EQUIV-LTL-63)
      (:TYPE-PRESCRIPTION BOOLEAN-LISTP)
      (:TYPE-PRESCRIPTION TRUEP))
Warnings: Non-rec
Time: 0.01 seconds (prove: 0.00, print: 0.00, other: 0.00)
Prover steps counted: 94
 LTL-63
LTL !>
```

6.5 Always-Eventually

For (91) (un-named) here is the statement of the theorem in ACL2.

```
;;;
;;;
;;; (91) LTL-91
;;;
;;; ACL2 proved this.
;;;
;;;
(defthm ltl-91
 (acl2::implies (acl2::and (boolean-listp p)
                  (boolean-listp q)
                  (equal (len p) (len q)))
      (truep (implies (and (eventually (always p))
                  (always (eventually q)))
          (always (eventually (and p q))))))
 :hints (("Goal" :in-theory (e/d ()
      (LTL-50-EVENTUALLY-IN-TERMS-OF-ALWAYS)))))
```

And here is the proof summary output from ACL2 indicating the proof attempt was successful.

```
Q.E.D.

The storage of LTL-91 depends upon the
:type-prescription rule TRUEP.

Summary
Form: ( DEFTHM LTL-91 ...)
Rules: ((:COMPOUND-RECOGNIZER ACL2::NATP-COMPOUND-RECOGNIZER)
     (:COMPOUND-RECOGNIZER ACL2::ZP-COMPOUND-RECOGNIZER)
     (:DEFINITION ALWAYS)
     (:DEFINITION AND)
     (:DEFINITION BOOLEAN-LISTP)
     (:DEFINITION EVENTUALLY)
     (:DEFINITION FALSEP)
     (:DEFINITION IMPLIES)
     (:DEFINITION LAST)
```

```
(:DEFINITION LEN)
(:DEFINITION MIN)
(:DEFINITION COMMON-LISP::NOT)
(:DEFINITION NOT)
(:DEFINITION SYNP)
(:DEFINITION TRUE)
(:DEFINITION TRUEP)
(:ELIM CAR-CDR-ELIM)
(:EXECUTABLE-COUNTERPART <)
(:EXECUTABLE-COUNTERPART ALWAYS)
(:EXECUTABLE-COUNTERPART AND)
(:EXECUTABLE-COUNTERPART BINARY-+)
(:EXECUTABLE-COUNTERPART BOOLEAN-LISTP)
(:EXECUTABLE-COUNTERPART BOOLEANP)
(:EXECUTABLE-COUNTERPART CAR)
(:EXECUTABLE-COUNTERPART CDR)
(:EXECUTABLE-COUNTERPART CONS)
(:EXECUTABLE-COUNTERPART CONSP)
(:EXECUTABLE-COUNTERPART EQUAL)
(:EXECUTABLE-COUNTERPART EVENTUALLY)
(:EXECUTABLE-COUNTERPART FALSEP)
(:EXECUTABLE-COUNTERPART IMPLIES)
(:EXECUTABLE-COUNTERPART LEN)
(:EXECUTABLE-COUNTERPART COMMON-LISP::NOT)
(:EXECUTABLE-COUNTERPART NOT)
(:EXECUTABLE-COUNTERPART TAU-SYSTEM)
(:EXECUTABLE-COUNTERPART TRUE)
(:EXECUTABLE-COUNTERPART TRUEP)
(:FAKE-RUNE-FOR-LINEAR NIL)
(:FAKE-RUNE-FOR-TYPE-SET NIL)
(:FORWARD-CHAINING BOOLEAN-LISTP-FORWARD-TO-SYMBOL-LISTP)
(:FORWARD-CHAINING SYMBOL-LISTP-FORWARD-TO-TRUE-LISTP)
(:INDUCTION ALWAYS)
(:INDUCTION AND)
(:INDUCTION BOOLEAN-LISTP)
(:INDUCTION EVENTUALLY)
(:INDUCTION LEN)
(:INDUCTION TRUEP)
(:REWRITE ACL2::|(+ 0 x)|)
(:REWRITE ACL2::|(+ c (+ d x))|)
(:REWRITE ACL2::|(+ x (if a b c))|)
(:REWRITE ACL2::|(+ y (+ x z))|)
(:REWRITE ACL2::|(< x (if a b c))|)
(:REWRITE ALWAYS-IS-BOOLEAN-LISTP)
(:REWRITE AND-IS-CONSP)
(:REWRITE ASSOCIATIVITY-OF-OR)
(:REWRITE BOOLEAN-LISTP-CONS)
(:REWRITE CAR-CONS)
(:REWRITE CAR-EVENTUALLY-2)
(:REWRITE CAR-OF-TRUE-IS-T)
(:REWRITE CDR-CONS)
(:REWRITE CDR-OF-AND)
(:REWRITE CDR-OF-TRUE-IS-TRUEP)
(:REWRITE DEFN-OF-FALSE)
(:REWRITE DEMORGAN-3-47-A)
```

```
(:REWRITE EA-FOR-P-ENDING-IN-T)
(:REWRITE EQUAL-ALWAYS-FALSE)
(:REWRITE EQUAL-ALWAYS-TRUE)
(:REWRITE EQUAL-FALSE-ALWAYS-LEMMA)
(:REWRITE EVENTUALLY-IS-BOOLEAN-LISTP)
(:REWRITE IDENTITY-OF-AND)
(:REWRITE IDENTITY-OF-AND-2)
(:REWRITE IDENTITY-OF-AND-LEMMA)
(:REWRITE IMPLIES-3-74-LEMMA)
(:REWRITE IMPLIES-3-75-LEMMA)
(:REWRITE IMPLIES-3-76B)
(:REWRITE LEFT-IDENTITY-OF-IMPLIES)
(:REWRITE LEFT-IDENTITY-OF-IMPLIES-LEMMA)
(:REWRITE LEN-OF-ALWAYS)
(:REWRITE LEN-OF-AND)
(:REWRITE LEN-OF-CDR)
(:REWRITE LEN-OF-EVENTUALLY)
(:REWRITE LEN-OF-NOT)
(:REWRITE LEN-OF-OR)
(:REWRITE LEN-OF-TRUE)
(:REWRITE LTL-32-TRUTH-OF-EVENTUALLY)
(:REWRITE LTL-52-DUAL-OF-EVENTUALLY-2)
(:REWRITE LTL-82-METATHEOREM)
(:REWRITE LTL-88-LEMMA-1)
(:REWRITE LTL-NOT-DOUBLE-NEGATION)
(:REWRITE NOT-IS-BOOLEAN-LISTP)
(:REWRITE NOT-IS-CONSP)
(:REWRITE OR-IS-BOOLEAN-LISTP)
(:REWRITE P-OR-NOT-P)
(:REWRITE ACL2::REDUCE-ADDITIVE-CONSTANT-EQUAL)
(:REWRITE REFLEXIVITY-OF-IMPLIES-2)
(:REWRITE ACL2::REMOVE-WEAK-INEQUALITIES)
(:REWRITE RIGHT-ZERO-OF-IMPLIES-LEMMA)
(:REWRITE SYMMETRY-OF-AND)
(:REWRITE SYMMETRY-OF-OR)
(:REWRITE TRUE-IS-BOOLEAN-LISTP)
(:REWRITE TRUE-IS-CONSP)
(:REWRITE TRUE-IS-TRUEP)
(:REWRITE TRUEP-IMPLIES-2)
(:REWRITE TRUEP-NOT)
(:REWRITE TRUEP-OR)
(:REWRITE TRUTH-OF-EVENTUALLY-TRUE-N)
(:REWRITE ZERO-OF-AND-LEMMA)
(:REWRITE ZERO-OF-OR-LEMMA)
(:REWRITE ZERO-OF-OR-LEMMA-COMMUTED)
(:TYPE-PRESCRIPTION ALWAYS)
(:TYPE-PRESCRIPTION AND)
(:TYPE-PRESCRIPTION BOOLEAN-LISTP)
(:TYPE-PRESCRIPTION EVENTUALLY)
(:TYPE-PRESCRIPTION FALSEP)
(:TYPE-PRESCRIPTION IMPLIES)
(:TYPE-PRESCRIPTION LEN)
(:TYPE-PRESCRIPTION MIN)
(:TYPE-PRESCRIPTION NOT)
(:TYPE-PRESCRIPTION SYMBOL-LISTP)
```

```
         (:TYPE-PRESCRIPTION TRUE)
         (:TYPE-PRESCRIPTION TRUEP))
Splitter rules (see :DOC splitter):
 if-intro: ((:DEFINITION ALWAYS)
            (:DEFINITION AND)
            (:DEFINITION BOOLEAN-LISTP)
            (:DEFINITION EVENTUALLY)
            (:DEFINITION FALSEP)
            (:DEFINITION IMPLIES)
            (:DEFINITION LEN)
            (:DEFINITION MIN)
            (:DEFINITION TRUEP))
Time: 22.16 seconds (prove: 22.05, print: 0.11, other: 0.00)
Prover steps counted: 3260591
 LTL-91
 LTL !>
```

6.6 Wait

For (96) Distributivity of Not over Until here is the statement of the theorem in ACL2.

```
;;;
;;; (96) Distributivity of Not over Until
;;;
;;; ACL2 proved this.
;;;
(defthm ltl-96-distributivity-of-not-over-until
 (acl2::implies (acl2::and (boolean-listp p)
                           (boolean-listp q)
                           (equal (len p) (len q)))
        (equal (not (until p q))
               (wait (not q) (and (not p) (not q)))))
 :hints (("Goal" :in-theory (e/d () ()))))
```

And here is the proof summary output from ACL2 indicating the proof attempt was successful.

```
Q.E.D.

Summary
Form: ( DEFTHM LTL-96-DISTRIBUTIVITY-OF-NOT-OVER-UNTIL ...)
Rules: ((:CONGRUENCE IFF-IMPLIES-EQUAL-NOT)
        (:DEFINITION AND)
        (:DEFINITION BOOLEAN-LISTP)
        (:DEFINITION ENDP)
        (:DEFINITION FALSEP)
        (:DEFINITION LEN)
        (:DEFINITION MIN)
        (:DEFINITION COMMON-LISP::NOT)
```

```
(:DEFINITION NOT)
(:DEFINITION SYNP)
(:DEFINITION TRUEP)
(:DEFINITION UNTIL)
(:DEFINITION WAIT)
(:ELIM CAR-CDR-ELIM)
(:EXECUTABLE-COUNTERPART <)
(:EXECUTABLE-COUNTERPART AND)
(:EXECUTABLE-COUNTERPART BINARY-+)
(:EXECUTABLE-COUNTERPART BOOLEAN-LISTP)
(:EXECUTABLE-COUNTERPART BOOLEANP)
(:EXECUTABLE-COUNTERPART CAR)
(:EXECUTABLE-COUNTERPART CDR)
(:EXECUTABLE-COUNTERPART CONS)
(:EXECUTABLE-COUNTERPART CONSP)
(:EXECUTABLE-COUNTERPART EQUAL)
(:EXECUTABLE-COUNTERPART FALSEP)
(:EXECUTABLE-COUNTERPART LEN)
(:EXECUTABLE-COUNTERPART MIN)
(:EXECUTABLE-COUNTERPART COMMON-LISP::NOT)
(:EXECUTABLE-COUNTERPART NOT)
(:EXECUTABLE-COUNTERPART TRUEP)
(:EXECUTABLE-COUNTERPART UNARY-/)
(:EXECUTABLE-COUNTERPART UNTIL)
(:EXECUTABLE-COUNTERPART WAIT)
(:FAKE-RUNE-FOR-LINEAR NIL)
(:FAKE-RUNE-FOR-TYPE-SET NIL)
(:FORWARD-CHAINING BOOLEAN-LISTP-FORWARD-TO-SYMBOL-LISTP)
(:FORWARD-CHAINING SYMBOL-LISTP-FORWARD-TO-TRUE-LISTP)
(:INDUCTION BOOLEAN-LISTP)
(:INDUCTION LEN)
(:INDUCTION NOT)
(:INDUCTION UNTIL)
(:REWRITE ACL2::|(* 0 x)|)
(:REWRITE ACL2::|(* a (/ a))|)
(:REWRITE ACL2::|(* y x)|)
(:REWRITE ACL2::|(+ 0 x)|)
(:REWRITE ACL2::|(+ c (+ d x))|)
(:REWRITE ACL2::|(+ y (+ x z))|)
(:REWRITE ACL2::|(/ (/ x))|)
(:REWRITE ACL2::|(< 0 (/ x))|)
(:REWRITE ACL2::|(equal (/ x) c)|)
(:REWRITE AND-IS-BOOLEAN-LISTP)
(:REWRITE BOOLEAN-LISTP-CONS)
(:REWRITE CAR-CONS)
(:REWRITE CAR-OF-AND)
(:REWRITE CAR-OF-NOT)
(:REWRITE CAR-UNTIL)
(:REWRITE CDR-CONS)
(:REWRITE CDR-OF-NOT)
(:REWRITE CDR-UNTIL)
(:REWRITE CONS-EQUAL)
(:REWRITE DEFAULT-CDR)
(:REWRITE IDENTITY-OF-AND-LEMMA)
(:REWRITE LEN-OF-AND)
```

```
            (:REWRITE LEN-OF-CDR)
            (:REWRITE LEN-OF-NOT)
            (:REWRITE LTL-19-IDEMPOTENCY-OF-UNTIL)
            (:REWRITE LTL-20-RIGHT-ZERO-OF-UNTIL-LEMMA)
            (:REWRITE LTL-21-LEFT-IDENTITY-OF-UNTIL-LEMMA)
            (:REWRITE NOT-IS-BOOLEAN-LISTP)
            (:REWRITE NOT-IS-CONSP)
            (:REWRITE ACL2::REDUCE-ADDITIVE-CONSTANT-EQUAL)
            (:REWRITE ACL2::SIMPLIFY-PRODUCTS-GATHER-EXPONENTS-<)
            (:REWRITE SYMMETRY-OF-AND)
            (:REWRITE TRUEP-NOT)
            (:REWRITE ZERO-OF-AND-LEMMA)
            (:TYPE-PRESCRIPTION AND)
            (:TYPE-PRESCRIPTION BOOLEAN-LISTP)
            (:TYPE-PRESCRIPTION LEN)
            (:TYPE-PRESCRIPTION NOT)
            (:TYPE-PRESCRIPTION SYMBOL-LISTP)
            (:TYPE-PRESCRIPTION TRUEP)
            (:TYPE-PRESCRIPTION UNTIL)
            (:TYPE-PRESCRIPTION WAIT))
Splitter rules (see :DOC splitter):
 case-split: ((:REWRITE ACL2::SIMPLIFY-PRODUCTS-GATHER-EXPONENTS-<))
 if-intro: ((:DEFINITION BOOLEAN-LISTP)
            (:DEFINITION LEN)
            (:DEFINITION COMMON-LISP::NOT)
            (:DEFINITION NOT)
            (:DEFINITION TRUEP)
            (:DEFINITION WAIT)
            (:REWRITE CAR-OF-AND)
            (:REWRITE CAR-OF-NOT)
            (:REWRITE CAR-UNTIL)
            (:REWRITE CONS-EQUAL))
Time: 1.81 seconds (prove: 1.79, print: 0.01, other: 0.00)
Prover steps counted: 329421
LTL-96-DISTRIBUTIVITY-OF-NOT-OVER-UNTIL
LTL !>
```

7 CDS4LTL Proofs

All the theorems of CDS4LTL, at this point in time, have been proved in ACL2. To reproduce these results, first perform all the steps described earlier in section 5 of this paper on page 228. These propositional calculus theorems are required for the proofs in CDS4LTL. After all those steps have been accomplished, load the following files into ACL2. They must be loaded in the order shown below. The process of proving all the theorems may take up to four hours depending on your particular computer.

```
;;;
;;; LTL Operators and Theorems Files
;;;
```

```
(ld "~/Warford Copy/ltl-functions/next-theorems.lisp")
(ld "~/Warford Copy/ltl-functions/until-theorems.lisp")
(ld "~/Warford Copy/ltl-functions/eventually-theorems.lisp")
(ld "~/Warford Copy/ltl-functions/always-theorems.lisp")
(ld "~/Warford Copy/ltl-functions/proof-metatheorems.lisp")
(ld "~/Warford Copy/ltl-functions/always-eventually-theorems.lisp")
(ld "~/Warford Copy/ltl-functions/wait-theorems.lisp")
```

8 Summary

The purpose of this work was to use ACL2 to model the operators of LTL, and use the theorem prover to confirm that all theorems stated in the CDS4LTL paper [3] are in fact theorems. This work has now been completed. There are 154 axioms, definitions and theorems in the paper as of this writing (December 2015). All 18 axioms and definitions have ACL2 confirmation proofs. So the foundations of the system have been rigorously checked. In fact, as a result of this work one of the original axioms, called Weak Associativity of Until, was found to not be a theorem and was removed from the system. Of the 136 theorems in CDS4LTL, all of them now have ACL2 confirmation proofs. The last 11 theorems to be proven with ACL2 are listed in Table 1.

Theorem Number	Theorem Name	Comments
(98)	Distributivity of And over Wait	Proved on 7 Dec 2015
(107)	Right Distributivity of Wait over And	Proved on 10 Dec 2015
(115)	Conjunction Rule of Wait	Proved on 23 Nov 2015
(144)	No name	Proved on 4 Dec 2015
(145)	No name	Proved on 4 Dec 2015
(146)	Right Monotonicity of Until	Proved on 11 Dec 2015
(150)	Absorption	Proved on 3 Dec 2015
(151)	Absorption	Proved on 19 Dec 2015 ~~Not yet proved manually~~ (proved manually 11/24/15)
(152)	Ordering	Proved on 19 Dec 2015 ~~Not yet proved manually~~ (proved manually 11/27/15)
(153)	Ordering	Proved on 14 Dec 2015
(154)	Ordering	Proved on 22 Dec 2015

Table 1: Record of the last eleven CDS4LTL theorems to be proved by ACL2.

9 Acknowledgements

I would like to thank Professor Warford of Pepperdine University, first for teaching me formal methods through his iTunes University courses, and second for inviting me to work with him and David Vega on CDS4LTL. I also thank David Vega for inviting me to join the team. It has been a challenging and rewarding experience.

10 References

[1] ACL2 Community. *ACL2 System and Libraries on GitHub*. 2017.

[2] Sandip Ray. *Scalable Techniques for Formal Verification*. New York: Springer, 2010, p. 237.

[3] J. Stanley Warford, David Vega, and Scott M. Staley. "A Calculational Deductive System for Linear Temporal Logic (Complete)". In: (2019).

[4] J. Stanley Warford, David Vega, and Scott M. Staley. *Theorems from CDS4LTL*. Pepperdine University Research Report, Natural Science Division, 2018.

Book Acknowledgements

This book's publication process, including LaTeX typesetting of article three, combining the papers into a complete volume, working out the details of using a print on demand supplier and developing the cover art concept and content has been led by Julee Staley.

CDS4LTL Research Team

J. Stanley Warford is a Professor of Computer Science at Pepperdine University in Malibu, California. He is also a graduate of Pepperdine having been awarded a BS in Mathematics in 1966. He has a Ph.D. in engineering from UCLA in 1984. He joined the faculty of Pepperdine in 1975, where he has taught in the Seaver College for over 40 years. In 2008 Professor Warford received the Howard A. White Award for teaching excellence.

David Vega is a Member of the Technical Staff at The Aerospace Corporation. He is a graduate of Pepperdine having been awarded a BS in Mathematics and Computer Science in 2010. He also received a MS in Computer Science from Washington University in St. Louis in 2012. He was awarded a Pepperdine Dean's Scholarship, a Natural Science Division Scholarship, the Litton Endowed Scholarship (twice), and the Tooma Undergraduate Research Fellowship (twice).

Scott M. Staley is a Chief Engineer (retired) of the Ford Motor Company Research Labs in Dearborn, Michigan where he had a 23 year career in advanced technology development. Prior to that he was an Assistant Professor of Mechanical Engineering at the University of Miami and Florida International University. He has a BS (1975) and Ph.D. (1984) in Mechanical Engineering from Purdue University, and a MS (1979) in Mechanical Engineering from the University of Connecticut. In 2012 he was recognized with the Outstanding Mechanical Engineer Award by Purdue University.

www.ingramcontent.com/pod-product-compliance
Lightning Source LLC
Chambersburg PA
CBHW081425220526
45466CB00008B/2270